Phytate destruction – consequences for precision animal nutrition

Phytate destruction
–
consequences for precision animal nutrition

edited by:
C.L. Walk,
I. Kühn,
H.H. Stein,
M.T. Kidd
and
M. Rodehutscord

Wageningen Academic
Publishers

EAN: 9789086862900
e-EAN: 9789086868360
ISBN: 978-90-8686-290-0
e-ISBN: 978-90-8686-836-0
DOI: 10.3920/978-90-8686-836-0

Cover design:
Tom Ward of Tom Ward Studio

First published, 2016

Table of contents

Foreword

Dear Reader,

The costs associated with the anti-nutritional influence of phytate on animal growth performance, feed efficiency and nutrient utilisation may be as much as \$2 billion/year. This can have a significant impact on an industry facing increasing consumer and economic pressures. When re-evaluating nutrition strategies to meet current and future challenges, the impact of phytate on feed efficiency, nutrient utilisation, sustainability and ultimately profitability, cannot be ignored.

While not an exhaustive review, some topics in this book have been extensively studied, e.g. phytate as an anti-nutrient, and this book aims to provide a source of information as well as novel perspectives on superdoses of phytase and phytate destruction. The book provides an update on phosphorus and calcium digestibility methods, factors to consider when using digestibility assays and ideas proposed by the authors for future directions in nutrient evaluations. While other topics presented in this book, such as nutrient requirements for trace minerals, might be less of a focus with regards to research interests in recent years, new information presented in this book suggests this should be reignited due to significant effects of phytase on trace minerals and benefits in meat quality and growth performance under stressed conditions. Topics such as digestible calcium requirements for broilers and pigs or the role of inositol polyphosphates are considerably novel information for the monogastric nutritionist or physiologist. The book also aims to delve into research on a cellular and microbiological level, to better understand nutrient requirements and recommendations in normal or deficient conditions. Yet the book also provides commercial perspectives on current nutrient or dietary recommendations from the pig, poultry and catfish industry, as examples of how science can be put into practice.

I hope you find the information in this book thought provoking, maybe sometimes controversial and in some cases hypothetical. The book was developed to harness the information and ideas presented at the 3rd International Phytate Summit held in November 2016. The theme of the conference was *The Value Chain of Phytate Destruction* and this book aims to provide you with a scientific, peer-reviewed source of the current information and future directions regarding phytate, phytate esters, *myo*-inositol, calcium and phosphorus, trace minerals, and amino acids in pig, poultry and aqua nutrition.

The contributors to the chapters are experts in their field and it is an honour to have worked with them to create this book. It should be said that the authors were given a very 'optimistic' time frame to submit their chapters, the editors were given a very short time frame to return comments, and everyone rose to the challenge and surpassed expectations in content, perspectives and even the minimum number of pages!

Happy reading.

Carrie Walk
Senior Research Manager, AB Vista

Chapter 1
Update on phytate degradation pattern in the gastrointestinal tract of pigs and broiler chickens

M. Rodehutscord and P. Rosenfelder*
Institute of Animal Science, University of Hohenheim, Emil-Wolff-Str. 10, 70599 Stuttgart, Germany; inst450@uni-hohenheim.de

Abstract

Plant raw materials used in pig and poultry feedstuffs contain *myo*-inositol 1,2,3,4,5,6-hexakis dihydrogen phosphate ($InsP_6$) and its salts (phytate) as the main storage form of phosphorus (P). The extent to which P is released from $InsP_6$ largely determines the digestibility of P from plant raw materials and the amount of mineral phosphates needed to meet the animal's P requirement. Phytases and other phosphatases catalyse $InsP_6$ hydrolysis in the digestive tract, and enzymes involved originate from an unknown mix of dietary and endogenous origin, the latter produced by the intestinal epithelium and the microbiota colonizing the digestive tract. The extent of $InsP_6$ degradation and appearance of degradation products in the digestive tract are distinctly different between pigs and broiler chickens. It appears that plant intrinsic phytase activity is a relevant factor for $InsP_6$ degradation and P digestibility in pigs, whereas in broilers the activities of endogenous enzymes play the major role. The potential to utilise P from $InsP_6$ is much higher in broilers than pigs. However, $InsP_6$ degradation is easy to manipulate by supplements of P, calcium (Ca) and exogenous phytases. While good progress has been made in identifying factors that cause variation in $InsP_6$ degradation, multiple interactions are not yet understood well enough to be considered in feed formulation in the industry. A better understanding of these interactions is of utmost importance in conditions of finite global rock phosphate stores, environmental problems due to soil P accumulation in certain regions and high costs of feeding.

Keywords: inositol phosphates, phosphatases, phosphorus, microbes, mucosa

1.1 Introduction

All living organisms depend upon a continuous supply of phosphates. Amounts of phosphorus (P) required by livestock have been estimated and dietary recommendations for pigs and poultry were published by different scientific committees (GfE, 1999, 2006; NRC, 1994, 2012; Rostagno, 2011). The P supply along the food chain is maintained by the application of fertilizer and feed phosphates, which are produced to a large extent from rock phosphates. The global rock phosphate stores are limited, and they may be depleted within a century (Cordell *et al.*, 2009). This limitation, together with the accumulation of these stores in only a few countries, is considered one of the greatest challenges facing sustainable food production (Gross, 2010; Neset and Cordell, 2012). Specifically in regards

DOI 10.3920/978-90-8686-836-0_1, © Wageningen Academic Publishers 2016

to the farm animal sector, there are two main approaches to address these challenges; first, exploring possibilities for reduction in feed phosphate application, and second, improving the efficacy of phytase supplements in non-ruminant feeding (Rodehutscord, 2008). These challenges are related to phytic acid (*myo*-inositol 1,2,3,4,5,6-hexakis dihydrogen phosphate; $InsP_6$) and their salts (phytate) and have initiated an increase in research regarding the presence and role of inositol phosphates with variable degree of phosphorylation ($InsP_x$) in the gastrointestinal tract of pigs and poultry. In this contribution, the extent of $InsP_6$ degradation and the appearance of $InsP_x$ with lower degree of phosphorylation including influencing factors will be addressed.

1.2 Inositol phosphates in feedstuffs

In plant seeds, $InsP_6$ is the primary stored form of P (Eeckhout and De Paepe, 1994; Ravindran *et al.*, 1994; Rodehutscord *et al.*, 2016). This, together with the dominant role of plant seeds and their processing products (e.g. oilseed meals produced from soybeans and rapeseed) in animal feeding, makes $InsP_6$ the most important source of organic P in diets for pigs and poultry. Pig and poultry diets contain about 0.2-0.3% of $InsP_6$-P, depending on what raw materials are used and their growing and processing conditions. Of note, substantial variation exists in $InsP_6$ concentration within one feed raw material. This has been extensively investigated for different genotypes of cereal grains (Figure 1.1), but soil type or agronomic details such as fertilisation also affect $InsP_6$ concentration in plant seeds.

Inositol pentakisphosphates ($InsP_5$) and other $InsP_x$ were found to be contained in cereal grains only at marginal levels (Rodehutscord *et al.*, 2016). However, in oilseed meals, especially rapeseed meal, up to one third of the total InsP-P can be contained as $InsP_5$-P

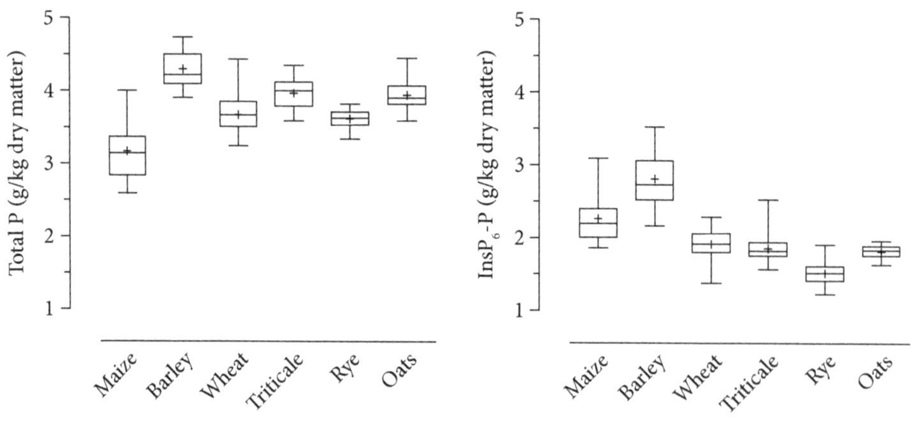

Figure 1.1. *Variation of total P and $InsP_6$-P in cereal grains. Genotypes of barley (n=21), wheat (n=29), triticale (n=21), rye (n=22) and oats (n=14) were grown using the same agronomic conditions and maize (n=27) using similar agronomic conditions (Rodehutscord* et al.*, 2016).*

(Pontoppidan *et al.*, 2007). This indicates degradation of $InsP_6$ may occur during thermal processing of oilseeds. It is unknown yet whether variable proportions of $InsP_5$ isomers in the feed are related to differences in the efficiency of P release from the inositol ring in the digestive tract. Uncertainty also exists whether or not technological processing of complete diets (e.g. pelleting or extrusion) can modulate the $InsP_x$ pattern of the diet. A 7% $InsP_6$ reduction has been observed after pelleting a diet based on barley, rapeseed cake and peas (Skoglund *et al.*, 1997), and extrusion of a diet based on barley, wheat, rye and soybean meal resulted in marginal effects on dietary $InsP_6$ and lower $InsP_x$ concentration (Schlemmer *et al.*, 2001). This area deserves further experimentation.

The observation of different $InsP_x$ to exist in the feed (for whatever reason) points to the need for considering the assay that was used for $InsP_6$ or phytate analysis. Different assays are in use and may or may not be able to detect inositol phosphates other than $InsP_6$.

Utilization of P from $InsP_6$ and lower $InsP_x$ by the animal requires stepwise cleavage of phosphate groups from the inositol ring. This hydrolysis is catalysed by phytases (Sandberg and Andlid, 2002) and other phosphatases. Phytases are not specific for $InsP_6$ and can further catalyse hydrolysis of lower $InsP_x$. Lower $InsP_x$ can also be catalysed by other phosphatases that cannot hydrolyse $InsP_6$. The sequence of cleavage yields $InsP_x$ with varying degrees of phosphorylation as intermediate products and may also yield *myo*-inositol as an end product. Nature has evolved specific strategies for different organisms to cleave phosphate from $InsP_6$ under various conditions (Mullaney and Ullah, 2007). Hence, 'phytase' refers to a group of enzymes that differ in catalytic properties, structures, sizes and origins (Mullaney and Ullah, 2003). Their activity results in the appearance of different $InsP_x$ in the digestive tract, and will be considered in a subsequent section of this chapter.

1.3 Disappearance of $InsP_6$ in the digestive tract of pigs and poultry

1.3.1 Pigs

Studies using pigs fed diets with low intrinsic plant phytase activity and mainly based on maize and soybean meal showed differences in intestinal $InsP_6$ breakdown (Table 1.1). $InsP_6$ disappearance ranged between 7 and 22% in the duodenum and 9 and 60% in the ileum, with only one study reaching a value or more than 40% in the ileum (Baumgärtel *et al.*, 2008; Jongbloed *et al.*, 1992; Kemme *et al.*, 1999, 2006; Rapp *et al.*, 2001; Rutherfurd *et al.*, 2014). Different levels of calcium (Ca) and P in the diets likely contributed to this variation. Other reasons for differences in $InsP_6$ disappearance might be the method of analysis of $InsP_x$, ingredient composition, adaption time and age of the animals (Dersjant-Li *et al.*, 2015), as well as sampling time after feeding (Blaabjerg *et al.*, 2011), the definition of sections samples were taken from and sample processing. In the stomach of pigs about one third of $InsP_6$ in a maize-soybean meal diet containing some wheat bran was hydrolysed while any further hydrolysis in the small intestine is not relevant (Figure 1.2; Rutherfurd *et al.*, 2014). However, gastric $InsP_6$ degradation depends very much on dietary phytase activity. In a study of Schlemmer *et al.* (2001) inactivation of dietary phytase activity by

Table 1.1. *Literature values on $InsP_6$ disappearance in the gastrointestinal tract of pigs fed low-calcium, low-phosphorus diets mainly based on maize and soybean meal.*

Reference	Ca (g/kg)	Total P (g/kg)	Sampling location	$InsP_6$ disappearance (%)
Kemme *et al.* (1999)	5.5	3.0	ileum	35
Rutherfurd *et al.* (2014)	3.9	4.0	stomach	37
			jejunum	38
			ileum	39
			faeces	87
Jongbloed *et al.* (1992)	5.2	3.3	duodenum	22
			ileum	10
Rapp *et al.* (2001)	5.6	4.3	duodenum	9
			ileum	19
Baumgärtel *et al.* (2008)	9.1	2.9	ileum	60
Zeng *et al.* (2014)	4.1	3.8	ileum	11
Kemme *et al.* (2006)	5.8	5.0	duodenum	7
			ileum	27

extrusion resulted in only 16% of total InsP appearing as $InsP_2$ – $InsP_5$ in the stomach of pigs. Furthermore, $InsP_6$ degradation depends on the passage rate through the stomach with still 73% $InsP_6$ remaining 5 hours after feeding when compared to a measurement directly after feeding (Blaabjerg *et al.*, 2011).

Regardless of the magnitude of $InsP_6$ disappearance measured in the terminal ileum, post-ileal $InsP_6$ disappearance was nearly complete in pigs (Baumgärtel *et al.*, 2008; Sandberg *et al.*, 1993; Schlemmer *et al.*, 2001; Seynaeve *et al.*, 2000). Since P absorption posterior to the ileum does not seem to be relevant, phosphate released from $InsP_6$ in the hindgut remains unavailable to the animal. Consequently, P digestibility in pigs, determined according to a standard protocol and using mash feed (GfE, 1994), hardly exceeded 30% in low-phytase feedstuffs such as maize or oilseed meals (Düngelhoef *et al.*, 1994; Rodehutscord *et al.*, 1996), whereas it was about 45% in barley, 52% in triticale, 68-73% in rye and 43-74% in wheat (Düngelhoef *et al.*, 1994; Hovenjürgen *et al.*, 2003; Rodehutscord *et al.*, 1996), all cereals known to contain intrinsic plant phytase activity.

1.3.2 Poultry

In contrast to pigs, several studies conducted with broiler chickens have reported that precaecal (pc) $InsP_6$ disappearance ranged between 62 and 89% (Table 1.2). All of these studies used diets that were mainly based on maize and soybean meal and thus were very low in plant intrinsic phytase activity. Additionally, none of the studies used diets that contained any mineral P supplement, and all were low in Ca content. This factor combination is a relatively artificial situation and not representative for practical-type diets. Nevertheless, it demonstrates the high biological potential the broiler chicken has to

Table 1.2. *Literature values on $InsP_6$ disappearance in broiler chickens fed low-calcium, low-phosphorus diets.*

Main ingredients of the diet	Ca (g/kg)	Total P (g/kg)	Sampling location	$InsP_6$ disappearance (%)	Reference
Maize-SBM[1]	1.8	4.1	ileum	69	Tamim and Angel (2003)
Maize-SBM	3.6	3.7	ileum	75	Applegate *et al.* (2003)
Maize-SBM	1.8	4.0	ileum	69	Tamim *et al.* (2004)
Maize-SBM	4.3	3.0	ileum	62	Shastak *et al.* (2014)
Maize	1.3	3.1	ileum	89	Leytem *et al.* (2008)
Maize-SBM	6.7	4.7	ileum	74	Zeller *et al.* (2015b)
Maize-SBM	5.4	4.0	ileum	67	Zeller *et al.* (2015c)
Maize-SBM	6.0	2.9	excreta	70	Shastak *et al.* (2014)
Maize-SBM	5.0	5.0	excreta	65	Mohammed *et al.* (1991)
Maize-SBM	6.3	4.3	excreta	63	Mitchell and Edwards Jr. (1996)

[1] SBM = soybean meal.

hydrolyse $InsP_6$ in the digestive tract. To what extent factors such as the Ca and P mineral supplementation can suppress the expression of this potential will be discussed in Chapter 10 (Rodehutscord, 2016).

When using digesta from the intestine of laying hens which had been fed a maize-soybean meal-based diet, substantial breakdown of sodium phytate *in vitro* was observed (Marounek *et al.*, 2008), especially when content from the caeca was used. In the crop of broiler chickens, $InsP_6$ disappearance was shown to be very low when diets did not contain phytase (Figure 1.2). However, following passage of digesta through the acidic environment of the proventriculus/gizzard, $InsP_6$ disappearance measured in the duodenum/jejunum reached levels of more than 50%. $InsP_6$ disappearance further increased with passage of

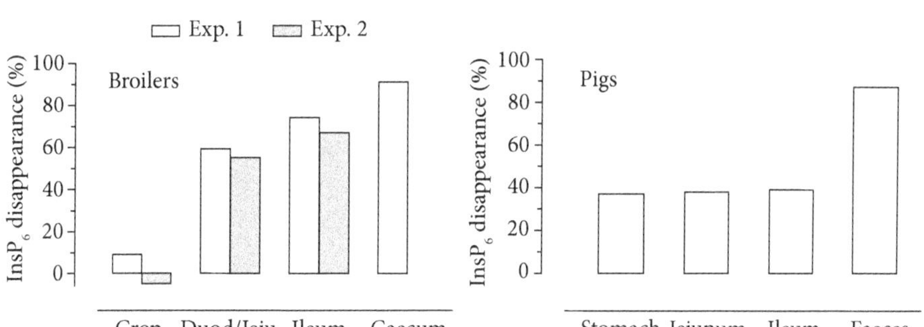

Figure 1.2. *$InsP_6$ disappearance measured in sections of the digestive tract of broiler chickens (left panel) and pigs (right panel) fed maize-soybean meal-based diets without a phytase supplement; data from Rutherfurd* et al. *(2014) and Zeller* et al. *(2015b,c, 2016).*

the digesta to the terminal ileum, and it was about 90% in the caeca of broilers. Of note, these high values of $InsP_6$ disappearance in broilers were achieved without intrinsic plant phytase or supplemented phytase in the diets.

1.4 Role of endogenous mucosal phytases

The disappearance of $InsP_6$ in the digestive tract upon feeding diets devoid of $InsP_6$ degrading enzymes needs other enzyme sources to be discussed for their potential contribution to $InsP_6$ hydrolysis. Whereas it is often assumed that animals do not produce endogenous phytase in intestinal epithelia, there is a growing body of evidence that this assumption is not correct.

Information on the relevance of epithelial phytase is scarcer in pigs than in poultry. One study reported hydrolytic enzyme activity of the piglet's mucosa towards $InsP_3$, $InsP_4$, $InsP_5$ and $InsP_6$ *in vitro*, and this activity was higher in mucosa obtained from the jejunum compared to duodenum and ileum (Hu *et al.*, 1996). In poultry, studies mainly conducted with broiler chickens but also with laying hens revealed some phytase activity when using purified brush border membrane vesicles from different sections of the small intestine (Huber *et al.*, 2015; Maenz and Classen, 1998; Onyango *et al.*, 2006). Epithelial phytase was found highest in preparations of the duodenum and it decreased in the more posterior parts of the small intestine (Maenz and Classen, 1998). There was also some indication that epithelial phytase secretion is reduced at higher inorganic phosphate (P_i) concentrations in the intestinal lumen (Huber *et al.*, 2015). Epithelial phytase secretion was also reduced when diets containing 0.9% Ca compared to 0.4% Ca were fed (Applegate *et al.*, 2003). However, it was stated that the regulation of endogenous phytase secretion is not well examined, and that the quantitative contribution of endogenous phytase to intestinal $InsP_6$ breakdown cannot be assessed at this time (Huber *et al.*, 2015). Because of their localization, brush border membrane-associated phytase may only have a restricted relevance for hydrolysis occurring in the gut lumen. Furthermore, endogenous enzymes may express different behaviour in the environment of the digestive tract than under the well-standardised *in vitro* assay conditions. The extent of $InsP_6$ breakdown in the duodenum/jejunum of broilers after feeding diets with different supplementary levels of a mineral phosphate and phytase (Zeller *et al.*, 2015c) did not correlate with the endogenous epithelial phytase activity measured in the companion work (Huber *et al.*, 2015). In three experiments using different dietary Ca and vitamin D_3 combinations as well as different bird strains (Applegate *et al.*, 2003), a significant correlation between pc $InsP_6$ disappearance and V_{max} of phytase activity of the brush border membrane vesicles was observed in one, but not in the other two experiments. Hence, while brush border membrane vesicles clearly show that substantial enzyme activity is there, its quantitative relevance for $InsP_x$ degradation in the complex situation present in the lumen of the digestive tract still needs to be elucidated.

1.5 Role of endogenous microbial phytases

Another potential source of luminal phytase activity is the microbiota colonizing the digestive tract. With regard to pigs, Humer *et al.* (2015) reviewed that contents of the stomach and intestine have negligible phytase activity. Likewise, the extent of phytate hydrolysis and liberation of InsP-P generated by microbial phytase in the gastrointestinal tract of pigs, proximal to the large intestine, is critical (Pagano *et al.*, 2007; Selle and Ravindran, 2008). When studying phytase activity along the gastrointestinal tract of pigs fed a diet without added phytase, high activity was found in the colon but not in the anterior sections (Pagano *et al.*, 2007). Correspondingly, extensive post-ileal $InsP_6$ disappearance has been found (Schlemmer *et al.*, 2001; Seynaeve *et al.*, 2000). As absorption of P occurs primarily in the upper parts of the small intestine, $InsP_6$-P cleaved in the colon is irrelevant for P supply to the pig.

Gnotobiotic broiler chickens compared with conventional non-gnotobiotic chickens had much higher $InsP_6$ levels in the caecal content than conventional broiler chickens (Kerr *et al.*, 2000). In laying hens, the specific phytase activity in caeca content was higher by a factor of more than 10 when compared with content from crop, stomach, and small intestine (Marounek *et al.*, 2010). These results indicate the substantial involvement of microbes in $InsP_6$ breakdown in the intestine. *In vitro* studies have shown the $InsP_6$ degrading activity of various bacteria (Konietzny and Greiner, 2002; Vats and Banerjee, 2004). Additionally, lactic acid producing bacteria isolated from the chicken intestine were identified as possible $InsP_6$ degrading candidates (Raghavendra and Halami, 2009). Among the bacteria in the small intestine of broiler chickens, lactobacilli are the most common ones (Rehman *et al.*, 2007). Interestingly, broiler chickens fed a diet supplemented with *Lactobacillus* species had increased P retention (Angel *et al.*, 2005). When using deep sequencing techniques, Witzig *et al.* (2015) recently found lactobacilli to be the dominating bacteria in the crop and small intestine of broiler chickens, but the phylotypes of lactobacilli differed between the crop and the small intestine. In addition to the dominant phylotypes, genes for $InsP_6$ phosphatases were identified in *Bacteroides* spp., *Burkholderia* spp. and three species of the genus *Bifidobacterium,* which are all also present in the digestive tract of chickens (Stentz *et al.*, 2014; Tamayo-Ramos *et al.*, 2012). This strongly supports the hypothesis of the chicken's gut microbiome contributing to $InsP_6$ breakdown. However, as for epithelial phytases, it is not possible to quantify the quantitative relevance of microbial phytase at this time.

Probably the crop microbiota is important for $InsP_6$ breakdown further down the digestive tract. Lactobacilli colonize there and produce enzymes. Hydration and temperature favour bacterial growth and enzyme activity (Svihus *et al.*, 2002). Short retention time of digesta in the crop may cause enzymes produced there to be active mainly in the subsequent sections, especially in the gizzard. Later in the small intestine, additional phytase and other phosphatases from the epithelium get involved, all together contributing to the development of $InsP_6$ disappearance shown in Figure 1.2. This hypothesis still needs to be tested in further experiments, but the hypothesis can be extended towards the lack of a 'pre-fermentation' section in the digestive tract of pigs being the reason for the differences observed in $InsP_6$ disappearance between pigs and poultry.

The microbial population in the caeca of broilers has a very high diversity, including bacteria known to be capable of $InsP_6$ breakdown (Rehman *et al.*, 2007; Witzig *et al.*, 2015). Correspondingly, *in vitro* phytase activity in caeca content of laying hens was much higher than in anterior sections of the digestive tract after feeding a maize-soybean meal-based diet (Marounek *et al.*, 2008). The $InsP_6$ content in the caeca was higher in gnotobiotic compared to conventional broilers (Kerr *et al.*, 2000). This demonstrated a high impact of microorganisms on $InsP_6$ breakdown in the caeca. In fact, less than10% of dietary $InsP_6$ were recovered in the caeca of broilers when fed a diet without phytase (Zeller *et al.*, 2015b). However, the relevance of caecal $InsP_6$ hydrolysis for the birds is still unclear. Based on the results of Son *et al.* (2002) it can be assumed, as an approximation, that not more than one quarter of ileal digesta enters the caeca for fermentation. Post-ileal absorption of P has not been reported to exist. Hence, P released from $InsP_x$ in the caeca is excreted as phosphate or bound in other forms. In pigs, the low faecal recovery of $InsP_6$ together with the relatively low pc $InsP_6$ disappearance also indicates extensive microbial $InsP_6$ degradation in the post-ileal compartment of the digestive tract.

1.6 Relevance of intrinsic plant phytase activity

Some plant seeds and feedstuffs obtained from them contain intrinsic phytase activity (Eeckhout and De Paepe, 1994; Rodehutscord *et al.*, 2016). Intrinsic phytase activity is high in cereal grains such as wheat, rye and triticale, while it is lower in legume seeds. Phytase is hardly detectable in oilseed meals and maize. Intrinsic plant phytase – assuming it is not deactivated during processing of the feed – can, in principle, contribute to $InsP_6$ breakdown in the digestive tract. Again, differences in the relevance of this $InsP_6$ hydrolysis seem to exist between pigs and broilers.

1.6.1 Pigs

Schlemmer *et al.* (2001) measured a 90% reduction in phytase activity in the stomach digesta of pigs when diets with high intrinsic phytase activity were fed. Nevertheless, the contribution of $InsP_6$ to total $InsP_x$ in stomach and small intestine was lower in the untreated diet when compared to the extruded diet (Figure 1.3). Peptic digestion as well as unfavourable pH for intrinsic plant phytases in the duodenum may cause a strong inactivation of these phytases. The pH in the stomach and small intestine is favourable for proteolytic enzymes which may break down the plant phytases making them undetectable in the lower small intestine (Yi and Kornegay, 1996). Along this line, Schlemmer *et al.* (2009) concluded that plant phytases might not be involved in relevant hydrolysis of InsP in the small and large intestine. Inactivation of the intrinsic phytase of a wheat-barley-soybean meal diet through steam-pelleting at about 90 °C resulted in pc $InsP_6$-P degradation of only 29% when compared to the untreated diet (52% $InsP_6$-P degradation) (Blaabjerg *et al.*, 2010). Studies that investigated P digestibility of different raw materials in pigs using standardised trial conditions including low dietary P showed that P digestibility was highest in wheat and rye (high phytase activity) followed by barley (medium phytase activity) and maize or oilseed meals (low/no phytase activity) (Düngelhoef *et al.*, 1994; Hovenjürgen *et al.*, 2003; Rodehutscord *et al.*, 1996). This demonstrates that intrinsic

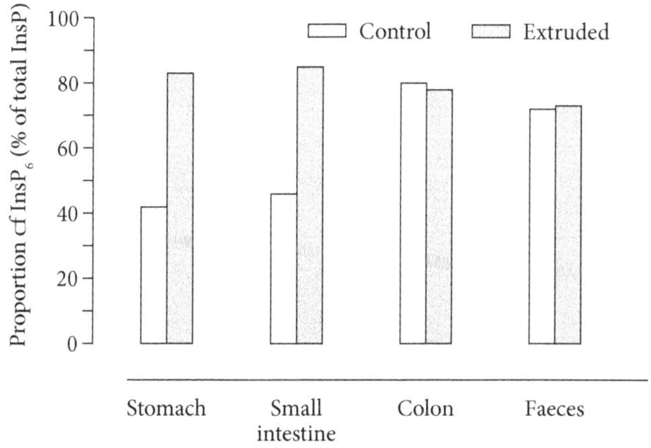

Figure 1.3. *$InsP_6$ content (% of total InsP in the respective section) measured in different sections of the digestive tract and faeces of pigs fed diets based on cereal grains and soybean meal and not extruded (control; 43.1 mU/mg protein phytase activity) or extruded (0.2 mU/mg protein phytase activity) (Schlemmer* et al., *2001).*

plant phytase is a relevant factor for P digestibility in the pig if not destroyed during feed processing.

1.6.2 Poultry

In broilers, plant phytase activity would primarily take place in the crop prior to initiation of proteolysis in the gizzard. However, it is a matter of debate how relevant the intrinsic plant phytase activity can be for $InsP_6$ breakdown in the intestine. Phosphorus retention studies involving wheat-based diets found some positive relationship between the intrinsic phytase activity of the wheat batches that were used and P retention in broiler chickens (Barrier-Guillot *et al.*, 1996; Oloffs *et al.*, 2000). Phosphorus retention in broiler chickens was also lower when extruded wheat was fed instead of a non-extruded wheat (Oloffs *et al.*, 1998). Other studies using different grains with different phytase activity did not indicate a relationship between the intrinsic plant phytase activity of the diet and $InsP_6$ disappearance in broiler chickens (Juanpere *et al.*, 2004; Leytem *et al.*, 2008). Considering these divergent reports, it was of interest to investigate intrinsic phytase effects along the digestive tract. Following inclusion of microwave-treated wheat instead of untreated wheat in the diet, both intrinsic phytase activity and $InsP_6$ disappearance in the crop of broiler chickens were substantially reduced (Zeller *et al.*, 2016) (Figure 1.4). However, with further passage through the digestive tract, the differences in $InsP_6$ breakdown between the diets containing microwave-treated or untreated wheat disappeared. These differences indicate that other phytase sources had compensated for the lack in intrinsic plant phytase activity (Zeller *et al.*, 2015a). Other authors also concluded that in comparison to endogenous epithelial and microbial phytase, intrinsic feed phytase seems to contribute very little to $InsP_6$ degradation when measured in the terminal ileum of broiler chickens (Leytem *et al.*, 2008; Shastak *et al.*, 2014). When a maize-based diet with phytase below the detection limit was compared with a wheat-based diet with 660 U/kg phytase, pc $InsP_6$ disappearance in broiler chickens was approximately 61%, and there was no difference between the two

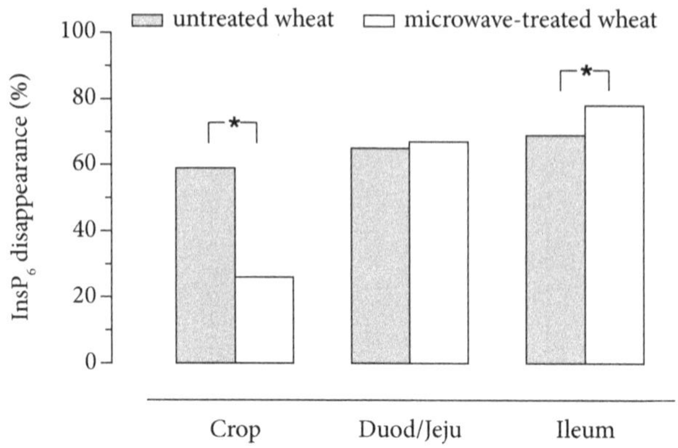

Figure 1.4. *$InsP_6$ disappearance measured in different sections of the digestive tract of broiler chickens fed diets that contained untreated wheat (632 U/kg phytase) or microwave-treated wheat (121 U/kg phytase) (Zeller* et al.*, 2015a, 2016)*

diets (Shastak *et al.*, 2014). Overall, the discrepancies in the evaluation of intrinsic plant phytase relevance that are obvious from this chapter cannot yet be explained. It is possible that they are related to P and Ca mineral inclusion levels of the diets.

1.7 Presence and potential relevance of $InsP_x$ isomers

Disappearance of $InsP_6$ in the digestive process does not mean that *myo*-inositol is completely de-phosphorylated. Lower $InsP_x$ isomers carrying different numbers of phosphate groups are widely present in the gastrointestinal tract, and their concentrations are variable. This is of interest for different reasons. First, part of the InsP-P remains unavailable to the animal because it is still bound to the inositol ring. Second, some specific $InsP_x$ isomers are known to be able to express specific cellular signalling and other physiological functions. This topic is discussed in Chapter 4 (Huber, 2016).

One reason for the variation in $InsP_x$ pattern is the origin and specificity of the enzyme. Based on stereo specificity of $InsP_6$ hydrolysis, different types of phytases have been recognized by IUPAC-IUB: 3-phytases (E.C.3.1.3.8), 4-phytases (E.C.3.1.3.26) and 5-phytases (E.C.3.1.3.72). This classification is based on the suggestion to use the numbering based on D-configuration for *myo*-inositol (IUPAC, 1976). Bacterial phytases, such as *Escherichia coli* phytase, which preferentially generate D-Ins(1,2,3,4,5)P_5 are classified as 6-phytases (Greiner *et al.*, 2000). Phytases that generate D-Ins(1,2,3,5,6)P_5, such as the majority of plant phytases, should be called 4-phytases based on the configuration introduced by IUPAC, but are conventionally also called 6-phytases (based on L-configuration). As the major $InsP_5$ generated by phytases from *Selenomonas cerevisiae* and *Aspergillus niger* has been identified to be D-Ins(1,2,4,5,6)P_5, these phytases are called 3-phytases according to the D-configuration (Konietzny and Greiner, 2002). In principle, all enzymes belonging to this group can be active in the digestive tract of animals and contribute to $InsP_6$ degradation. By investigating the concentration of degradation products in the digestive tract it is possible to gain insight into the contribution that different enzymes have to the stepwise degradation of $InsP_6$.

However, and in contrast to $InsP_6$, conclusions from the measurement of lower $InsP_x$ are not of quantitative nature and disappearance rates cannot be calculated. First, the concentration of a specific degradation product results as an equilibrium or two dephosphorylation steps (the one that generates $InsP_x$ from $InsP_{x+1}$ and the one that generates $InsP_{x-1}$ from $InsP_x$). Second, the different phytases mentioned before are not unique in their degradation pathways. While they have a specific priority relative to the initial phosphate group to attack and the degradation pathway, they also have minor pathways leading to co-appearance of other $InsP_5$ isomers. Hence, simple configuration of phytases based on the initial attack of the susceptible phospho-ester bond might be hasty (Greiner *et al.*, 2000). However, looking at dominant and less dominant isomers in the digestive tract gives indication of differences in enzymes involved.

1.7.1 $InsP_5$ isomers

Only very few studies have determined the appearance of $InsP_x$ isomers in the pig. In one study, Ins(1,2,3,4,5)P_5 was the dominant $InsP_5$ isomer in the stomach and small intestine when a diet with intrinsic plant phytase activity was fed (Figure 1.5; Schlemmer *et al.*, 2001). Upon destruction of this phytase by extrusion of the diet this $InsP_5$ isomer was no longer dominating. The $InsP_5$ pattern determined in the colon was only slightly affected by extrusion of the diet.

In broilers, upon feeding a maize-soybean meal-diet without a P supplement, Ins(1,2,4,5,6)P_5 and Ins(1,2,3,4,5)P_5 were the dominant $InsP_5$ isomers analysed in the crop (Figure 1.6). Passage through the digestive tract led to a shift towards a decrease in the proportion of Ins(1,2,4,5,6)P_5 and increase in Ins(1,2,3,4,5)P_5, which points towards a relatively high relevance of 6-phytases in gizzard and small intestine. Whether this 6-phytases

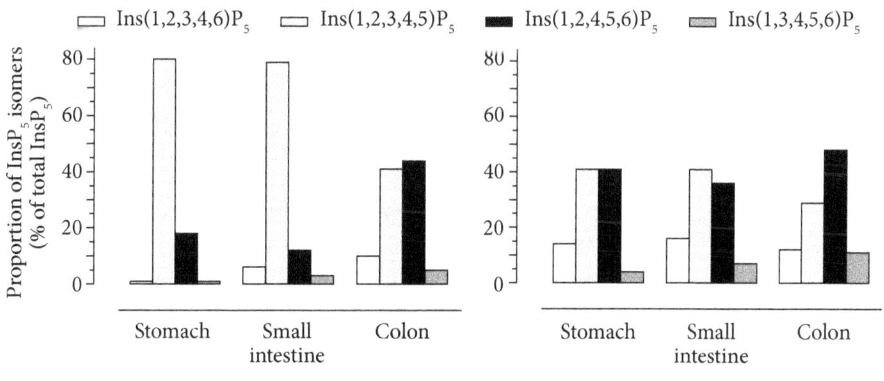

Figure 1.5. *Differences in the pattern of $InsP_5$ isomers (% of total $InsP_5$) measured in different sections of the digestive tract of pigs after feeding a cereal grain-soybean meal diet that was not extruded (left panel) or extruded (right panel) to inactivate intrinsic plant phytase activity (Schlemmer* et al.*, 2001).*

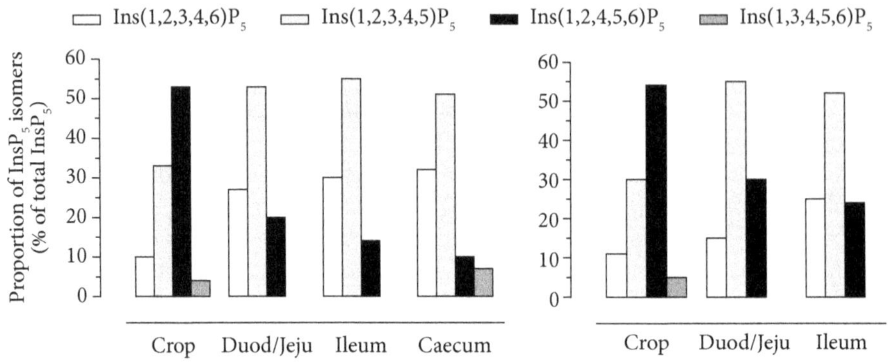

Figure 1.6. *Pattern of* $InsP_5$ *isomers detected in different sections of the gastrointestinal tract after feeding low-P maize-soybean meal diets without added phytase in two experiments with broiler chickens (Zeller* et al., *2015b,c, 2016).*

are produced by the microbiota or the birds' mucosa cannot yet be distinguished, but lactobacilli likewise play a role as supported by Borda-Molina *et al.* (2016). Ins(1,2,3,4,6) P_5 also appeared in the intestine, which was unexpected because a 5-phytase has been described only for lily pollen, *Bifidobacterium pseudocatenulatum* and *S. ruminantium* ssp. *lactilytica* (Haros *et al.*, 2009; Puhl *et al.*, 2008). However, recent studies found *B. pseudocatenulatum* in the small intestine of chickens and human faeces (Alegría *et al.*, 2014; Ruiz *et al.*, 2015).

Upon supplementation of MCP to a low-P basal diet, proportions of Ins(1,2,3,4,5)P_5 decreased and Ins(1,2,4,5,6)P_5 increased in duodenum/jejunum of broilers, while the effect was absent or in the opposite direction in the terminal ileum (Zeller *et al.*, 2015c). In another broiler study MCP inclusion in the diet led to an increase in the concentration of total $InsP_5$ in the terminal ileum, with the biggest relative increase found for Ins(1,2,4,5,6) P_5 (Shastak *et al.*, 2014). This demonstrates that MCP modulates formation or degradation (or both) of $InsP_5$ isomers due to differently affecting specific phytase activity.

When microbial phytase is added to the diet, not only $InsP_6$ disappearance but also the $InsP_5$ pattern is strongly affected. In pigs, the inclusion of 150 U/kg diet of a microbial 3-phytase to a maize-soybean meal diet increased the formation of Ins(1,2,4,5,6)P_5 in the duodenum and ileum whereas the presence of other $InsP_5$ was reduced (Kemme *et al.*, 2006). In broilers, effects of added phytase were already detectable in the crop. Supplementation of 3- and 6-phytase resulted in clear dominance of either Ins(1,2,4,5,6) P_5 or Ins(1,2,3,4,5)P_5, respectively, in the crop (Zeller *et al.*, 2015b,c), and these differences continued to exist until the end of ileum. These authors showed for the first time that the main degradation pathways of 3- and 6-phytases known before only from *in vitro* studies can be found also in the complex environment of the digestive tract of poultry.

1.7.2 $InsP_4$ isomers

Among inositol tetrakisphosphates ($InsP_4$) in the digestive tract, Ins(1,2,5,6)P_4 and Ins(1,2,3,4)P_4 were detectable in pigs and broilers. Their relative contributions varied. For pigs fed an extruded diet with very low intrinsic phytase activity and no added phytase, only about 1% of total InsP were present as Ins(1,2,5,6)P_4 and Ins(1,2,3,4)P_4 in the stomach, small intestine and colon (Schlemmer *et al.*, 2001). However, feeding the non-extruded diet that contained intrinsic phytase activity, concentrations of Ins(1,2,3,4)P_4 and Ins(1,2,5,6)P_4 amounted to 6 and 17% of total InsP, respectively, in stomach and small intestine. The supplementation of a 3-phytase increased the concentrations of $InsP_4$ in the duodenum and ileum of pigs when compared to the un-supplemented diet (Kemme *et al.*, 2006). In broilers, upon feeding diets without added phytase, Ins(1,2,5,6)P_4 was present in crop but not in the small intestine (Zeller *et al.*, 2015b,c, 2016). Ins(1,2,3,4)P_4 instead was found in the duodenum/jejunum and ileum. Supplements of 3- and 6-phytase caused a characteristic pattern of $InsP_4$, similar to $InsP_5$. Supplementation of an *E. coli*-derived phytase caused a high increase of Ins(1,2,5,6)P_4 in the crop and proventriculus/gizzard, and this isomer also appeared in the small intestine. Accumulation of this isomer upon *E. coli*-derived phytase supplementation and slower hydrolysis to $InsP_3$ confirmed similar findings obtained *in vitro* and using phytases from *E. coli* (Greiner *et al.*, 1993). Instead, supplementation of an *A. niger* phytase did not cause a big accumulation of $InsP_4$ but rather an accumulation of $InsP_3$ isomers in the crop (Zeller *et al.*, 2015b).

1.7.3 *Myo*-inositol release

Upon addition of phytase at 500 and 1000 U/kg to a maize-soybean meal diet the concentration of *myo*-inositol in the gizzard content of broilers was found to be significantly increased (Walk *et al.*, 2014). In line with this result, we found a significantly increased *myo*-inositol concentration following phytase addition also in the ileum of broilers, but this phytase effect was partly compensated by simultaneous supplementation of a mineral phosphate (Sommerfeld *et al.*, 2016). Significantly increased *myo*-inositol concentrations were also found in the peripheral blood plasma of broilers with phytase dosages from 1000-3,000 U/kg of diet (Cowieson *et al.*, 2015). Hence, complete and fast destruction of $InsP_6$ in principle is possible in the digestive tract of broilers when diets are supplemented with phytase. The quantitative relevance for the supply of metabolisable *myo*-inositol can yet not be calculated.

1.8 Conclusions

1. Better understanding of the $InsP_6$ degradation process in the digestive tract and factors that influence the degradation process is the key towards a more sustainable use of P sources and phytases in the pig and poultry industry.
2. Big differences exist between pigs and poultry in the extent of $InsP_6$ degradation from intrinsic plant-based, endogenous and microbial phytases.
3. The potential to degrade $InsP_6$ is high in broiler chickens when challenged by low P and Ca concentration of the diet.

4. A better understanding of the quantitative contribution of endogenous phytases produced by microbes in different feeding and management conditions is required.

References

Alegría, Á., Delgado, S., Guadamuro, L., Flórez, A.B., Felis, G.E., Torriani, S. and Mayo, B., 2014. The genome of *Bifidobacterium pseudocatenulatum* IPLA 36007, a human intestinal strain with isoflavone-activation activity. Gut Pathogens 6:1-6.

Angel, R., Dalloul, R.A. and Doerr, J., 2005. Performance of broiler chickens fed diets supplemented with a direct-fed microbial. Poultry Science 84:1222-1231.

Applegate, T.J., Angel, R. and Classen, H.L., 2003. Effect of dietary calcium, 25-hydroxycholecalciferol, or bird strain on small intestinal phytase activity in broiler chickens. Poultry Science 82:1140-1148.

Barrier-Guillot, B., Casado, P., Maupetit, P., Jondreville, C., Gatel, F. and Larbier, M., 1996. Wheat phosphorus availability: 2 – *In vivo* study in broilers and pigs; relationship with endogenous phytasic activity and phytic phosphorus content in wheat. Journal of the Science of Food and Agriculture 70:69-74.

Baumgärtel, T., Metzler, B.U., Mosenthin, R., Greiner, R. and Rodehutscord, M., 2008. Precaecal and postileal metabolism of P, Ca and N in pigs as affected by different carbohydrate sources fed at low level of P intake. Archives of Animal Nutrition 62:169-181.

Blaabjerg, K., Jørgensen, H., Tauson, A.-H. and Poulsen, H.D., 2011. The presence of inositol phosphates in gastric pig digesta is affected by time after feeding a nonfermented or fermented liquid wheat- and barley-based diet. Journal of Animal Science 89:3153-3162.

Blaabjerg, K., Jørgensen, H., Tauson, A.H. and Poulsen, H.D., 2010. Heat-treatment, phytase and fermented liquid feeding affect the presence of inositol phosphates in ileal digesta and phosphorus digestibility in pigs fed a wheat and barley diet. Animal 4:876-885.

Borda-Molina, D., Vital, M., Sommerfeld, V., Rodehutscord, M. and Camarinha-Silva, A., 2016. Spatial variation in the gut microbiota of broiler chickens fed diets supplemented or not with calcium, phosphorus and phytase. Proceedings of the 10th Joint Symposium INRA-ROWETT: Gut microbiology. Clermont-Ferrand, France (abstr.)

Cordell, D., Drangert, J.-O. and White, S., 2009. The story of phosphorus: global food security and food for thought. Global Environmental Change 19:292-305.

Cowieson, A.J., Aureli, R., Guggenbuhl, P. and Fru-Nji, F., 2015. Possible involvement of *myo*-inositol in the physiological response of broilers to high doses of microbial phytase. Animal Production Science 55:710-719.

Dersjant-Li, Y., Awati, A., Schulze, H. and Partridge, G., 2015. Phytase in non-ruminant animal nutrition: a critical review on phytase activities in the gastrointestinal tract and influencing factors. Journal of the Science of Food and Agriculture 95:878-896.

Düngelhoef, M., Rodehutscord, M., Spiekers, H. and Pfeffer, E., 1994. Effects of supplemental microbial phytase on availability of phosphorus contained in maize, wheat and triticale to pigs. Animal Feed Science and Technology 49:1-10.

Eeckhout, W. and De Paepe, M., 1994. Total phosphorus, phytate-phosphorus and phytase activity in plant feedstuffs. Animal Feed Science and Technology 47:19-29.

GfE, 1994. Mitteilungen des Ausschusses für Bedarfsnormen der Gesellschaft für Ernährungsphysiologie: Die Bestimmung des verdaulichen Phosphors beim Schwein. Proc. Soc. Nutr. Physiol. 2:113-119.

GfE, 1999. Energie- und Nährstoffbedarf landwirtschaftlicher Nutztiere. 7. Empfehlungen zur Energie- und Nährstoffversorgung der Legehennen und Masthühner (Broiler). DLG-Verlag, Frankfurt am Main, Germany.

GfE, 2006. Empfehlungen zur Energie- und Nährstoffversorgung von Schweinen. DLG-Verlag, Frankfurt am Main, Germany.

Greiner, R., Carlsson, N.-G. and Alminger, M.L., 2000. Stereospecificity of *myo*-inositol hexakisphosphate dephosphorylation by a phytate-degrading enzyme of *Escherichia coli*. Journal of Biotechnology 84:53-62.

Greiner, R., Konietzny, U. and Jany, K.D., 1993. Purification and characterization of two phytases from *Escherichia coli*. Archives of Biochemistry and Biophysics 303:107-113.

Gross, M., 2010. Fears over phosphorus supplies. Current Biology 20:R386.

Haros, M., Carlsson, N.-G., Almgren, A., Larsson-Alminger, M., Sandberg, A.-S. and Andlid, T., 2009. Phytate degradation by human gut isolated *Bifidobacterium pseudocatenulatum* ATCC27919 and its probiotic potential. International Journal of Food Microbiology 135:7-14.

Hovenjürgen, M., Rodehutscord, M. and Pfeffer, E., 2003. Effect of fertilization and variety on digestibility of phosphorus from plant feedstuffs in pigs. Journal of Animal and Feed Sciences 12:83-93.

Hu, H.L., Wise, A. and Henderson, C., 1996. Hydrolysis of phytate and inositol tri-, tetra-, and penta-phosphates by the intestinal mucosa of the pig. Nutrition Research 16:781-787.

Huber, K., 2016. Cellular *myo*-inositol metabolism. In: Walk, C.L., Kühn, I., Stein, H.H., Kidd, M.T. and Rodehutscord, M. (eds.) Phytate destruction – consequences for precision animal nutrition. Wageningen Academic Publishers, Wageningen, the Netherlands, pp. 53-60.

Huber, K., Zeller, E. and Rodehutscord, M., 2015. Modulation of small intestinal phosphate transporter by dietary supplements of mineral phosphorus and phytase in broilers. Poultry Science 94:1009-1017.

Humer, E., Schwarz, C. and Schedle, K., 2015. Phytate in pig and poultry nutrition. Journal of Animal Physiology and Animal Nutrition 99:605-625.

IUPAC [Commission on the Nomenclature of Organic Chemistry (CNOC) and IUPAC-IUB Commission on Biochemical Nomenclature (CBN)], 1976. Nomenclature of cyclitols. Recommendations, 1973. Biochemical Journal 153:23-31.

Jongbloed, A.W., Mroz, Z. and Kemme, P.A., 1992. The effect of supplementary *Aspergillus niger* phytase in diets for pigs on concentration and apparent digestibility of dry matter, total phosphorus, and phytic acid in different sections of the alimentary tract. Journal of Animal Science 70:1159-1168.

Juanpere, J., Pérez-Vendrell, A.M. and Brufau, J., 2004. Effect of microbial phytase on broilers fed barley-based diets in the presence or not of endogenous phytase. Animal Feed Science and Technology 115:265-279.

Kemme, P.A., Jongbloed, A.W., Mroz, Z., Kogut, J. and Beynen, A.C., 1999. Digestibility of nutrients in growing-finishing pigs is affected by *Aspergillus niger* phytase, phytate and lactic acid levels. 2. Apparent total tract digestibility of phosphorus, calcium and magnesium and ileal degradation of phytic acid. Livestock Production Science 58:119-127.

Kemme, P.A., Schlemmer, U., Mroz, Z. and Jongbloed, A.W., 2006. Monitoring the stepwise phytate degradation in the upper gastrointestinal tract of pigs. Journal of the Science of Food and Agriculture 86:612-622.

Kerr, M.J., Classen, H.L. and Newkirk, R.W., 2000. The effects of gastrointestinal tract micro-flora and dietary phytase on inositol hexaphosphate hydrolysis in the chicken. Poultry Science 79 (Supplement 1):11 (abs.).

Konietzny, U. and Greiner, R., 2002. Molecular and catalytic properties of phytate-degrading enzymes (phytases). International Journal of Food Science & Technology 37:791-812.

Leytem, A.B., Willing, B.P. and Thacker, P.A., 2008. Phytate utilization and phosphorus excretion by broiler chickens fed diets containing cereal grains varying in phytate and phytase content. Animal Feed Science and Technology 146:160-168.

Maenz, D.D. and Classen, H.L., 1998. Phytase activity in the small intestinal brush border membrane of the chicken. Poultry Science 77:557-563.

Marounek, M., Skřivan, M., Dlouhá, G. and Břeňová, N., 2008. Availability of phytate phosphorus and endogenous phytase activity in the digestive tract of laying hens 20 and 47 weeks old. Animal Feed Science and Technology 146:353-359.

Marounek, M., Skřivan, M., Rosero, O. and Rop, O., 2010. Intestinal and total tract phytate digestibility and phytase activity in the digestive tract of hens fed a wheat-maize-soyabean diet. Journal of Animal and Feed Sciences 19:430-439.

Mitchell, R.D. and Edwards Jr., H.M., 1996. Effect of phytase and 1,25-dihydroxycholecalciferol on phytate utilization and the quantitative requirement for calcium and phosphorus in young broiler chickens. Poultry Science 75:95-110.

Mohammed, A., Gibney, M.J. and Taylor, T.G., 1991. The effects of dietary levels of inorganic phosphorus, calcium and cholecalciferol on the digestibility of phytate-P by the chick. British Journal of Nutrition 66:251-259.

Mullaney, E.J. and Ullah, A.H.J., 2003. The term phytase comprises several different classes of enzymes. Biochemical and Biophysical Research Communications 312:179-184.

Mullaney, E.J. and Ullah, A.H.J., 2007. Phytases: attributes, catalytic mechanisms and applications. In: Turner, B.L., Richardson, A.E. and Mullaney, E.J. (eds.) Inositol phosphates linking agriculture and the environment. CABI Publishing, Wallingford, UK, pp. 97-110.

Neset, T.-S.S. and Cordell, D., 2012. Global phosphorus scarcity: identifying synergies for a sustainable future. Journal of the Science of Food and Agriculture 92:2-6.

NRC, 1994. Nutrient requirements of poultry. National Academy Press, Washington, D.C., USA.

NRC, 2012. Nutrient requirements of swine. National Academy Press, Washington, D.C., USA.

Oloffs, K., Cossa, J. and Jeroch, H., 2000. Die Bedeutung der korneigenen (nativen) Phytaseaktivität im Weizen für die Phosphor-Verwertung bei Broilern und Legehennen. Archiv für Geflügelkunde 64:157-161.

Oloffs, K., Dolbusin, A. and Jeroch, H., 1998. Einfluß von mikrobieller und nativer Weizenphytase auf die Phosphor-Verwertung bei Broilern. Archiv für Geflügelkunde 62:260-263.

Onyango, E.M., Asem, E.K. and Adeola, O., 2006. Dietary cholecalciferol and phosphorus influence intestinal mucosa phytase activity in broiler chicks. British Poultry Science 47:632-639.

Pagano, A.R., Roneker, K.R. and Lei, X.G., 2007. Distribution of supplemental *Escherichia coli* AppA2 phytase activity in digesta of various gastrointestinal segments of young pigs. Journal of Animal Science 85:1444-1452.

Pontoppidan, K., Pettersson, D. and Sandberg, A.-S., 2007. The type of thermal feed treatment influences the inositol phosphate composition. Animal Feed Science and Technology 132:137-147.

Puhl, A.A., Greiner, R. and Selinger, L.B., 2008. A protein tyrosine phosphatase-like inositol polyphosphatase from *Selenomonas ruminantium* subsp. *lactilytica* has specificity for the 5-phosphate of *myo*-inositol hexa*kis*phosphate. The International Journal of Biochemistry & Cell Biology 40:2053-2064.

Raghavendra, P. and Halami, P.M., 2009. Screening, selection and characterization of phytic acid degrading lactic acid bacteria from chicken intestine. International Journal of Food Microbiology 133:129-134.

Rapp, C., Lantzsch, H.-J. and Drochner, W., 2001. Hydrolysis of phytic acid by intrinsic plant or supplemented microbial phytase (*Aspergillus niger*) in the stomach and small intestine of minipigs fitted with re-entrant cannulas. 3. Hydrolysis of phytic acid (IP_6) and occurence of hydrolysis products (IP_5, IP_4, IP_3 and IP_2). Journal of Animal Physiology and Animal Nutrition 85:420-430.

Ravindran, V., Ravindran, G. and Sivalogan, S., 1994. Total and phytate phosphorus contents of various foods and feedstuffs of plant origin. Food Chemistry 50:133-136.

Rehman, H.U., Vahjen, W., Awad, W.A. and Zentek, J., 2007. Indigenous bacteria and bacterial metabolic products in the gastrointestinal tract of broiler chickens. Archives of Animal Nutrition 61:319-335.

Rodehutscord, M., 2016. Interactions between minerals and phytate degradation in poultry – challenges for phosphorus digestibility assays. In: Walk, C.L., Kühn, I., Stein, H.H., Kidd, M.T. and Rodehutscord, M. (eds.) Phytate destruction – consequences for precision animal nutrition. Wageningen Academic Publishers, Wageningen, the Netherlands, pp. 167-177.

Rodehutscord, M., 2008. Ansatzpunkte zur Schonung der begrenzten Phosphorressourcen. Archiv für Tierzucht 51:39-48.

Rodehutscord, M., Faust, M. and Lorenz, H., 1996. Digestibility of phosphorus contained in soybean meal, barley, and different varieties of wheat, without and with supplemental phytase fed to pigs and additivity of digestibility in a wheat-soybean-meal diet. Journal of Animal Physiology and Animal Nutrition 75:40-48.

Rodehutscord, M., Rückert, C., Maurer, H.P., Schenkel, H., Schipprack, W., Bach Knudsen, B.E., Schollenberger, M., Laux, M., Eklund, M., Siegert, W. and Mosenthin, R., 2016. Variation in chemical composition and physical characteristics of cereal grains from different genotypes. Archives of Animal Nutrition 70:87-107.

Rostagno, H.S. (Ed.), 2011. Brazilian tables for poultry and swine: composition of feedstuffs and nutritional requirements. 3rd edition. Universidade Federal de Vicosa, Departamento de Zootecnia, Brazil.

Ruiz, R., Peinado, M.J., Aranda-Olmedo, I., Abecia, L., Suárez-Pereira, E., Ortiz Mellet, C., García Fernández, J.M. and Rubio, L.A., 2015. Effects of feed additives on ileal mucosa-associated microbiota composition of broiler chickens. Journal of Animal Science 93:3410-3420.

Rutherfurd, S.M., Chung, T.K. and Moughan, P.J., 2014. Effect of microbial phytase on phytate P degradation and apparent digestibility of total P and Ca throughout the gastrointestinal tract of the growing pig. Journal of Animal Science 92:189-197.

Sandberg, A.-S. and Andlid, T., 2002. Phytogenic and microbial phytases in human nutrition. International Journal of Food Science & Technology 37:823-833.

Sandberg, A.-S., Larsen, T. and Sandström, B., 1993. High dietary calcium level decreases colonic phytate degradation in pigs fed a rapeseed diet. Journal of Nutrition 123:559-566.

Schlemmer, U., Frølich, W., Prieto, R.M. and Grases, F., 2009. Phytate in foods and significance for humans: Food sources, intake, processing, bioavailability, protective role and analysis. Molecular Nutrition & Food Research 53:S330-S375.

Schlemmer, U., Jany, K.-D., Berk, A., Schulz, E. and Rechkemmer, G., 2001. Degradation of phytate in the gut of pigs – pathway of gastro-intestinal inositol phosphate hydrolysis and enzymes involved. Archives of Animal Nutrition 55:255-280.

Selle, P.H. and Ravindran, V., 2008. Phytate-degrading enzymes in pig nutrition. Livestock Science 113:99-122.

Seynaeve, M., Janssens, G., Hesta, M., Nevel, C.V. and Wilde, R.O.D., 2000. Effects of dietary Ca/P ratio, P level and microbial phytase supplementation on nutrient digestibilities in growing pigs: breakdown of phytic acid, partition of P and phytase activity along the intestinal tract. Journal of Animal Physiology and Animal Nutrition 83:193-204.

Shastak, Y., Zeller, E., Witzig, M., Schollenberger, M. and Rodehutscord, M., 2014. Effects of the composition of the basal diet on the evaluation of mineral phosphorus sources and interactions with phytate hydrolysis in broilers. Poultry Science 93:2548-2559.

Skoglund, E., Larsen, T. and Sandberg, A.-S., 1997. Comparison between steeping and pelleting a mixed diet at different calcium levels on phytate degradation in pigs. Canadian Journal of Animal Science 77:471-477.

Sommerfeld, V., Schollenberger, M., Kühn, I. and Rodehutscord, M., 2016. Interactive effects of P, Ca, and phytase supplements on $InsP_6$ degradation and *myo*-inositol release in broiler chickens. Proc. PSA Annual Meeting, New Orleans, USA, July 11-14, 2016. Available at: http://tinyurl.com/godm46u.

Son, J.H., Ragland, D. and Adeola, O., 2002. Quantification of digesta flow into the caeca. British Poultry Science 43:322-324.

Stentz, R., Osborne, S., Horn, N., Li, A.W.H., Hautefort, I., Bongaerts, R., Rouyer, M., Bailey, P., Shears, S.B., Hemmings, A.M., Brearley, C.A. and Carding, S.R., 2014. A bacterial homolog of a eukaryotic inositol phosphate signaling enzyme mediates cross-kingdom dialog in the mammalian gut. Cell Reports 6:646-656.

Svihus, B., Hetland, H., Choct, M. and Sundby, F., 2002. Passage rate through the anterior digestive tract of broiler chickens fed on diets with ground and whole wheat. British Poultry Science 43:662-668.

Tamayo-Ramos, J.A., Sanz-Penella, J.M., Yebra, M.J., Monedero, V. and Haros, M., 2012. Novel phytases from *Bifidobacterium pseudocatenulatum* ATCC 27919 and *Bifidobacterium longum* subsp. *infantis* ATCC 15697. Applied and Environmental Microbiology 78:5013-5015.

Tamim, N.M. and Angel, R., 2003. Phytate phosphorus hydrolysis as influenced by dietary calcium and micro-mineral source in broiler diets. Journal of Agricultural and Food Chemistry 51:4687-4693.

Tamim, N.M., Angel, R. and Christman, M., 2004. Influence of dietary calcium and phytase on phytate phosphorus hydrolysis in broiler chickens. Poultry Science 83:1358-1367.

Vats, P. and Banerjee, U.C., 2004. Production studies and catalytic properties of phytases (*myo*-inositolhexakisphosphate phosphohydrolases): an overview. Enzyme and Microbial Technology 35:3-14.

Walk, C.L., Santos, T.T. and Bedford, M.R., 2014. Influence of superdoses of a novel microbial phytase on growth performance, tibia ash, and gizzard phytate and inositol in young broilers. Poultry Science 93:1172-1177.

Witzig, M., Camarinha-Silva, A., Green-Engert, R., Hoelzle, K., Zeller, E., Seifert, J., Hoelzle, L.E. and Rodehutscord, M., 2015. Spatial variation of the gut microbiota in broiler chickens as affected by dietary available phosphorus and assessed by T-RFLP analysis and 454 pyrosequencing. PLoS ONE 10:e0143442.

Yi, Z. and Kornegay, E.T., 1996. Sites of phytase activity in the gastrointestinal tract of young pigs. Animal Feed Science and Technology 61:361-368.

Zeller, E., Schollenberger, M., Kühn, I. and Rodehutscord, M., 2015a. Effect of diets containing enzyme supplements and microwave-treated or untreated wheat on inositol phosphates in the small intestine of broilers. Animal Feed Science and Technology 204:42-51.

Zeller, E., Schollenberger, M., Kühn, I. and Rodehutscord, M., 2015b. Hydrolysis of phytate and formation of inositol phosphate isomers without or with supplemented phytases in different segments of the digestive tract of broilers. Journal of Nutritional Science 4:e1.

Zeller, E., Schollenberger, M., Kühn, I. and Rodehutscord, M., 2016. Dietary effects on inositol phosphate breakdown in the crop of broilers. Archives of Animal Nutrition 70:57-71.

Zeller, E., Schollenberger, M., Witzig, M., Shastak, Y., Kühn, I., Hoelzle, L.E. and Rodehutscord, M., 2015c. Interactions between supplemented mineral phosphorus and phytase on phytate hydrolysis and inositol phosphates in the small intestine of broilers. Poultry Science 94:1018-1029.

Zeng, Z.K., Wang, D., Piao, X.S., Li, P.F., Zhang, H.Y., Shi, C.X. and Yu, S.K., 2014. Effects of adding super dose phytase to the phosphorus-deficient diets of young pigs on growth performance, bone quality, minerals and amino acids digestibilities. Asian-Australasian Journal of Animal Science 27:237-246.

Chapter 2
Relevance of gut microbiome and endogenous phosphatases on phytate breakdown in poultry – implications on phosphorus utilisation

D. Józefiak[1*], *R.M. Engberg*[2] *and A. Ptak*[3]
[1]*Department of Animal Nutrition and Feed Management, Poznan University of Life Sciences, Wołyńska 33, 60-637 Poznań, Poland;* [2]*Department of Animal Science, Aarhus University, Blichers Allé 20, 8830 Tjele, Denmark;* [3]*PIAST GROUP, Research and Development Center, Lewkowiec 50A, 63-400 Ostrów Wielkopolski, Poland; damjo@up.poznan.pl*

Abstract

Gastrointestinal microbiota of the poultry is still relatively unexplored area. Most of the research has been conducted on broiler chickens with very few studies showing phytate-microbiome interactions. Incorporation of novel techniques in microbiology particularly, pyrosequencing should clarify our knowledge about the potential role of the microbiota in phytate breakdown. Different microbiota populations colonising the poultry gastrointestinal tract (GIT) may metabolise nutrients that are unavailable for the host-bird, including phytate. The end products of microbial fermentation are usually short-chain fatty acids ethanol and different gasses. For today it is clear that all major groups of the gastrointestinal bacteria belonging to *Clostridium coccoides* cluster, *Clostridium leptum* cluster, *Bacteroides-Prevotella-Porphyrmonas group*, *Enterobacteriaceae* and *Lactobacillus* sp. are indirectly or directly affected by the presence of the phytate as well as phytase supplementation. However, it is not clear which bacteria play a key role in phytate degradation and how significant for microbiota is dietary phytase, even though proliferation of some important pathogens as *Clostridium perfringens* seem to inhibited. In conclusion, poultry GIT microbiota can utilize 10-25% of dietary phytate. Phytase supplementation significantly affects the microbiome but the mode of action is mainly indirect, through changes in the buffering capacity of the digesta and release of IP_6 isomers as well as *myo*-inositol.

Keywords: microbiota, phytase, phytic acid, gastrointestinal tract

2.1 Microbiota of the poultry gastrointestinal tract

With respect to the composition of the gastrointestinal bacterial community, the most information is available from broiler chickens (Stanley *et al.*, 2014). The introduction of culture independent molecular techniques based on 16S rRNA gene sequencing and high-throughput techniques allow for the handling of large sample numbers for the identification of gastrointestinal tract (GIT) associated bacteria with a taxonomic

DOI 10.3920/978-90-8686-836-0_2, © Wageningen Academic Publishers 2016

resolution (species, genus, family, etc.), depending on size (read length) and quality of the obtained sequences (Corrigan *et al.*, 2012; Gong *et al.*, 2002a,b, 2007; Lu *et al.*, 2003a,b, 2008; Singh *et al.*, 2012; Stanley *et al.*, 2012, 2013; Torok *et al.*, 2009, 2013). In the near future, we can expect an increasing number of metagenomic and RNA based studies, which will provide further insight into the function of the avian microbiota (Stanley *et al.*, 2012, 2013). In crop contents, the dominant bacteria belong to the genus *Lactobacillus*, and the species *L. salivarius, L. acidophilus, L. reuteri, L. johnsonii, L. crispatus, L. gallinarum, L. amylovorus,* and *L. gasseri,* have been identified in the broiler chicken (Guan *et al.*, 2003). The dominant bacteria in the crop of ducks and turkeys are also lactic acid bacteria (Kierończyk *et al.*, in press; Kurzak *et al.*, 1998). In the small intestine of broiler chickens, *Lactobacilli* are dominant as well, and may account for almost 70% of the microbiota, based on sequencing data (Bjerrum *et al.*, 2006; Gong *et al.*, 2007; Lu *et al.*, 2003a; Van Der Hoeven-Hangoor *et al.*, 2013). Other bacteria, belonging to the *Clostridiaceae* (11%), *Streptococcus* (6.5%), and *Enterococcus* (6.5%), are present in the ileum (Lu *et al.*, 2003a). In turkeys, lactic acid bacteria, including *Lactobacilli* and *Enterococci*, are dominant in the small intestine (Naqi *et al.*, 1970).

As compared to the small intestine, the microbial community in the caeca of chickens is much more diverse, showing high variability both at flock and individual levels (Stanley *et al.*, 2013). At this location, the *Clostridiaceae* are the most abundant bacteria accounting for approximately 65% of the microbiota (sequence-based data). Other frequently detected species are those related to *Fusobacterium* (14%), *Lactobacillus* (8%) and *Bacteroides* (5%). Members of *Bacillus, Streptococaceae, Enterococcaceae, Actinobacteria, Proteobacteria and Flavobacteriaceas* typically account for smaller amounts of the caecal microbiota (Józefiak *et al.*, 2010b; Lu *et al.*, 2003a). Besides bacteria, the chicken caeca also harbour methanogenic Archaea, (Saengkerdsub *et al.*, 2007). Further, molecular identification and gastrointestinal metagenomic analyses revealed the presence of various fungi, protozoa and viruses in the chicken GIT (Hume *et al.*, 2012). In the caecal content of geese, *Clostridiales, Bacteroidetes* as well as *Erysipelotrichia* have been demonstrated to be highly abundant (Liu *et al.*, 2011). In ducks likewise, *Clostridiales* and *Bacteroidetes* make up 90% of the caecal microbiota (Vasaï *et al.*, 2014). The bacterial community structure of caecal content from turkeys (Corrigan *et al.*, 2012) reveals the phylum *Firmicutes*, including *Clostridiaceae* and *Ruminococcaceae*, to be dominant (more than 57% of all sequences) and sequences related to *Bacillus* and *Lactobacillus* are also frequently identified. The second most abundant phylum is the *Bacteroidetes*, including *Bacteroides* spp, *Alistepes* spp. and *Barnsiella* spp., whereas *Proteobacteria*, including *Escherichia*, is the least abundant phylum in caecal contents of turkeys. Generally, bacterial numbers are highest in crop and caeca. Due to the acidic environment in the proventriculus and gizzard, bacteria numbers are reduced compared with other parts of the gastro-intestinal tract. Lactic acid bacteria are dominant, particularly in crop, gizzard and small intestine. Also in caecal content, lactic acid bacteria are present in high concentrations (log 9/g content). However, the numbers of total anaerobic bacteria other than lactic acid bacteria are highest in the ceca. The numbers of *Escherichia coli* increase along with the digestive tract with highest counts in caecal content (>log 8/g faeces). The *Lactobacillus* numbers in crop contents are in the range of 10^8-10^9 (Rehman *et al.*, 2007).

Although new molecular methods and tools for metagenomic approaches are available now, our understanding of the complex function of the intestinal microorganisms in relation to dietary conditions is sparse. To obtain detailed information about microbial function in terms of preferred substrates, metabolic pathways and possible beneficial properties, which may be exploited commercially for detoxification of mycotoxins (Yu *et al.*, 2010), culturing and isolation of intestinal microorganisms remains an important research objective. The GIT microbiome should be considered a complete micro-ecosystem including all intestinal segments from the crop to the rectum. In most of the work conducted on chickens and other poultry species, only the lower GIT segments have been considered. This incomplete picture can cause wrong perceptions of the micro-ecology and functions of the poultry GIT. To summarise, the available information on the composition of the microbiota at various locations of the poultry GIT indicates that the upper part, including crop, gizzard and small intestine, is colonised dominantly by lactic acid bacteria. In the caeca, the microbiota is much more diverse and dominated by strictly anaerobe bacteria with a variety of different fermentation properties.

2.2 Phytase, phytate and microbiota

Poultry diets are primarily based on raw materials of plant origin, which contain high quantities of phytic acid. The unique structure of this anti-nutritional factor determines its ability to chelate cations such as calcium (Ca), magnesium, zinc, copper, iron, potassium (K), and to form insoluble salts (Selle and Ravindran, 2007, 2008). In addition, phytate may increase endogenous losses of sodium (Na), and cause a sodium deficiency, which may impact the activity of Na–K-ATPases in the GIT that are involved in absorption of nutrients (Dersjant-Li *et al.*, 2015; Liu *et al.*, 2008). Indeed, ingestion of phytic acid by broiler chickens and pigs may reduce the activity of Na-K-ATPase in the GIT (Dersjant-Li *et al.*, 2015; Liu *et al.*, 2008). Furthermore, lipid and Ca-phytate may be involved in the formation of metallic soaps in the gut lumen, resulting in reduced digestion of saturated fats (Ravindran *et al.*, 2000; Zaefarian *et al.*, 2013). Therefore, presence of phytate in the GIT may impair the availability of many nutrients for the host, and simultaneously increase fermentable substrates for GIT microbiota. Moreover, phytate hydrolysis through exogenous phytase supplementation may induce changes in the intestinal microbiome, particularly in the lower parts of the GIT.

Different microbiota populations colonising the poultry GIT may metabolise nutrients that are unavailable for the host-bird, including phytate. The end products of microbial fermentation are usually short-chain fatty acids (SCFAs) ethanol and different gasses (Józefiak *et al.*; 2004). It is suggested that in pigs, endogenous phytase activity in the lower GIT does not improve phosphorus (P) utilisation because although phytate bound P may be released in the large intestine, no P is absorbed in this part of the intestinal tract, and the released P is, therefore, excreted in the faeces (Dersjant-Li *et al.*, 2015). While phytate creates many different complexes with different nutrients, it cannot be excluded that as a consequence of phytate hydrolysis, other fermentation end products may be generated. This may affect intestinal morphology, digestion and absorption processes. Moreover, the gut microbiome plays an important role in immune responses and creation of protective barriers against colonisation of pathogens. However, no information about possible interactions between phytate concentration and microbiome responses is available.

In both poultry and pigs, the main site of endogenous phytase activity is the lower parts of the GIT, i.e. caeca and colon, respectively (Dersjant-Li *et al.*, 2015). Results of experiments conducted with laying hens and broiler chickens (Table 2.1 and 2.2) indicate that there is low activity of endogenous phytase in the poultry GIT, demonstrating that up to 25% of P can be utilized without exogenous phytase supplementation.

Addition of exogenous phytases to diets for poultry improves performance, mineral retention, AMEn, and amino acid digestibility (Augspurger *et al.*, 2003; Cowieson and Adeola, 2005; Newkirk and Classen, 2001; Ravindran *et al.*, 1999, Rutherfurd *et al.*, 2002; Żyła *et al.*, 2013). However, incorporation of exogenous phytase in the matrix also leads to significant changes in the utilisation of dietary limestone and phosphates, and these feed

Table 2.1. *Phytase activity (μmol phytic acid h^{-1}) in the digestive tract of laying hens fed wheat-corn-soybean meal-based diet without microbial phytase supplementation (Marounek* et al., *2010).*[1]

Segment	Specific (per g digesta)	Total (per segment)
Crop	10.2[a]	98[a]
Stomach	9.2[a]	97[a]
Small intestine	14.6[a]	359[b]
Small intestinal mucosa	11.5[a]	227[ab]
Sum pre-caecal		781
Caeca	135.4[b]	663[c]
Total		1,444

[1] Means within a column not sharing a common letter differ significantly ($P<0.05$).

Table 2.2. *Phytase activity in diet and digesta of broiler chickens fed diets with or without added microbial phytase from 8 to 22 days of age; activity measured at day 22 (Onyango* et al., *2005).*[1]

	NC: low-P diet	NC +1000 FTU *Escherichia coli* phytase/kg	NC +1000 FTU *Peniophora lycii* phytase/kg
Feed (FTU/kg)	14	825	1,152
Digesta (FTU/kg DM intake)			
Crop	67[c]	649[a]	404[b]
Proventriculus and gizzard	28[b]	406[a]	63[b]
Jejunum	29[b]	554[a]	25[b]
Ileum	16[b]	91[a]	6[b]

[1] Means within a row not sharing a common letter differ significantly ($P<0.05$).

ingredients have high acid-binding/buffering capacity and contribute to an increase in digesta pH (Lawlor *et al.*, 2005). This will indirectly affect the GIT environment, improving proliferation of many types of microbiota particularly Gram-positive strains which do not tolerate lower pH values.

Hydrogen ion concentration varied significantly among specific sections of the digestive tract of the chicken and between individuals as well (Sturkie, 1976). The pH of the digestive tract can be influenced by many factors including Ca and P concentration of the diet, and high levels of Ca may increase pH in certain parts of the digestive tract (Guinotte *et al.*, 1995; Shafey *et al.*, 1991; Smulikowska *et al.*, 2010; Walk *et al.*, 2012). Shafey *et al.* (1991) reported that increasing dietary Ca from 1.07 to 2.53% increased crop pH from 4.89 to 5.32 and pH in the ileum from 6.62 to 7.39. Guinotte *et al.* (1995) found that an increase in dietary Ca from 1.0 to 3.6% increased pH from 2.76 to 3.82 in the gizzards of pullets. Similarly, Walk *et al.* (2012) observed a reduction in gizzard and ileal pH values in broilers fed 0.64% Ca as compared with those fed 1.03% Ca. Also, McDonald and Solvyns (1964) increased digesta pH from 5.6 to 6.1 along the entire length of the small intestine by increasing the concentration of dietary Ca, and this measure was negatively correlated with broiler chicken growth performance.

An excess in buffering capacity may interfere with digestion rate and extent, particularly during the gastric phase, and may also force additional commitment of digestive resources (HCl and pepsin) if digestion is to be maintained. Application of phytase is, therefore, expected to ease the constraints on gastric digestion such that inputs of HCl and pepsin can be reduced without increasing pH and thus compromising digestion. However, the literature is not consistent in this regard and there are many conflicting reports regarding the effects of phytase and Ca:P levels on the gastrointestinal pH in monogastric animals. Applying phytase at 1000 FTU/kg to diets with high and low levels of P, Smulikowska *et al.* (2010) observed an increase in the pH of jejunal digesta, while pH of the crop, stomach, ileal and caecal digesta were not affected. Furthermore, the application of a microbial phytase did not influence gastrointestinal pH (in the crop, gizzard, duodenum, jejunum, ileum) in 42-d-old broilers (Nourmohammadi *et al.*, 2011) or stomach pH values in weanling pigs (Radcliffe *et al.*, 1998). Supplemental phytase, however, increased faecal pH but did not affect ileal pH in pigs (Metzler *et al.*, 2008). Also Akyurek *et al.* (2011) observed an increase in caecal pH but not in ileal pH in 21-d-old broiler chickens when microbial phytase was incorporated at 500 FTU/kg of a reduced P diet. Walk *et al.* (2012) reported that feeding 5,000 FTU/kg significantly increased gizzard, duodenum, jejunum and ileum pHs in 16-d-old broilers, and that this effect was consistent in a low and high Ca diets. Furthermore, Walk *et al.* (2012) suggested that the increase in the gastrointestinal pH created by high levels of supplemental phytase may directly arise from the breakdown of the phytate molecule, release of Ca, and from a reduction of the acidogenic activity of phytate.

The amorphous Ca phosphate complex can also influence fermentation processes in the gastrointestinal tract. In rats and ruminants, Ca and P are important modulators of microbial fermentation (Bovee-Oudenhoven *et al.*, 1997; Durand and Komisarczuk, 1988). For example, there is a minimal P requirement for the fermentation of carbohydrates by rumen bacteria (Komisarczuk *et al.*, 1987). According to *in vitro* studies, depletion of P reduced the fermentative activity of rumen bacteria, resulting in decreased production of SCFAs and reduced bacterial ATP concentrations (Komisarczuk *et al.*, 1987). Besides antimicrobial effects, SCFAs play an essential role as bioregulators and mucosal growth promoters via direct or indirect mechanisms in the gut as well as influencing metabolism systemically once absorbed into the blood (Lan *et al.*, 2005; Mroz *et al.*, 2006). Smulikowska *et al.* (2010) reported that total SCFAs as well as acetic acid, propionic acid and butyric acid concentrations in the caecal digesta of 29-day-old chickens were increased with increasing dietary P content. Smulikowska *et al.* (2010) also observed that phytase supplementation increased acetic acid and butyric acid concentrations but only in the diet supplemented with the lower and not the higher monocalcium phosphate inclusion level, indicating a maximum threshold was reached. Ptak *et al.* (2015), reported an interaction among dietary concentrations of phytase and dietary Ca and digestible P which was demonstrated by the fact that reducing the Ca:P content of the diet reduced total SCFA, DL-lactate and acetic acid in the ileum and phytase increased concentrations of these acids only in broilers fed the diet with reduced Ca and digestible P. In contrast, in the study of Metzler *et al.* (2008), dietary monocalcium phosphate did not significantly increase ileal or faecal SCFA concentrations in pigs, but it did significantly reduce ileal DL-lactate concentrations. Further, Metzler-Zebeli *et al.* (2010) demonstrated that high-dietary monocalcium phosphate decreased total lactate and L-lactate concentrations in the gastric digesta of pigs. In the same study, however, phytase supplementation did not affect ileal concentrations of SCFA and lactate, but reduced faecal concentration of n-butyrate and its molar proportion. However, in later work, Metzler-Zebeli *et al.* (2011) observed no effect of dietary monocalcium phosphate on ileal SCFA concentrations. In this regard, the data relating SCFA concentrations in the GIT of monogastrics to phosphate supply are equivocal, likely for the same reasons as discussed for the pH data above.

The changes in SCFAs concentrations indicate that shifts in the gut microbiome may take place as a consequence of addition of phytase to the diets. High dietary Ca and P content in diets fed to rats increased intestinal proliferation of lactic acid bacteria (Ten Bruggencate *et al.*, 2004), and a diet rich in dietary Ca and P resulted in decreased numbers of *Enterococcus* spp., *Clostridium cluster* IV and Lactobacilli species in the ileum of growing pigs (Metzler-Zebeli *et al.*, 2010). Reduced counts of *Streptococcus* spp. in the stomach of pigs fed high Ca and P diets has also been reported (Metzler-Zebeli *et al.*, 2011), but Clostridium cluster XIVa numbers were increased by the high Ca:P diet (Metzler-Zebeli *et al.*, 2011). These findings are in agreement with the results reported by Ptak *et al.* (2015), in which the diets deficient in Ca and digestible P reduced butyrate but increased lactate producing bacterial groups. Additionally, a reduction in *Clostridium perfringens* and *Enterobacteriaceae* counts was also observed with low Ca:P diets. Calcium can influence bacterial growth by a range of factors, such as counteracting the adverse effects of acidic fermentation and precipitating cytotoxic components within the intestinal lumen (Ten Bruggencate *et al.*, 2004), induction of increased gastric acid secretion or interference with bacterial adhesion to the intestinal mucosal walls (Geesey *et al.*, 2000; Larsen *et al.*, 2007). Supplementation of phytase also increased the total numbers of bacteria as well as *Lactobacillus* sp. and *Enterococcus* sp. (Ptak *et al.*, 2015). Metzler-Zebeli *et al.* (2010) reported that addition of phytase increased the ileal numbers of strictly anaerobic bacteria, such as the *Clostridium coccoides* cluster, the *Clostridium leptum* cluster, and the *Bacteroides-Prevotella-Porphyrmonas group*, and tended to enhance those of *Enterobacteriaceae*. The authors stated that these observations might be associated with the greater availability of P in the lumen of the gastrointestinal tract.

Phosphorus supply may indeed play a controlling role for the total numbers of bacteria as suggested by the significant Ca/P and phytase effects, which were independent of one another, and tended to be additive (Ptak *et al.*, 2015). Phytase seems to have an important role to play in modulation of the microbiome, but its effects are influenced by Ca and P content of the diet. It should be also noted that phytase can increase the pool of *myo*-inositol derivates in the intestinal lumen (Cowieson *et al.*, 2011, 2013; Józefiak *et al.*, 2010a). *Myo*-inositol is an essential constituent of phospholipids, and thus is used for building the cell walls of many organisms (Michell, 2008). This is the reason *myo*-inositol is present in many types of media, used for *in vitro* culturing techniques. However, the significance, distribution and utilisation of *myo*-inositol and its derivatives varies across the three biological kingdoms (Michell, 2008). It is used by most archaea and all eukaryotic organisms, but by relatively few bacteria (Michell, 2008; Reynolds, 2009). Most of the bacteria, besides the Actinomycetes, do not use *myo*-inositol as a component of phospholipids in their lipid membranes or in redox reactions. However, for *Bacillus subtilis, Aerobacter aerogenes, Rhizobium leguminosarum, Sinorhizobium meliloti, Corynebacterium glutamicum*, and the *Lactobacillus casei, myo*-inositol is a carbon source (Reynolds, 2009). In addition, the assumption of the potential effects of *myo*-inositol on the microbiological status of digestive systems is partially confirmed by a study conducted on juvenile carp (Jiang *et al.*, 2009). These authors observed a reduction in the number of *Aeromonas hydrophila* and *E. coli*, and increase the population of the Lactobacillus after addition of *myo*-inositol to the diet.

2.3 Conclusions

1. In poultry, gut microbiota can utilize 10-25% of dietary phytate.
2. Phosphorus supply appears to play a controlling role for the different taxonomic groups of bacteria.
3. Phytase supplementation significantly affects microbiome but mode of action is indirect, through changes in the buffering capacity of the digesta and release of IP_6 isomers as well as *myo*-inositol.

References

Akyurek, H., Ozduven, M.L., Okur, A.A., Koc, F. and Samli, H.E., 2011. The effects of supplementing an organic acid blend and/or microbial phytase to a corn-soybean based diet fed to broiler chickens. African Journal of Agricultural Research 6:642-649.

Augspurger, N.R., Webel, D.M., Lei, X.G. and Baker, D.H., 2003. Efficacy of an *E. coli* phytase expressed in yeast for releasing phytate-bound phosphorus in young chicks and pigs. Journal of Animal Science 81:474-483.

Bjerrum, L., Engberg, R.M., Leser, T., Jensen, B., Finster, K. and Pedersen, K., 2006. Microbial community composition of the ileum and cecum of broiler chickens as revealed by molecular and culture-based techniques. Poultry Science 85:1151-1164.

Bovee-Oudenhoven, I.M.J., Termont, D.S.M.L., Heldt, P.J. and Van der Meer, R., 1997. Increasing the intestinal resistance of rats to the invasive pathogen *Salmonella enteritidis*: additive effects of dietary lactulose and calcium. Gut 40:497-504.

Corrigan, A., Horgan, K., Clipson, N. and Murphy, R.A., 2012. Effect of dietary prebiotic (mannan-oligosaccharide) supplementation on the caecal bacterial community structure of turkeys. Microbial Ecology 64:826-836.

Cowieson, A.J. and Adeola, O., 2005. Carbohydrases, protease, and phytase have an additive beneficial effect in nutritionally marginal diets for broiler chicks. Poultry Science 84:1860-1867.

Cowieson, A.J., Ptak, A., Maćkowiak, P., Sassem, M., Pruszyńska-Oszmałek, E., Żyła, K., Świątkiewicz, S., Kaczmarek, S. and Jozefiak, D., 2013. The effect of microbial phytase and *myo*-inositol on performance and blood biochemistry of broiler chickens fed wheat/corn-based diets. Poultry Science 92:2124-2134.

Cowieson, A.J., Wilcock, P. and Bedford, M.R., 2011. Super-dosing effects of phytase in poultry and other monogastrics. World's Poultry Science Journal 67:225-235.

Dersjant-Li, Y., Awati, A., Schulze, H. and Partridge, G., 2015. Phytase in non-ruminant animal nutrition: a critical review on phytase activities in the gastrointestinal tract and influencing factors. Journal of the Science of Food and Agriculture 95:878-896.

Durand, M. and Komisarczuk, S., 1988. Influence of major minerals on rumen microbiota. Journal of Nutrition 118:249-260.

Geesey, G.G., Wigglesworth-Cooksey, B. and Cooksey, K.E., 2000. Influence of calcium and other cations on surface adhesion of bacteria and diatoms: a review. Biofouling 14:195-205.

Gong, J., Forster, R.J., Yu, H., Chambers, J.R., Sabour, P.M., Wheatcroft, R. and Chen, S., 2002a. Diversity and phylogenetic analysis of bacteria in the mucosa of chicken ceca and comparison with bacteria in the cecal lumen. FEMS Microbiology Letters 208:1-7.

Gong, J.H., Forster, R.J., Yu, H., Chambers, J.R., Wheatcroft, R., Sabour, P.M. and Chen, S., 2002b. Molecular analysis of bacterial populations in the ileum of broiler chickens and comparison with bacteria in the cecum. FEMS Microbiology Ecology 41:171-179.

Gong, J., Si, W., Forster, R.J., Huang, R., Yu, H., Yin, Y., Yang, C. and Han, Y., 2007. 16S rRNA gene-based analysis of mucosa-associated bacterial community and phylogeny in the chicken gastrointestinal tracts: from crops to ceca. FEMS Microbiology Ecology 59:147-157.

Guan, L.L., Hagen, K.E., Tannock, G.W., Korver, D.R., Fasenko, G.M. and Allison, G.E., 2003. Detection and identification of *Lactobacillus* species in crops of broilers of different ages by using PCR-denaturing gradient gel electrophoresis and amplified ribosomal DNA restriction analysis. Applied and Environmental Microbiology 69:6750-6757.

Guinotte, F., Gautron, J., Nys, Y. and Soumarmon, A., 1995. Calcium solubilisation and retention in the gastrointestinal tract in chicks (*Gallus domesticus*) as a function of gastric acid secretion inhibition and of calcium carbonate particle size. British Journal of Nutrition 73:125-139.

Hume, M.E., Hernandez, C.A., Barbosa, N.A., Sakomura, N.K., Dowd, S.E. and Oviedo-Rondón, E.O., 2012. Molecular identification and characterization of ileal and cecal fungus communities in broilers given probiotics, specific essential oil blends, and under mixed Eimeria infection. Foodborne Pathogens and Disease 9:853-860.

Jiang, W.D., Feng, L., Liu, Y., Jiang, J. and Zhou, X.Q., 2009. Growth, digestive capacity and intestinal microflora of juvenile Jian carp (*Cyprinus carpio* var. Jian) fed graded levels of dietary inositol. Aquaculture Research 40:955-962.

Józefiak, D., Rutkowski, A., Martin, S., 2004. Carbohydrate fermentation in the avian ceca: a review. Animal Feed Science and Technology 113:1-15.

Józefiak, D., Ptak, A., Kaczmarek, S., Maćkowiak, P., Sassem, M. and Słomiński, B.A., 2010a. Multi-carbohydrase and phytase supplementation improves growth performance and liver insulin receptor sensitivity in broiler chickens fed diets containing full-fat rapeseed. Poultry Science 89:1939-1946.

Józefiak, D., Rutkowski, A., Kaczmarek, S., Jensen, B.B., Engberg, R.M. and Højberg, O., 2010b. Effect of β-glucanase and xylanase supplementation of barley- and rye-based diets on caecal microbiota of broiler chickens. British Poultry Science 51:546-557.

Komisarczuk, S., Merry, R.J. and McAllan, A.B., 1987. Effect of different levels of phosphorus on rumen microbial fermentation and synthesis determined using a continuous culture technique. British Journal of Nutrition 57:279-290.

Kierończyk, K., Rawski, M., Długosz, J., Świątkiewicz, S. and Józefiak, D., in press. Avian crop function – a review. Annals of Animal Science. DOI: http://doi.org/10.1515/aoas-2016-0032.

Kurzak, P., Ehrmann, M.A. and Vogel, R.E., 1998. Diversity of lactic acid bacteria associated with ducks. Systematic and Applied Microbiology 21:588-592.

Lan, Y., Verstegen, M.W.A., Tamminga, S. and Williams, B.A., 2005. The role of the commensal gut microbial community in broiler chickens. World's Poultry Science Journal 61:95-104.

Larsen, N., Nissan, P. and Willats, W.G., 2007. The effect of calcium ions on adhesion and competitive exclusion of *Lactobacillus* ssp. and *E. coli* O138. International Journal of Food Microbiology 114:113-119.

Lawlor, P.G., Lynch, P.B., Patrick, J. Caffrey, P.J., O'Reilly, J.J. and O'Connell, M.K., 2005. Measurements of the acid-binding capacity of ingredients used in pig diets. Irish Veterinary Journal 58:447-452.

Liu, B.Y., Wang, Z.Y., Wang, H.R., Hu, P., Xu, D. and Wang, Q., 2011. Molecular profiling of bacterial species in the caecum of geese. Czech Journal of Animal Science 56:192-203.

Liu, N., Ru, Y.J., Li, F.D. and Cowieson, A.J., 2008. Effect of diet containing phytate and phytase on the activity and messenger ribonucleic acid expression of carbohydrase and transporter in chickens. Journal of Animal Science 86:3432-3439.

Lu, J., Hofacre, C., Smith, F. and Lee, M.D., 2008. Effects of feed additives on the development on the ileal bacterial community of the broiler chicken. Animal 2:669-676.

Lu, J., Idris, U., Harmon, B., Hofacre, C., Maurer, J.J. and Lee, M.D., 2003a. Diversity and succession of the intestinal bacterial community of the maturing broiler chicken. Applied and Environmental Microbiology 69:6816-6824.

Lu, J., Sanchez, S., Hofacre, C., Maurer, J.J., Harmon, B.G. and Lee, M.D., 2003b. Evaluation of broiler litter with reference to the microbial composition as assessed by using 16S rRNA and functional gene markers. Applied and Environmental Microbiology 69:901-908.

Marounek, M., Skřivan, M., Rostro, O. and Rop, O., 2010. Intestinal and total tract phytate digestibility and phytase activity in the digestive tract of hens fed a wheat-maize-soyabean diet. Journal of Animal and Feed Sciences 19:433-442.

McDonald, M.W. and Solvyns, A., 1964. Dietary calcium levels and chicken growth. In: Australasian Poultry Science Convention. Australian Branch, WPSA, Proceedings of the Australasian Poultry Science Convention. Surfers Paradise, Qld. Merck Sharp and Dohme, pp.112-116.

Metzler, B.U., Mosenthin, R., Baumgärtel, T. and Rodehutscord, M., 2008. Effects of phosphorus level, phytase and pectin on the chemical composition and carbohydrase activity of fecal bacteria and microbial metabolites in the gastrointestinal tract of pigs. Journal of Animal Science 86:1544-1555.

Metzler-Zebeli, B.U., Vahjen, W., Baumgärtel, T., Rodehutscord, M. and Mosenthin, R., 2010. Ileal microbiota of growing pigs fed different dietary calcium phosphate levels and phytase content and subjected to ileal pectin infusion. Journal of Animal Science 88:147-158.

Metzler-Zebeli, B.U., Zijlstra, R.T., Mosenthin, R. and Gänzle, M.G., 2011. Dietary calcium phosphate content and oat â-glucan influence gastrointestinal microbiota, butyrate-producing bacteria and butyrate fermentation in weaned pigs. FEMS Microbiology Ecology 75:402-413.
Michell, R.H., 2008. Inositol derivatives: evolution and functions. Nature Reviews Molecular Cell Biology 9:151-161.
Mroz, Z., Koopmans, S.J., Bannink, A., Partanen, K., Krasucki, W., Řverland, M. and Radcliffe, S., 2006. Carboxylic acids as bioregulators and gut growth promoters in nonruminants. In: Mosenthin, R., Zentek, J. and Żebrowska, T. (eds.) Biology of nutrition in growing animals. Elsevier, Edinburgh, UK, pp. 81-133.
Naqi, S., Lewis, D. and Hall, C., 1970. The intestinal microflora of turkeys. Avian Diseases 14:620-625.
Newkirk, R.W. and Classen, H.L., 2001. The non-mineral nutritional impact of phytate in canola meal fed to broiler chicks. Animal Feed Science and Technology 91:115-128.
Nourmohammadi, R., Hosseini, S.M., Saraee, H. and Arab, A., 2011. Plasma thyroid hormone concentrations and pH values of some GI-tract segments of broilers fed on different dietary citric acid and microbial phytase levels. Journal of Animal and Veterinary Advances 6:1-6.
Onyango, E.M., Bedford, M.R. and Adeola O., 2005. Efficacy of an evolved *Escherichia coli* phytase in diets for broiler chicks. Poultry Science 84:248-255.
Ptak, A., Bedford, M.R., Świątkiewicz, S., Żyła, K. and Józefiak, D., 2015. Phytase modulates ileal microbiota and enhances growth performance of the broiler chickens. PLoS ONE 10: e0119770. DOI: http://doi.org/10.1371/journal.pone.0119770.
Radcliffe, J.S., Zhang, Z. and Kornegay, E.T., 1998. The effects of microbial phytase, citric acid, and their interaction in a corn-soybean meal-based diet for weanling pigs. Journal of Animal Science 76:1880-1886.
Ravindran, V., Cabahug, S., Ravindran, G. and Bryden, W.L., 1999. Influence of microbial phytase on apparent ileal amino acid digestibility of feedstuffs for broilers. Poultry Science 78:699-706.
Ravindran, V., Cabahug, S., Ravindran, G., Selle, P.H. and Bryden W.L., 2000. Response of broiler chickens to microbial phytase supplementation as influenced by dietary citric acid and non-phytate phosphorus levels. II. Effects on apparent metabolisable energy, nutrient digestibility and nutrient retention. British Poultry Science 41:193-200.
Rehman, H.U., Vahjen, W., Awad, W.A. and Zentek, J., 2007. Indigenous bacteria and bacterial metabolic products in the gastrointestinal tract of broiler chickens. Archives of Animal Nutrition 61:319-335.
Reynolds, T.B., 2009. Strategies for acquiring the phospholipid metabolite inositol in pathogenic bacteria, fungi and protozoa: making it and taking it. Microbiology 155:1386-1396.
Rutherfurd, S.M., Chung, T.K. and Moughan, P.J., 2002. The effect of microbial phytase on ileal phosphorus and amino acid digestibility in the broiler chicken. British Poultry Science 43:598-606.
Saengkerdsub, S., Anderson, R.C., Wilkinson, H.H., Kim, W.-K., Nisbet, D.J. and Ricke, S.C., 2007. Identification and quantification of methanogenic archaea in adult chicken ceca. Applied and Environmental Microbiology 73:353-356.
Selle, P.H. and Ravindran, V., 2007. Microbial phytase in poultry nutrition. Animal Feed Science and Technology 135:1-41.
Selle, P.H. and Ravindran, V., 2008. Phytate-degrading enzymes in pig nutrition. Livestock Science 113:99-122.
Shafey, T.M., McDonald, M.W. and Single, J.G., 1991. Effects of dietary calcium and available phosphorus on digesta pH and on the availability of calcium, iron, magnesium and zinc from the intestinal contents of meat chicken. British Poultry Science 32:185-194.
Singh, K.M., Shah, T., Deshpande, S., Jakhesara, S.J., Koringa, P.G., Rank, D.N. and Joshi, C.G., 2012. High through put 16S rRNA gene-based pyrosequencing analysis of the fecal microbiota of high FCR and low FCR broiler growers. Molecular Biology Reports 39:10595-10602.

Smulikowska, S., Czerwiński, J. and Mieczkowska, A., 2010. Effect of an organic acid blend and phytase added to a rapeseed cake-containing diet on performance, intestinal morphology, caecal microflora activity and thyroid status of broiler chickens. Journal of Animal Physiology and Animal Nutrition 94:15-23.

Stanley, D., Denman, S.E., Hughes, R.J., Geier, M.S., Crowley, T.M., Chen, H., Haring, V.R. and Moore, R.J., 2012. Intestinal microbiota associated with differential feed conversion efficiency in chickens. Applied Microbiology and Biotechnology 96:1361-1369.

Stanley, D., Geier, M.S., Hughes, R.J., Denman, S.E. and Moore, R.J., 2013. Highly variable microbiota development in the chicken gastrointestinal tract. PLoS ONE 8:e84290. DOI: http://doi.org/10.1371/journal.pone.0084290.

Stanley, D., Hughes, R.J. and Moore, R.J., 2014. Microbiota of the chicken gastrointestinal tract: influence on health, productivity and disease. Applied Microbiology and Biotechnology 98:4301-4310.

Sturkie, P.D., 1976. Secretion of gastric and pancreatic juice, pH of tract, digestion in alimentary canal, liver and bile, and absorption. In: Sturkie, P.D. (eds.) Avian physiology. 3rd ed. Spinger-Verlag, New York, NY, USA, pp. 196-209.

Ten Bruggencate, S.J.M., Bovee-Oudenhoven, I.M.J., Lettink-Wissink, M.L.G., Katan, M.B. and Van der Meer, R., 2004. Dietary fructooligosaccharides and inulin decrease resistance of rats to salmonella: protective role of calcium. Gut 53:530-535.

Torok, V., Hughes, R., Ophel-Keller, K. and Ali, M., 2009. Influence of different litter materials on cecal microbiota colonization in broilers. Poultry Science 88:2474-2481.

Torok, V.A., Dyson, C., Mckay, A. and Ophel-Keller, K., 2013. Quantitative molecular assays for evaluating changes in broiler gut microbiota linked with diet and performance. Animal Production Science 53:1260-1268.

Van Der Hoeven-Hangoor, E., Van Der Vossen, J.M.B.M., Schuren, F.H.J., Verstegen, M.W.A., De Oliveira, J.E., Montijn, R.C. and Hendriks, W.H., 2013. Ileal microbiota composition of broilers fed various commercial diet compositions. Poultry Science 92:2713-2723.

Vasaï, F., Ricaud, K.B., Bernadet, M.D., Cauquil, L., Bouchez, O., Combes, S. and Davail, S., 2014. Overfeeding and genetics effect the composition of intestinal microbiota in *Anas platyrhynchos* and *Carina moschata* (Muscovy) ducks. FEMS Microbiology Ecology 87:204-216.

Walk, C.L., Bedford, M.R. and McElroy, A.P., 2012. Influence of limestone and phytase on broiler performance, gastrointestinal pH, and apparent ileal nutrient digestibility. Poultry Science 91:1371-1378.

Yu, H., Zhou, T., Gong, J., Young, C., Su, X., Li, X.-Z., Zhu, H., Tsao, R. and Yang, R., 2010. Isolation of deoxynivalenol-transforming bacteria from the chicken intestines using the approach of PCR-DGGE guided microbial selection. BMC Microbiology 10:182. DOI: http://doi.org/10.1186/1471-2180-10-182.

Zaefarian, F., Romero, J.F. and Ravindran, V., 2013. Influence of a microbial phytase on the performance and the utilisation of energy, crude protein and fatty acids of young broilers fed on phosphorus-adequate maize- and wheat-based diets. British Poultry Science 54:653-660.

Żyła, K., Duliński, R., Pierzchalska, M., Grabacka, M., Józefiak, D. and Świątkiewicz, S., 2013. Phytases and *myo*-inositol modulate performance, bone mineralization and alter lipid fractions in serum of broilers. Journal of Animal and Feed Sciences 22:56-62.

Chapter 3
Reduction of phytate to tetrakisphosphate (IP_4) to trisphosphate (IP_3), or perhaps even lower, does not remove its antinutritive properties

M.R. Bedford[*] *and C.L. Walk*
AB Vista, 3 Woodstock Court, Blenheim Road, Marlborough, Wiltshire SN8 4AN, United Kingdom; mike.bedford@abvista.com

Abstract

Phytase has traditionally been used as a means to recover phosphorus (P) from phytic acid (IP_6) and thus reduce feed costs by enabling reduction of inorganic phosphate sources. More recent work has indicated that benefits beyond provision of P are achievable if sufficient IP_6 is hydrolysed. Initial work focussed on simply the destruction of IP_6, and inverse correlations between animal performance and dietary IP_6 levels noted. More recent work has investigated the kinetics of phytase activity *in vivo* and *in vitro* and both show that use of industry standard levels of phytase significantly reduce IP_6 levels as expected, but increase levels of tetrakisphosphate (IP_4) and trisphospate (IP_3) *in vivo*. The latter was not expected. If IP_4 and IP_3 do possess antinutritive capabilities, and evidence is provided here which suggests this may be the case, then in order to successfully implement an 'extraphosphoric' effect (superdosing) with a phytase, the target should be removal of all IP esters in the intestinal contents. There are large differences between phytases with regards to their kinetics not only on IP_6, but also pentakisphosphate (IP_5), IP_4, IP_3 and diphosphate (IP_2). It is unlikely, therefore, that the dose required to 'superdose', where approximately 85% IP_6 hydrolysis to IP_1 is targeted, is a simple and consistent multiple of that required to release 0.10% available P, where only 40% hydrolysis is necessary. Successful superdosing could significantly improve amino acid digestibility and hence knowledge of the source of the phytase is essential if the correct dosage is to be employed to unlock these benefits.

Keywords: superdoses, pepsin, amino acids, phytic acid

3.1 Introduction

Phytic acid was discovered in 1903 and at the time was viewed as being a rich source of phosphorus (P) in plants, and indeed in 1904 it was even suggested that phytin itself should be viewed as a rich source of P for humans (Andrews and Herrarte, 1951). Subsequent work through the 1920s to the 1950s changed this view of phytin from a beneficial source of P to an inert compound because of the poor availability of phytin P, and then to detrimental as a result of its ability to bind calcium (Ca) and precipitate rickets in marginal Ca diets (Andrews and Herrarte, 1951). Although it was recognised that humans did possess the ability to degrade phytin (of ingested phytin only 20-60% appeared in the faeces in a 1935

DOI 10.3920/978-90-8686-836-0_3, © Wageningen Academic Publishers 2016

study (McCance and Widdowson, 1935), it was hypothesised by Andrews and Herrarte (1951) that much of this hydrolysis takes place 'so low a point as to render impossible any absorption of the products'. Thus, phytic acid (IP_6) per se is now seen as being a potent antinutrient with the consequential inference that its destruction will improve subsequent performance. Multiple strands of evidence have suggested that indeed IP_6 not only chelates minerals, reducing their availability (Zeng *et al.*, 2014) and potentially precipitating a deficiency, but also interferes with digestion in such a manner that protein and energy digestibility can be compromised as well. Initial work indicated that supplementation of broiler, duck and piglet diets with synthetic IP_6 resulted in reduced rates of protein digestion and in particular this was driven by significant increases in endogenous losses in the form of mucin and bile acids (Cowieson *et al.*, 2006; Onyango and Adeola, 2011; Pirgozliev *et al.*, 2011). The loss of endogenous proteins as a result of administration of IP_6 is hypothesised to be due to interference with gastric digestion in particular. This interference is likely the result of the fact that even at the low pH of the stomach, IP_6 is negatively charged whereas most dietary proteins will be positively charged and hence precipitate on formation of binary linkages with IP_6 (Morales *et al.*, 2011, 2013). In such form, these proteins are resistant to gastric digestion (Yu *et al.*, 2012). The response of the animal seems to be to increase secretion of acid whilst paradoxically reduce pepsin secretion. This is mostly documented as an increase in gastric pH (Amerah *et al.*, 2014; Morgan *et al.*, 2016; Walk *et al.*, 2012) and an increase in digesta pepsin in diets supplemented with phytase (Morgan *et al.*, 2016; Woyengo *et al.*, 2010). Addition of IP_6 may also reduce pepsin activity in the digesta (Woyengo *et al.*, 2010) and addition of phytase increases mucosal pepsin (Liu *et al.*, 2009), both of which indicate that there is proportionally less gastric proteolytic digestion capacity with increasing IP_6 concentration. This may be due to either precipitation of the substrate or of pepsin or both. Precipitation of the substrate will not only reduce the efficacy of pepsin, but may also reduce the signal for release of pepsin. Either way, the rate at which gastric digestion proceeds is compromised in a dose dependent manner with increments of dietary IP_6 as indicated by duodenal digestibility data (Liu *et al.*, 2015; Truong *et al.*, 2015). Gastric digestion is designed to unfold and degrade the hydrophobic regions of dietary proteins (Moran Jr., in press) and any significant compromise in this process will likely not be overcome by increased pancreatic protease secretion. The protein solubility of eight feed ingredient sources (soybean meal, peas, beans, chickpea protein isolate, lupin meal, canola meal, wheat midds and wheat) was shown to be poorest between pH 3 and pH 4 in the presence of IP_6, the exact pH of lowest solubility being protein source dependent (Morales *et al.*, 2013). If the pH was reduced below 3, the solubility of all protein/IP_6 complexes increased which may be why there is such a consistent reduction in gastric pH with incremental dietary IP_6 or reduced phytase activity. Such a response may present the gastric environment with more substrate for pepsin although this comes at a cost. Reducing gastric pH increases the potential for damage to the cornified epithelium layer of the proventriculus/gizzard, which is mitigated by secretion of additional gastric mucin. Hence there is an association between dietary IP_6 levels and losses of endogenous proteins (Cowieson *et al.*, 2006; Onyango and Adeola, 2011; Pirgozliev *et al.*, 2011), and thus there is an association between IP_6 and digestibility of cysteine, serine, threonine, glycine and aspartic acid which are all highly represented in mucin.

Until recently, the focus has been almost exclusively on IP_6 and losses of amino acids and minerals. Indeed, earlier publications suggested that the most a phytase needed to achieve was reduction of IP_6 to tetrakisphosphate (IP_4), the endogenous phosphatases (mucosal) effecting the reduction from IP_4 to *myo*-inositol (IP_0). Results of recent work, however, indicate that not only are the endogenous phosphatases ineffective in hydrolysing diphosphate (IP_2) or higher esters (unpublished data), but the lower esters are also not as inert as they were purported to be. The next section focusses specifically on this topic, and makes reference to the role of these esters only in the 'extra-phosphoric' effects of phytases in animals. If the animals were P deficient then the benefit of such extra-phosphoric effects would not be realised.

3.2 Importance of IP_4-IP_1 in animal nutrition

A brief review of the literature highlights a tenet that has been often repeated and accepted as a result of significant *in vitro* data coupled with some *in vivo* support, namely that the antinutritive effects of IP_6 dwarf those of IP_4 and lower esters, and thus, partial dephosphorylation is all that is needed to remove the constraints that IP_6 place on digestion. *In vitro* data abounds to suggest that IP_6 and pentakisphosphate (IP_5) are more effective chelators of minerals than IP_4 or triphosphate (IP_3) (Luttrell, 1992; Persson *et al.*, 1998), and because some minerals are required as co-factors for activation of endogenous enzymes, the rate of digestion of fat, starch, and particularly protein may be compromised by IP_6 and IP_5 activity to a somewhat greater extent than by the lower esters (Knuckles *et al.*, 1985, 1989). *In vitro* solubility of proteins are also compromised to a far greater extent by IP_6 than by lower esters (Kies *et al.*, 2006), and as a result, destruction of IP_6 is seen as the principal goal to prevent binary associations between this molecule and that of soluble, positively charged proteins. The relatively innocuous nature of IP_4 and lower esters is inferred in many papers, but signals of the potential damage of these esters can be inferred from the work of Schlemmer *et al.* (2001) where the contents of the stomach and intestinal contents of the pig were analysed for $IP_6 \rightarrow IP_2$ in both the solid and liquid phases. When pigs were fed a diet that was largely devoid of phytase (ingredients had been extruded to inactivate any native phytases), it was clear that the majority of feed and intestinal phytate was present as IP_6 with smaller amounts of IP_5, indicating limited hydrolysis even in the large intestine. Most of the IP_6 was in the liquid phase in the stomach, but in the small and large intestine this was completely reversed, with most IP_6 being in the solid phase, which is as expected because the soluble IP_6 in the stomach is expected to rapidly precipitate with any available cations as the pH rises in the small and large intestine. The most interesting part of this study, however, was the comparison of this diet, with an identical diet where the cereal components had not been extruded, resulting in a modicum of cereal derived phytase being present in the diet. The effect of this marginal level of phytase was significant, resulting in reduced IP_6 in the stomach, both in the solid and liquid phase with the effect being most pronounced in the liquid phase. The appearance of significant amounts of IP_5, IP_4, IP_3 and IP_2, particularly IP_4, was noted with the majority of all of these esters being present in the liquid phase. Indeed, the total reduction in IP_6 as a result of the presence of the phytase was quantitatively recovered in the form of IP_5, IP_4, IP_3 and IP_2, which indicates that the phytase was unsuccessful in hydrolysing much IP_6 all the way to monophosphate (IP_1). Of interest to this paper however, is the clear transition between the stomach and small intestine where the IP_5, IP_4 and IP_3 but not IP_2 moved from the soluble to the insoluble phase. The implications are that phytase produces soluble lower esters in the stomach, most of which subsequently precipitate in the small intestine, presumably with metal ions and/or proteins. If this is the case then the lower esters are not innocuous and likely interfere with digestion. Indeed, the work of Xu *et al.* (1992) demonstrated that even though the solubility of both zinc (Zn) and Ca were compromised to a far greater extent by IP_6 and IP_5 compared with IP_4 and IP_3 at pH 3 and 4, as the pH rose further, the difference between these esters was much less and in the case of Zn, the differences disappeared at pH 6. Both Zn and Ca are co-factors for pancreatic enzyme activity, and their removal from the intestinal milieu may consequently reduce proteolytic efficiency.

In vitro evidence that IP_4 and IP_3 compromise pepsin activity was provided by Yu *et al.* (2012). Incubation of pepsin with an azocasein substrate resulted in almost no proteolysis when IP_6 was present, presumably as a result of IP_6 interacting with both the substrate and the pepsin. When a phytase was added to the incubation, the rate of proteolysis increased over time, but it was not maximised until all IP_4 and IP_3 were hydrolysed. Indeed, there was limited effect on proteolysis with complete removal of IP_6 and IP_5, indicating that IP_4 and IP_3 were as effective in inactivating pepsin as IP_6 and IP_5.

Thus IP_4 and IP_3 have the potential to compromise gastric and small intestinal proteolysis. *In vivo* data are limited, however, except for the recent work by Beeson *et al.* (2014a,b) in which a significant and negative relationship between the ileal contents of IP_4 and IP_3 and the digestibility of energy and nitrogen, as well as that of many mineral cations, was observed over a range of phytase doses. Indeed, the correlations between nutrient digestibilities and ileal IP_6 and IP_5 contents were less significant than that with the lower esters, indicating that when phytases are employed, generation of IP_4 and IP_3 may be as detrimental to performance as the IP_6 from which it is derived. Pontopiddan *et al.* (2012) reported that when 500 FTU/kg of a *Citrobacter brakii* phytase was fed to pigs there was a significant reduction in IP_6, but a significant accumulation of IP_4 and IP_3 in all sections of the gastro-intestinal tract, particularly in the terminal ileum. Results of additional work have indicated that increased dosages of phytase limit this accumulation (Walk *et al.*, 2014; Zeller *et al.*, 2015, 2016), and thus the benefits noted in animal performance from use of 'superdoses' of phytase may be due, in part, to the reduction in IP_4 and IP_3 levels in the intestine, which are the result of use of low levels of phytase.

In essence, the use of traditional levels of phytase may be removing one antinutrient in the guise of IP_6, only to produce others in the form of IP_4 and IP_3. If P recovery is the main goal, and extraphosphoric effects not the target, then such an event will be irrelevant. However if the focus is improved performance, unrelated to P recovery, then this could be significant. Moreover, it may not simply be a question of increasing the dose of phytase to reduce the accumulation of IP_4 as the ability of the given phytase to hydrolyse IP_4 may be poorly related to its ability to hydrolyse IP_6. The activity of a phytase is determined on IP_6 and yet the same enzyme has to work on four further substrates if it is to successfully reduce IP_6 to IP_1. The kinetics on IP_4 are poorer than that on IP_6 and IP_5 as is evident from the *in vivo* accumulation of IP_4. Results of recent *in vitro* work (Menezes-Blackburn *et al.*, 2015) confirm that IP_4 is a poor substrate for all commercial phytases used at this time, but there are differences between phytases, which are not reflected in their kinetics on IP_6. Thus, the ranking of phytases on their ability to hydrolyse IP_4 is quite different from the ranking based on the ability to hydrolyse IP_6 and IP_5. If removal of IP_4 and IP_3 *in vivo* is key to the success in enhancing animal performance, then dose and source of phytase needs to be considered together.

Recent *in vitro* work, (unpublished) demonstrates that the rate at which different phytases hydrolyse lower esters differs in comparison to their hydrolysis of IP_6 and IP_5. In this method, 200 µm of phytic acid at pH 2.5 is incubated with a fixed number of units of each phytase (units determined as initial velocity at pH 2.5 and 200 µm phytate). For the first few minutes the rate of P release is identical for all phytases tested, as is expected because the majority of the P released is from IP_6 and each phytase is dosed at the same number of units. However, as the reaction progresses, the rate of P release diverges between phytases, with some slowing significantly compared with others, particularly as approximately 33% or more of the phytate P is released. These differences could be due to a number of factors including end product inhibition (either P or a specific phytate ester) or simply poorer kinetics on the lower esters as suggested by Menezes-Blackburn *et al.* (2015). Regardless, the inference is that some phytases may be significantly better at completely dephytinising a diet than others.

3.3 Conclusions

1. Superdosing of phytase may improve animal performance, which has been demonstrated in the majority of cases where phytate hydrolysis is almost complete. Several factors influence the degree of hydrolysis with a given dose of phytase including Ca and P level of the diet as well as other factors such as age of the animal, fibre level, presence of microbial modulators and dietary fat concentration. Thus some degree in variation of response is expected.
2. Improvements in amino acid digestibility may be observed either as a result of reduced endogenous losses or improved efficiency of digestion of dietary proteins, and have been noted in most, but not all, trials where phytase was employed. Unfortunately, very few have coupled phytate ester digestibility with amino acid digestibility in an effort to determine if the phytase dose employed successfully removed the bulk of the IP_6 and lower esters. A disproportionate increase in amino acid digestibility was noted when IP_6 hydrolysis was increased from 80 to 90% at the ileal level (unpublished) indicating that even small, residual levels of phytate and presumably IP_4 and IP_3 may compromise digestive efficiency.
3. Further work is needed to establish that there is indeed a link between IP_4 and IP_3 digestibility and amino acid digestibility, and to identify those factors (eg dose, phytase identity, dietary Ca, P, fat, etc.) which influence this response. Such work could result in more consistent benefits being achieved in the field and consequently expedite industry uptake of this application of phytase.

References

Amerah, A.M., Plumstead, P.W., Barnard, L.P. and Kumar, A., 2014. Effect of calcium level and phytase addition on ileal phytate degradation and amino acid digestibility of broilers fed corn-based diets. Poultry Science 93:906-915.

Andrews, J.C. and Herrarte, E., 1951. Studies on phytin. Journal of the Elisha Mitchell Scientific Society 67:45-53.

Beeson, L., Walk, C.L. and Olukosi, O., 2014a. Growth performance and nutrient digestibility responses of broilers to high supplemental levels of phytase in diets adequate or deficient in non-phytate phosphorus and calcium. World's Poultry Science Association UK Spring meeting (abstract).

Beeson, L., Walk, C.L. and Olukosi, O., 2014b. The flow of inositol phosphate esters and phytate phosphorus in the proximal and distal parts of the digestive tract of broilers receiving diets adequate in available phosphorus and supplemented with high levels of phytase. Journal of Animal Science. 92:Supplement 2 223 (Abstr).

Cowieson, A.J., Acamovic, T. and Bedford, M.R., 2006. Phytic acid and phytase: implications for protein utilization by poultry. Poultry Science 85:878-885.

Kies, A.K., De Jonge, L.H., Kemme, P.A. and Jongbloed, A.W., 2006. Interaction between protein, phytate and microbial phytase. *In vitro* studies. Journal of Agricultural and Food Chemistry 54:1753-1758.

Knuckles, B.E., Kuzmicky, D.D. and Betschart, A.A., 1985. Effect of phytate and partially hydrolyzed phytate on *in vitro* protein digestibility. Journal of Food Science 50:1080-1082.

Knuckles, B.E., Kuzmicky, D.D., Gumbmann, M.R. and Betschart, A.A., 1989. Effect of *myo*-inositol phosphate esters on *in-vitro* and *in-vivo* digestion of protein. Journal of Food Science 54:1348-1349.

Liu, N., Ru, Y.J., Li, F.D., Wang, J.P. and Lei, X.Q., 2009. Effect of dietary phytate and phytase on proteolytic digestion and growth regulation of broilers. Archives of Animal Nutrition 63:292-303.

Liu, S.Y., Bold, R.M., Plumstead, P.W. and Selle, P.H., 2015. Effects of 500 and 1000 FTU/kg phytase supplementation of maize-based diets with two tiers of nutrient specifications on performance of broiler chickens. Animal Feed Science and Technology 207:159-167.

Luttrell, B.M., 1992. The biological relevance of the binding of calcium ions by inositol phosphates. Journal of Biological Chemistry 268:1521-1524.

McCance, R.A. and Widdowson, E.M., 1935. Phytin in human nutrition. Biochemical Journal 29:2694-2699.

Menezes-Blackburn, D., Gabler, S. and Greiner, R., 2015. Performance of seven commercial phytases in an *in vitro* simulation of poultry digestive tract. Journal of Agricultural and Food Chemistry 63:6142-6149.

Morales, G.A., Moyano, F.J. and Marquez, L., 2011. *In vitro* assesment of the effects of phytate and phytase on nitrogen and phosphorus bioaccessibility within fish digestive tract. Animal Feed Science and Technology 170:209-221.

Morales, G.A., Saenz de Rodriganez, M., Marquez, L., Diaz, M. and Moyano, F.J., 2013. Solubilisation of protein fractions induced by *Escherichia* coli phytase and its effects of *in vitro* fish digestion of plant proteins. Animal Feed Science and Technology 181:54-64.

Moran Jr., E.T., in press. Gastric digestion of protein through pancreozyme action optimizes intestinal forms for absorption, mucin formation and villus integrity. Animal Feed Science and Technology. DOI: http://dx.doi.org/10.1016/j.anifeedsci.2016.05.015.

Morgan, N.K., Walk, C.L., Bedford, M.R., Scholey, D.V. and Burton, E.J., 2016. Effect of feeding broilers diets differing in susceptible phytate content. Animal Nutrition 2(1):33-39.

Onyango, E.M. and Adeola, O., 2011. Inositol hexaphosphate increases mucin loss from the digestive tract of ducks. Journal of Animal Physiology and Animal Nutrition 96:416-420.

Persson, H., Turk, M., Nyman, M. and Sandberg, A.-S., 1998. Binding of Cu^{2+}, Zn^{2+} and Cd^{2+} to inositol tri-, tetra-, penta- and hexa-phosphates. Journal of Agriculture and Food Chemistry 46:3194-3200.

Pirgozliev, V., Bedford, M.R., Acamovic, T. and Allimehr, M., 2011. The effects of supplementary bacterial phytase on dietary true metabolisable energy, nutrient digestibility and endogenous losses in precision fed turkeys. British Poultry Science 52:214-220.

Pontoppidan, K., Glitsoe, V., Guggenbuhl, P., Quintana, A.P., Nunes, C.S., Pettersson, D. and Sandberg, A.S., 2012. *In vitro* and *in vivo* degradation of *myo*-inositol hexakisphosphate by a phytase from *Citrobacter braakii*. Archives of Animal Nutrition 66:431-444.

Schlemmer, U., Jany, K.-D., Berk, A., Schulz, E. and Rechkemmer, G., 2001. Degradation of phytate in the gut of pigs – pathway of gastro-intestinal inositol phosphate hydrolysis and enzymes involved. Archives of Animal Nutrition 55:255-280.

Truong, H.H., Bold, R.M., Liu, S.Y. and Selle, P.H., 2015. Standard phytase inclusion in maize based broiler diets enhances digestibility coefficients of starch, amino acids and sodium in four small intestinal segments and digestive dynamics of starch and protein. Animal Feed Science and Technology 209:240-248.

Walk, C.L., Bedford, M.R. and McElroy, A.P., 2012. Influence of limestone and phytase on broiler performance, gastrointestinal pH and apparent ileal nutrient digestibility. Poultry Science 91:1371-1378.

Walk, C.L., Santos, T.T. and Bedford, M.R., 2014. Influence of superdoses of a novel microbial phytase on growth performance, tibia ash, and gizzard phytate and inositol in young broilers. Poultry Science 93:1172-1177.

Woyengo, T.A., Adeola, O., Udenigwe, C.C. and Nyachoti, C.M., 2010. Gastro-intestinal digesta pH, pepsin activity and soluble mineral concentration responses to supplemental phytic acid and phytase in piglets. Livestock Science 134:91-93.

Xu, P., Price, J., Wise, A. and Aggett, P.J., 1992. Interaction of inositol phosphates with calcium, zinc and histidine. Journal of Inorganic Biochemistry 47:119-130.

Yu, S., Cowieson, A.J., Gilbert, C., Plumstead, P.W. and Dalsgaard, S., 2012. Interactions of phytate and *myo*-inositol phosphate esters (IP_{1-5}) including IP_5 isomers with dietary protein and iron and inhibition of pepsin. Journal of Animal Science 90:1824-1832.

Zeller, E., Schollenberger, M., Kuhn, I. and Rodehutscord, M., 2015. Effect of diets containing enzyme supplements and microwave-treated or untreated wheat on inositol phosphates in the small intestine of broilers. Animal Feed Science and Technology 204:42-51.

Zeller, E., Schollenberger, M., Kuhn, I. and Rodehutscord, M., 2016. Dietary effects on inositol breakdown in the crop of broilers. Archives of Animal Nutrition 70:57-71.

Zeng, Z.K., Wang, D., Piao, X.S., Li, P.F., Zhang, H.Y., Shi, C.X. and Yu, S.K., 2014. Effects of adding super dose phytase to the phosphorus-deficent diets of young pigs on growth performance, bone quality, minerals and amino acids digestibilities. Asian-Australasian Journal of Animal Sciences 27:237-246.

Chapter 4
Cellular *myo*-inositol metabolism

K. Huber
University of Hohenheim, Faculty of Agricultural Sciences, Institute of Animal Science, Fruwirthstraße 35, 70599 Stuttgart, Germany; korinna.huber@uni-hohenheim.de

Abstract

Myo-inositol (MI) is an important constituent of cells in animals and humans. Sources for MI are dietary MI absorbed from the intestine, the body-own synthesis of MI from glucose and the release of MI from cellular phospholipids. Therefore, MI is discussed to be a non-essential substance for mammalians. However, under certain metabolic conditions such as diabetes type 2, dietary MI supplementation improved insulin sensitivity and lipid metabolism. The underlying mechanisms are not clarified yet. Furthermore, MI is a structural component of phosphoinositides in plasma membranes as phosphatidylinositol-(4,5)bisphosphate (PIP_2) and phosphatidylinositol-(3,4,5) trisphosphate (PIP_3). Both, PIP_2 and PIP_3 are produced via phosphorylation of MI by phosphoinositide kinases. Located in the inner leaf of the plasma membrane of cells, these phosphoinositides modulate cell migration and endocytosis. PIP_2 is also the precursor for phospholipase C to produce inositol-(1,4,5)trisphosphate (IP_3) and diacylglycerol (DAG), both influencing the intracellular ionised calcium signalling. PIP_3 is a signalling molecule of the insulin signalling cascade activating protein kinase B. Thus, glucose metabolism of cells is influenced by MI. To conclude, MI is important for many metabolic pathways, endocrine signalling and cell architecture. Supplementing MI with the diet ameliorates metabolic disturbances implicating that MI is a good candidate as nutriceutical in animal and human nutrition.

Keywords: cellular transporter, phosphatidylinositol synthesis, physiological role

4.1 Introduction

Myo-inositol (MI), a six-fold alcohol of cyclohexane, is an essential factor which is involved in many cellular functions. Main sources of MI are the diet and cellular biosynthesis. The aim of this review was to describe the transport mechanisms by which MI could be taken up in the small intestine and by the cells, and to summarize knowledge about cellular MI metabolism. Its major physiological role is as a constituent of phosphoinositides which are phosphorylated at different sites of the MI ring. Due to the variation in nomenclature found in literature, Table 4.1 presents an overview about the terminology of the most important cellular phosphoinositides.

DOI 10.3920/978-90-8686-836-0_4, © Wageningen Academic Publishers 2016

Table 4.1. *Terminology of phosphoinositides.*

	Inositolphosphates[1]	Phosphatidylinositols	Common abbreviations
Ins P1	inositol-(3)phosphate	phosphatidylinositol-(3)phosphate	
	inositol-(4)phosphate	phosphatidylinositol-(4)phosphate	
	inostiol-(5)phosphate	phosphatidylinositol-(5)phosphate	
Ins P2	inositol-(3,4)bisphosphate	phosphatidylinositol-(3,4)bisphosphate	
	inositol-(3,5)bisphosphate	phosphatidylinositol-(3,5)bisphosphate	PIP_2
	inositol-(4,5)bisphosphate	phosphatidylinositol-(4,5)bisphosphate	
Ins P3	inositol-(1,4,5)trisphosphate	phosphatidylinositol-(1,4,5)trisphosphate	IP_3
	inositol-(3,4,5)trisphosphate	phosphatidylinositol-(3,4,5)trisphosphate	PIP_3
Ins P4	inositol-(1,2,3,4)tetraphosphate		
	all other combinations of phosphorylation sites		
Ins P5	inositol-(1,2,3,4,5)pentaphosphate		
	all other combinations of phosphorylation sites		
Ins P6	inositol-(1,2,3,4,5,6)hexakisphosphate		phytic acid
	any salt of phytic acid		phytate

[1] Inositol was used throughout the table instead of *myo*-inositol.

4.2 Transmembranal transport of *myo*-inositol

4.2.1 Transport in the small intestine

Dietary sources of MI can be free MI, phytate (*myo*-inositol-(1,2,3,4,5,6)hexakisphosphate and their salts) and phospholipids. Brush border membrane-associated endogenous phytases, phosphatases, pancreatic phospholipases and microbial phytases and phosphatases were degrading MI contained in feedstuffs resulting in free MI, at least in part (Holub, 1986). Of free MI, 99.8% was absorbed in human intestine (Croze and Soulage, 2013) suggesting a high capacity transport system in intestinal epithelium. In the small intestines of rats and rabbits, MI was transported by a secondary active, electrogenic and sodium-dependent process across the apical membrane of enterocytes (Aouameur *et al.*, 2007; Scalera *et al.*, 1991). The transport system was transferring MI together with sodium ions across the brush border membrane into the cell (Aouameur *et al.*, 2007; Scalera *et al.*, 1991). D-glucose and phloridzin were able to inhibit this transport but not in a competitive manner (Aouameur *et al.*, 2007). Glucose transport via sodium-linked glucose transporter 1 (SGLT1) was not influenced by MI (Aouameur *et al.*, 2007). For mammalians, sodium-dependent MI cotransporters SMIT1 (SLC5A3) and SMIT2 (SLC5A11) were detected (Schneider, 2015). SMIT2 was found in the intestine (Schneider, 2015), expressing a stoichiometry of 2 sodium (Na+):1 MI and a K_m value of about 0.14 mM (Aouameur *et al.*, 2007; Scalera *et al.*, 1991) and was proven to be responsible for intestinal MI absorption. High intestinal absorptive capacity for MI indicates the importance of this molecule for intermediary metabolism (Figure 4.1).

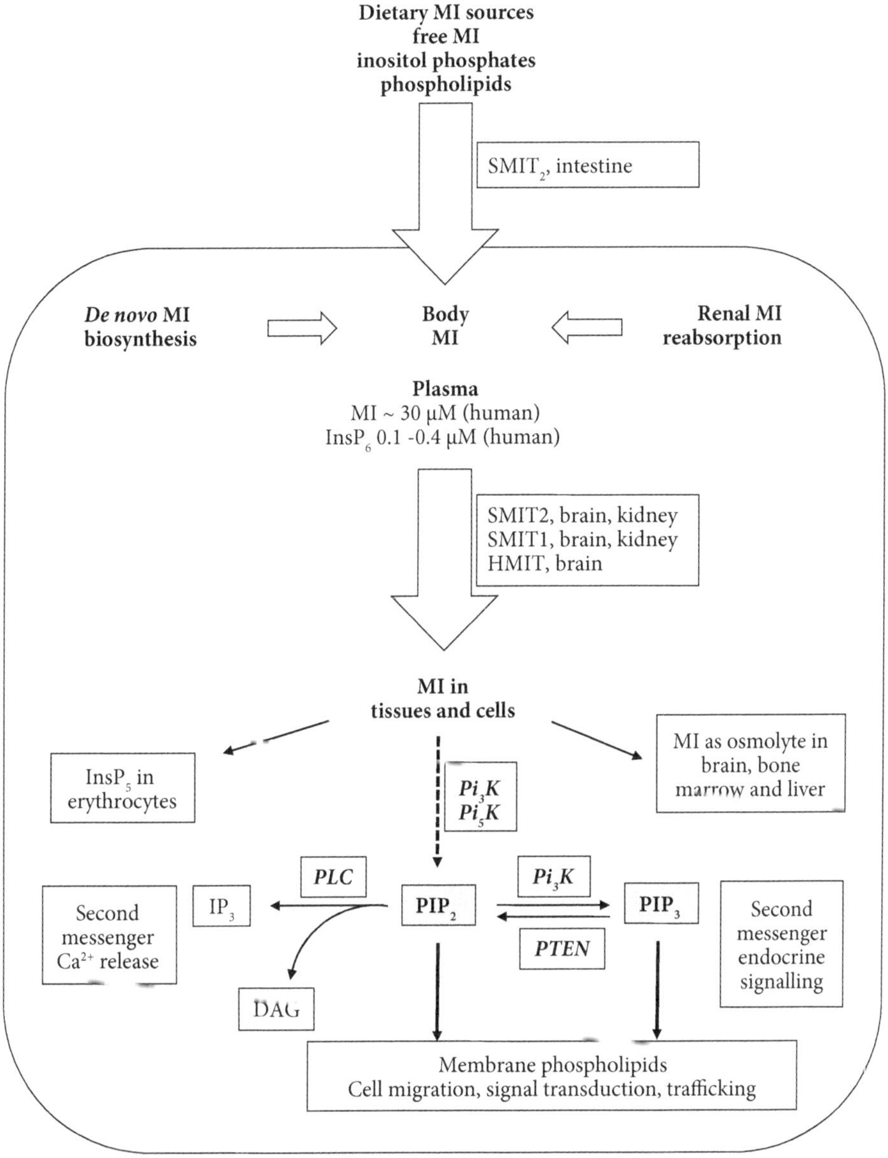

Figure 4.1. *Overview of major pathways of* myo-*inositol metabolism – Sources, cellular transport and utilisation of* myo-*inositol within the cell. Abbreviations: MI =* myo-*inositol; $InsP_6$ = inositol (1,2,3,4,5,6) hexakisphosphate; SMIT1/2 = sodium-linked* myo-*inositol transporter 1 and 2; HMIT = proton-linked* myo-*inositol transporter; PIP_2 = phosphatidylinositol-(4,5)bisphosphate; PIP_3 = phosphatidylinositol-(3,4,5) trisphosphate; IP_3 = phosphatidyl-inositol-(1,4,5)tris-phosphate; DAG = diacylglycerol; PLC = phospholipase C; Pi_3K = phosphoinositide-3 kinase; Pi_5K = phosphoinositide-5 kinase; PTEN = phosphatase and tensin homolog.*

4.2.2 Transport in tissue cells

SMIT1 was mainly expressed in kidney (basolateral membrane) and brain to release MI from the cell; furthermore, in brain, a proton-dependent MI transporter (HMIT; SLC2A13) was identified (Schneider, 2015) (Figure 4.1). In studies with application of radioactively labelled MI in male rats, thyroid, brain, liver, spleen and reproductive tract were immediately incorporating MI, while adipose tissue and muscle contained only very low levels of labelled MI (Croze and Soulage, 2013). Thus, expression of specific MI transporters appeared to be very variable between tissues. Muscle and adipose tissues may rely on tissue-own MI biosynthesis. In kidney (apical membrane), SMIT2 was expressed for reabsorption of MI from the primary urine. Since glucose was inhibiting SMIT2-mediated MI transport due to competition for Na ions, diabetic patients with hyperglycemia and glucosuria had decreased renal MI reabsorption resulting in MI excretion with the urine (inosituria) (Croze and Soulage, 2013).

4.3 Cellular *myo*-inositol metabolism

4.3.1 Cellular biosynthesis and metabolism

Besides by uptake across the plasma membrane, MI amount in the cell is determined by the *de novo* synthesis from glucose (Figure 4.1) as shown in rat testis, brain, liver and kidney, and by the release of MI from phosphatidylinositols (Croze and Soulage, 2013). For *de novo* synthesis, D-glucose is converted to glucose-6-phosphate by hexokinase, then transformed to MI-1-phosphate by MI phosphate synthase and finally dephosphorylated to MI by inositol monophosphatase (Croze and Soulage, 2013). By specific epimerases, MI can be converted to other isoforms but only to a very small proportion. The kidney is the sole organ for MI degradation by MI oxidase activity (Croze and Soulage, 2013). By degradation and reabsorption of MI, the kidney is an important regulator of plasma MI concentration (Holub, 1986). *Myo*-inositol deficient diets reduced the concentration of free MI in urine of rats and also in many body tissues except the brain (Burton *et al.*, 1976; Shepherd and Taylor, 1974). Maintaining its concentrations in brain at dietary MI depletion underlines the importance of MI for the central nervous system. In brain, liver and bone marrow MI was detected as osmolyte to ameliorate osmotic stress in these tissues (Croze and Soulage, 2013). Dietary MI supplementation led to an increase in plasma and tissue MI concentrations (Burton and Wells, 1976).

Myo-inositol is mainly used for intracellular synthesis of phosphoinositides. Phosphoinositide kinases of type 3 or 5 (Pi_3K, Pi_5K) were synthesising phosphatidylinositols phosphorylated at the 3, 4 or 5 sites of the MI ring. These PI were then further phosphorylated to build up phosphatidylinositol-bisphosphates with phosphate groups at positions 3,4, 3,5 and 4,5, respectively. All phosphorylations are reversible reactions. Phosphatase and tensin homolog (PTEN) is counteracting the activity of PiK and is able to dephosphorylate phosphatidyl-trisphosphates. Of special importance is phosphatidylinositol-(4,5)bisphosphate (PIP_2), which is a major precursor for phosphatidylinositol-triphosphates. Pi_3K is phosphorylating PIP_2 to create phosphatidylinositol-(3,4,5)trisphosphate (PIP_3), a key component in endocrine signalling pathways (Shewan *et al.*, 2011). Both pathways, PIP_2 and PIP_3, synthesis were strongly associated with the cellular plasma membrane. In the cytosol, phospholipase C (PLC) is irreversibly hydrolysing PIP_2 to phosphatidylinositol-(1,4,5) trisphosphate (IP_3) and diacylglycerol (DAG). IP_3 is involved in release of calcium (Ca^{2+}) from intracellular stores, thereby mediating a plethora of cellular pathways (Shewan *et al.*, 2011). Phosphatidyl-pentaphosphate ($InsP_5$) was found in avian erythrocytes; a role in regulation of oxygen affinity of hemoglobins was suggested by Holub (1986).

Furthermore, PIP_2 and PIP_3 were major components of the inner leaf of plasma membranes and of organelle-specific membranes (Figure 4.1). Phosphoinositides in membranes were contributing to regulation of the cytoskeleton dynamics, in membrane trafficking, immune cell functions, chemotaxis and specialised docking sites for effectors (Aouameur *et al.*, 2007). They are necessary to fine-tune the composition of the membrane-cytosol interface to control the signalling potential of the cells and the polarity of epithelial cells (Shewan *et al.*, 2011).

It is known that PIP_3 is required for the basolateral entity of a plasma membrane, while PIP_2 is defining the apical entity and the trailing edge of migrating cells (Shewan *et al.*, 2011; Tsujita and Itoh, 2015). Pi_3K, Pi_5K and PTEN are balancing the concentrations of PIP_2 and PIP_3 for adequate membrane composition; however, it is not fully understood so far how these processes are coordinated in regulating mammalian cell architecture.

4.3.2 Physiological role

Considering the contributions of MI and phosphoinositides to membrane functions and to signalling pathways it is obvious that the physiological role of MI must be immense. Generally, MI appeared to act after its transformation to a variety of phosphoinositides (see Table 4.1) on cellular pathways. Furthermore, a multitude of phosphoinositides functions could be explained by the high extent of their interactions with proteins (Hammon and Balla, 2015). An overview of important MI and phosphoinositide effects is given in Figure 4.2.

Myo-inositol and also to a higher extent, phytic acid increased triglyceride (TG) synthesis in 3T3-L1 cells, which metabolically resembles adipocyte physiology (Kim *et al.*, 2014). Higher TG synthesis was based on stimulated protein expression of key lipotropic components, fatty acid synthase, peroxisome-proliferator activator receptor (PPAR) gamma and sterol regulatory element-binding protein (SREBP1). Supporting higher cellular TG concentrations, catecholamine-induced lipolysis was inhibited by MI and phytic acid (Kim *et al.*, 2014).

Thus, stimulating adipocyte differentiation and enhancing adipocyte lipid storage capacity indicated that MI could improve lipid-associated metabolic disturbances. Furthermore, insulin-stimulated glucose uptake was enhanced by MI and this was associated with increased phosphorylation of insulin receptor substrate in 3T3-L1 adipose cells (Kim *et al.*, 2014). *Myo*-inositol mediated enhanced GLUT4 translocation into the plasma membrane of white adipose tissue cells (Croze and Soulage, 2013) and murine skeletal muscle (Dang

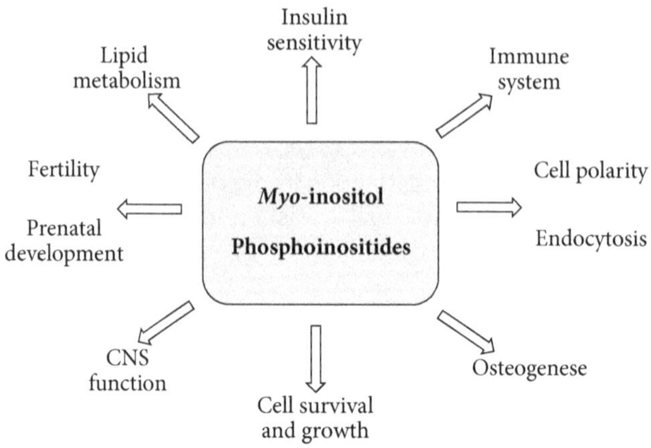

Figure 4.2. *Important physiological processes influenced by* myo-*inositol and phosphoinositides – By modelling the cytoskeleton, the plasma membranes and the intracellular signal cascades* myo-*inositol and its related phosphatidylinositols affect a plethora of cellular processes.*

et al., 2010) appeared to be the underlying pathways to improve insulin sensitivity. The cellular mechanisms by which MI was affecting the insulin-glucose and lipid metabolism are not clear yet.

Potentially, activation of Pi_3K by MI might be one of the pathways, which led to stimulation of the insulin signalling cascade. Activation of Pi_3K in the plasma membrane by insulin, but also by epidermal growth factor (EGF) and insulin-like growth factor (IGF1) resulted in increased concentrations of PIP_3 synthesized from MI. Phosphatidylinositol-(3,4,5) trisphosphate can be dephosphorylated at the 3′site by PTEN and at the 5′site by SH2-containing inositol 5-phosphatases (SHIP) to balance the intracellular content of this signal, which is very low abundance in non-stimulated cells (Manna and Jain, 2015). After binding to PKB/AKT, a major component of insulin signalling cascade, this kinase was phosphorylated by PIP_3-mediated conformational activation. Furthermore, auto-phosphorylation of atypical protein kinase zeta (PKCζ) was induced by PIP_3, which was inducing GLUT4 translocation into the cell membrane and thereby, increased cellular glucose uptake rate (Farese, 2001). In addition, many other PIP_3 targets were identified indicating that PIP_3 had a pivotal role in cell survival and growth, cell differentiation, endocytosis, chemotaxis, immune cell activation and cytokine production (Manna and Jain, 2015; Porciello *et al.*, 2016). PIP_3 signalling induced the mammalian target of rapamycin (mTOR) pathway, thereby stimulating mitochondrial genesis, protein synthesis and *de novo* lipogenesis (Manna and Jain, 2015). Furthermore, PIP_3 and PIP_2 were activating T cell metabolism and differentiation indicating a contribution of MI and related phosphoinositides to immune system function (Porciello *et al.*, 2016). PIP_2 produced by Pi_5K and also phosphatidylinositol-(3,4)bisphosphate were essential for clathrin-mediated endocytosis and intracellular trafficking of vesicles (Posor *et al.*, 2013). A therapeutical role of MI and its isomer, D-chiro-inositol (DCI) in fertility disorders was discussed in recent publications about polycystic ovary syndrome (PCOS) (Bevilacqua and Bizzarri, in press; Bizzarri *et al.*, 2016). PCOS of women is related to diabetes type 2 and obesity;

insulin resistance is commonly discussed as basal mechanism for this fertility disorder. However, it was suggested that MI and DCI were acting on cytoskeleton of the ovary cells, decreasing the cystic degeneration of follicles and thereby, ameliorating reproductive functions. PCOS-like syndromes were also described in dairy cows and broiler breeder hens (Ortega *et al.*, 2015; Walzem and Chen, 2014). Intrauterine development appeared also to be influenced by MI. In naturally occurring low birth weight piglets (intrauterine growth restricted due to larger litter sizes), a high MI and DCI concentration in plasma was detected by a metabolomics approach. *Myo*-inositol was discussed as biomarker of intrauterine metabolic programming (Nissen *et al.*, 2011).

4.4 Conclusions

Myo-inositol and its related phosphoinositides act on many significant biological functions of cells throughout the body. It is assessed to be an essential substrate, which is produced by body own metabolism and is intestinally absorbed from the diet. Increasing availability of MI by dietary supplementation appeared to be beneficial for living organisms, especially in situations of metabolic stress. Since no toxic effects were described, MI seems to be a safe potential nutriceutical substance.

1. *Myo*-inositol is absorbed by the intestinal epithelium efficiently in human and rats.
2. *Myo*-inositol improves insulin sensitivity and lipid metabolism.
3. *Myo*-inositol is an important regulator of cellular architecture.

References

Aouameur, R., Da Cal, S., Bissonnette, P., Coady, M.J. and Lapointe, J.Y., 2007. SMIT2 mediates all *myo*-inositol uptake in apical membranes of rat small intestine. American Journal of Physiology, Gastrointestinal and Liver Physiology 293:G1300-G1307.

Bevilacqua, A. and Bizzari, M., in press. Physiological role and clinical utility of inositols in polycystic ovary syndrome. Best Practice and Research Clinical Obstetrics and Gynaecology. DOI: http://dx.doi.org/10.1016/j.bpobgyn.2016.03.007.

Bizzari, M., Cucina, A., Dinicola, S., Harrath, A.H., Alwasel, S.H., Unfer, V. and Bevilacqua, A., 2016. Does *myo*-inositol effect on PCOS follicles involve cytoskeleton regulation? Medical Hypotheses 91:1-5.

Burton, L.E., Ray, R.E., Bradford, J.R., Orr, J.P., Nickerson, J.A., Wells, W.W., 1976. *Myo*-inositol metabolism in the neonatal and developing rat fed a *myo*-inositol free diet. Journal of Nutrition 106:1610-1616.

Burton, L.E. and Wells, W.W., 1976. Myo-inositol metabolism during lactation and development in the rat. The prevention of lactation-induced fatty liver by dietary *myo*-inositol. Journal of Nutrition 106:1617-1628.

Croze, M.L. and Soulage, C.O., 2013. Potential role and therapeutic interests of *myo*-inositol in metabolic diseases. Biochimie 95:1811-1827.

Dang, N.T., Mukai, R., Yoshida, K. and Ashida, H., 2010. D-pinitol and *myo*-inositol stimulate translocation of glucose transporter 4 in skeletal muscle of C57BL/6 mice. Bioscience, Biotechnology, and Biochemistry 74:1062-1067.

Farese, R.V., 2001. Insulin-sensitive phospholipid signaling systems and glucose transport. Update II. Experimental Biology and Medicine 226:283-295.

Hammon, G.R.V. and Balla, T., 2015. Polyphosphoinositide binding domains: key to inositol lipid biology. Biochimica and Biophysica Acta 1851:746-758.

Holub, B.J., 1986. Metabolism and function of *myo*-inositol and inositol phospholipids. Annual Reviews of Nutrition 6:563-597.

Kim, J.N., Han, S.N. and Kim, H.K., 2014. Phytic acid and *myo*-inositol support adipocyte differentiation and improve insulin sensitivity in 3T3-L1 cells. Nutrition Research 34:723-731.

Manna, P. and Jain, S.K., 2015. Phosphatidylinositol-3,4,5-triphosphate and cellular signalling: Implications for obesity and diabetes. Cell Physiology and Biochemistry 35:1253-1275.

Nissen, P.M., Nebel, C., Oksbjerd, N. and Bertram, H.C., 2011. Metabolomics reveals relationship between plasma inositols and birth weight: Possible markers for fetal programming of type 2 diabetes. Journal of Biomedicine and Biotechnology. DOI: http://dx.doi.org/10.1155/2011/378268.

Ortega, H.H., Marelli, B.E., Rey, F., Amweg, A.N., Diaz, P.U., Stangaferro, M.L. and Salvetti, N.R., 2015. Molecular aspects of bovine cystic ovarian disease pathogenesis. Reproduction 149:R251-R264.

Porciello, N., Kinkl, M., Viola, A. and Tuosto, L., 2016. Phosphatidyl 4-phosphate 5-kinases in the regulation of T cell activation. Frontiers in Immunology 7:186.

Posor, Y., Eichhorn-Gruenig, M., Puchkov, D., Schöneberg, J., Ullrich, A. Lampe, A., Müller, R., Zarbakhsh, S., Gulluni, F., Hirsch, E., Krauss, M., Schultz, C., Schmoranzer, J., Noe, F. and Haucke, V., 2013. Spatiotemporal control of endocytosis by phosphatidylinositol-3,4-bisphosphate. Nature 499:233-237.

Scalera, V., Natuzzi, D. and Prezioso, G., 1991. *Myo*-inositol transport in rat intestinal brush border membrane vesicles, and its inhibition by D-glucose. Biochimica and Biophysica Acta 1062:187-192.

Schneider, S., 2015. Inositol transport proteins. FEBS Letters 589:1049-1058.

Shepherd, N.D. and Taylor, T.G., 1974. A re-assessment of the status of *myo*-inositol as a vitamin. Proceedings of the Nutrition Society 33:63A-74A.

Shewan, A., Eastburn, D.J. and Mostov, K., 2011. Phosphoinositides in cell architecture. Cold Spring Harbor Perspectives in Biology 3:a004796.

Tsujita, K. and Itoh, T., 2015. Phosphoinositides in the regulation of actin cortex and cell migration. Biochimica and Biophysica Acta 1851:824-831.

Walzem, R.L. and Chen, S., 2014. Obesity-induced dysfunctions in female reproduction: lessons from birds and mammals. Advances in Nutrition 5:199-206.

Chapter 5
Inositol diphosphates: an expanding repertoire of functions and regulation

U. Padmanabhan[1], R.S. Kilari[2], L.J. Winward[3] and S.T. Safrany[4]*
[1]Haffkine Institute for Training, Research & Testing, Acharya Donde Marg, Parel, Mumbai 400012 India; [2]School of Pharmacy, University of Wolverhampton, Wolverhampton, WV1 1LY, United Kingdom; [3]School of Life Sciences, Dundee University, Dundee, DD1 5EH, United Kingdom; [4]Royal College of Surgeons in Ireland, Medical University of Bahrain, P.O. Box 15503, Adliya, Bahrain; ssafrany@rcsi-mub.com

Abstract

The first formal discovery of inositol diphosphates appeared in 1993. The intervening years have led us through a colourful and often turbulent journey. The most phosphorylated inositide, bis-diphosphoinositol tetrakisphosphate ($[PP]_2$-$InsP_4$ or $InsP_8$) continues to be coy; we have been unable to ascribe any function to it. We now understand the biochemical pathways involved in its synthesis. Inositol hexakisphosphate kinase (IHK) can produce inositol diphosphates containing 5-diphosphate groups whereas diphosphoinositol polyphosphate kinases (PP-$InsP_5$Ks) β-phosphorylate at the (D/L) 1-position. Both are required to synthesise $[PP]_2$-$InsP_4$. Functions for 5-diphosphates include regulation of insulin signalling, DNA repair, bone deposition, inorganic phosphate and inositol uptake, protein trafficking and ubiquitylation, apoptosis, immune signalling, psychiatric disorders and several other signalling pathways. Data using knockout mice for each IHK give a range of phenotypes, but we await publication of knockout mutants of PP-$InsP_5$Ks. Only then can we start establishing which phenotype can be assigned to each diphosphoinositol pentakisphosphate (PP-$InsP_5$) or $[PP]_2$-$InsP_4$. As binding partners exist for several of the kinases involved, care is required to determine whether these kinases act as scaffolds or localised delivery systems for inositol diphosphates which are then having their effects. Similarly, the kinases involved in inositol diphosphate synthesis are targets for kinase inhibitors. The low affinity of IHKs for ATP makes them particularly susceptible to ATP-competitive kinase inhibitors. Here we bring together results from genetic and pharmacological manipulation of inositol diphosphate levels in an attempt to further refine our current understanding and better determine what the future holds.

Keywords: DNA repair, hexakisphosphate, pyrophosphate

5.1 Introduction

The versatility of *myo*-inositol, containing one axial and five equatorial hydroxyl groups, is astonishing. Of the 63 possible combinations by which phosphates can replace these hydroxyls, over half are routinely found in mammalian cells. In addition to these,

C.L. Walk *et al.* (ed.) **Phytate destruction – consequences for precision animal nutrition**
DOI 10.3920/978-90-8686-836-0_5, © Wageningen Academic Publishers 2016

seven inositol phospholipids are interlinked. Should these molecules not offer enough complexity of signalling, inositol diphosphates (diphosphoinositol polyphosphates or inositol pyrophosphates) have also been found (Stephens *et al.*, 1993). Although levels and structures in mammals are greatly different to those found in *Entamoeba* and slime moulds, these diphosphates have been implicated in many cellular events. The need for further structural distinctions is not why these diphosphates have evolved. It is more likely that these molecules are being used for their ability to donate phosphates. Indeed, the energy released upon fission of the β-phosphate may exceed that available from ADP (Stephens *et al.*, 1993). This could be the reason for their existence – donation of phosphates to target molecules, as will be discussed later.

Under physiological situations, inositol diphosphates are created by the phosphorylation of either inositol 1,3,4,5,6-pentakisphosphate ($InsP_5$) to diphosphoinositol tetrakisphosphate (PP-$InsP_4$) and possibly further to bis-diphosphoinositol trisphosphate ($[PP]_2$-$InsP_3$), or inositol hexakisphosphate ($InsP_6$) to diphosphoinositol pentakisphosphate (PP-$InsP_5$) and bis-diphosphoinositol tetrakisphosphate ($[PP]_2$-$InsP_4$) (Draskovic *et al.*, 2008). It has taken several years and multiple false starts to determine the structures of the inositol diphosphates. Analysis of cell extracts has proved difficult due to the low levels found in mammalian cells. In contrast, other eukaryotes appeared to contain relatively high levels of diphosphates, so it was from these that the original analyses were performed. The first characterisation was from *Dictyostelium discoideum*, showing high levels of (D/L) 1-PP-$InsP_5$ and (D/L) 1,4-$[PP]_2$-$InsP_4$ (Stephens *et al.*, 1993). Synthesis of several PP-$InsP_5$ isomers followed by simple enzymatic analysis led us to believe that (D) 1-PP-$InsP_5$ was the most likely enantiomer present in mammalian samples (Falck *et al.*, 1995). In contrast, *Entamoeba histolytica* was proposed to contain 5-PP-$InsP_5$ (Martin *et al.*, 1993) or even 2-PP-*neo*-$InsP_5$ and 2,5-$[PP]_2$-*neo*-$InsP_4$ (Martin *et al.*, 2000). Further analysis using *D. discoideum* proposed 6-PP-$InsP_5$ as the major PP-$InsP_5$ and vicinal diphosphates in 5,6-$[PP]_2$-$InsP_4$ creating higher energies of hydrolysis due to steric repulsion (Laussmann *et al.*, 1997). The existence of 5-PP-$InsP_5$ and 1,5-$[PP]_2$-$InsP_4$ in *Polysphondilium pallidum* (Laussmann *et al.*, 1998) further clouded matters. Complex HPLC-metal dye detection analysis of mammalian cell samples also proposed the presence of 5-PP-$InsP_5$ (Albert *et al.*, 1997). Later characterisation of mammalian enzymes identified that the major product of $InsP_6$ phosphorylation by inositol hexakisphosphate kinases (IHKs) leads to the formation of 5-PP-$InsP_5$ (Draskovic *et al.*, 2008) and that the PP-$InsP_5$ kinase (PP-$InsP_5$K) phosphorylated both $InsP_6$ and PP-$InsP_5$ in the 1-position (Lin *et al.*, 2009), albeit phosphorylating $InsP_6$ at a much slower rate (Fridy *et al.*, 2007). Likewise, it appears that IHKs may be able to phosphorylate 1-PP-$InsP_5$ creating 1,5-$[PP]_2$-$InsP_4$ (Figure 5.1).

In vitro analysis of phosphorylation products proposes that IHKs can phosphorylate $InsP_5$ at both the 1- and 5-positions, although the favoured site for phosphorylation appears to be the 1-phosphate. Further phosphorylation was also proposed, allowing the production of triphosphates and beyond (Draskovic *et al.*, 2008). The *in vivo* relevance of such molecules remains uncertain.

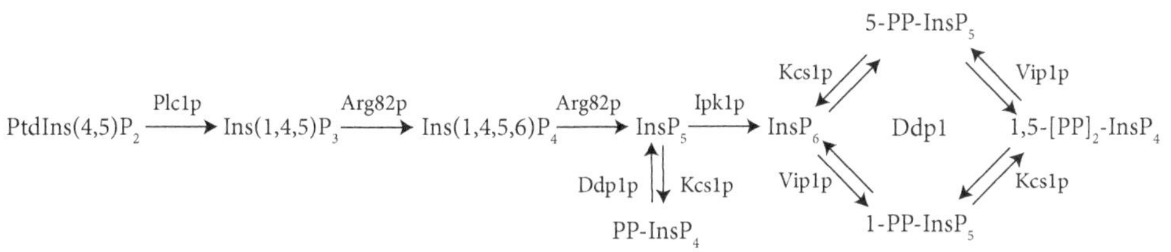

Figure 5.1. *Metabolic relationship of higher inositol phosphates. Yeast enzymes involved in each step are shown. Plc1p is a phospholipase C; Arg82p is an inositol tris/tetrakisphosphate kinase; Ipk1p is an inositol pentakisphosphate kinase; Kcs1p is an inositol hexakisphosphate kinase (IHK); Vip1p is a diphosphoinositol pentakisphosphate kinase (PP-InsP$_5$K); Ddp1p is a diphosphoinositol polyphosphate phosphohydrolase (DIPP).*

In this chapter, we will give an overview of the inositol diphosphates in three sections:
1. Effector proteins downstream of inositol diphosphates.
2. Biochemical regulation – enzymes directly involved in the formation, interconversion and breakdown of inositol diphosphates.
3. Pharmacological regulation – drugs and treatments affecting inositol diphosphate levels.

5.2 Effector proteins downstream of inositol diphosphates

Some 20% of InsP$_5$ and 50% of InsP$_6$ are converted into inositol diphosphates every hour (Menniti *et al.*, 1993). For the cell to expend so much energy producing and maintaining inositol diphosphate levels, it is reasonable to assume that these molecules have important roles to play. Before the formal discovery of inositol diphosphates, IHKs were already known due to their downstream consequences.

In *Saccharomyces cerevisiae*, the IHK knockout (*kcs1Δ*) had been studied before the enzyme activity of Kcs1p was known. Found to reverse the phenotype in a protein kinase C (pkc) mutant, Kinase C Suppressor 1 knockouts reversed the 73-fold increase in DNA recombination observed in a pkc mutant (Huang and Symington, 1995). Following identification that Kcs1p was the yeast IHK, Saiardi *et al.* (2000) showed that *kcs1Δ* did contain lower levels of inositol diphosphates, and that the yeast showed a range of phenotypes: reduced growth rate, temperature sensitivity and small, fragmented vacuoles. They suggested that altered protein trafficking may explain the vacuole phenotype. Further phenotypes included salt (NaCl)-sensitivity and growth impairment.

Soon after their discovery, several proteins with high *in vitro* affinity for inositol diphosphates were identified in pathways involved in protein trafficking. InsP$_6$ had been shown to bind AP-2, a clathrin assembly protein involved with potassium channel activity, although it did not regulate activity under the conditions tested (Timerman *et al.*, 1992). Cardiac AP-3 also bound inositol 1,4,5-trisphosphate (Ins(1,4,5)P$_3$) and inositol 1,3,4,5-tetrakisphosphate, which inhibited channel activity, and InsP$_6$ which did not (Kijima *et al.*, 1993). Synapse-specific AP-3 bound PP-InsP_5 with highest affinity, which blocked clathrin assembly (Ye *et al.*, 1995). Likewise, short-chain phosphatidylinositol 3,4,5-trisphosphate (PtdInsP_3) has been shown to inhibit clathrin assembly, binding the same site as InsP_6 (Gaidarov *et al.*, 1996), indicating these molecules may inhibit protein trafficking due to their inhibition of clathrin assembly. That short-chain PtdInsP_3 was able to inhibit a greater proportion of clathrin assembly than InsP_6 and PP-InsP_5 (Hao *et al.*, 1997) led many to believe that inositol lipids, not phosphates were the physiological regulators of these events. The need to use short-chained versions of inositol lipids that are water soluble could, in this instance, have caused misinterpretation regarding whether they are activators or inhibitors of these events. An alternative explanation could be that inositol phosphates hold the AP-2 in a disassembled, soluble form. Activation of phosphoinositide 3-kinase (PI 3-K) drives production of PtdInsP_3 and related lipids in the membrane, causing translocation to the membrane and conformational change to the assembled form (Gaidarov and Keen, 1999). Coatomer, a protein complex involved in vesicle transport with an ion channel activity, has high affinity for PP-InsP_5. Binding of PP-InsP_5 was found to close this potassium channel at physiologically-relevant doses. Channel activity was also blocked by InsP_6 (albeit at a higher dose) (Fleischer *et al.*, 1994). Similar observations were made using coatomer purified from *S. cerevisiae*, although here PP-InsP_5, $[PP]_2$-InsP_4 and InsP_6 were equipotent while PtdInsP_3 was found to bind weakly (Ali *et al.*, 1995). However, a later paper showed that α-COP, a component of coatomer, bound PtdInsP_3, although competition binding with PP-InsP_5 was not explored (Chaudhary *et al.*, 1998). This protein trafficking defect was observed in intact cells as well as receptor-mediated endocytosis was also impeded (Saiardi *et al.*, 2002).

The concept that inositol phosphates can bind inositol lipid-binding domains, and hence regulate proteins containing these domains, is well understood. Phospholipase C δ1 (PLCδ1) binds both phosphatidylinositol 4,5-bisphosphate (PtdIns(4,5)P_2) and the product of hydrolysis Ins(1,4,5)P_3 by virtue of its pleckstrin homology (PH) domain (Lemmon *et al.*, 1995). Using a green fluorescent protein (GFP) tag of this PH domain, it is possible to detect phospholipase C activation in live cells, as the Ins(1,4,5)P_3 produced binds the PLCδ1 PH domain, causing translocation from the plasma membrane (PtdIns(4,5)P_2-bound) to the cytoplasm (Ins(1,4,5)P_3-bound) (Stauffer *et al.*, 1998). The PH domain of *D. discoideum* cytosolic regulator of adenylyl cyclase (PH_{CRAC}) has high affinity for PP-InsP_5 as well as PtdInsP_3, which is required for directional chemotactic responses in *D. discoideum* (Luo *et al.*, 2003). Knockout of IHK sensitised cells to the extracellular messenger cyclic AMP, showing that PP-InsP_5 is acting as a negative regulator in the PtdInsP_3-mediated movement of *D. discoideum.*

The PH domains of 3-phosphoinositide-dependent protein kinase-1 (PDK1) and Akt (also known as protein kinase B) also appear to be regulated by inositol diphosphates (Komander *et al.*, 2004; Luo *et al.*, 2003). Under resting conditions, PDK1 and Akt are cytoplasmic, bound to these inositol phosphates. Stimulation of growth factor and insulin receptors activates PI 3-K, raising membrane PtdInsP_3 for which these proteins also have high affinity. PDK1 and Akt are brought together, allowing PDK1 to phosphorylate and activate Akt. Thus, the level of Akt activation is dependent on a balance between PI 3-K products and inositol phosphates. Mouse embryonic stem cells lacking IHK1 produce very little 5-PP-InsP_5 and show raised sensitivity to stimulation by insulin-like growth factor-1. Akt phosphorylation is enhanced, leading to increased activation of a series of downstream events (Bhandari *et al.*, 2008; Chakraborty *et al.*, 2010). Akt regulates both glucose uptake into muscle and fat by driving the translocation of the glucose transporter GLUT4 to the plasma membrane then glycogen synthesis by inactivating glycogen synthase kinases 3α and 3β (GSK3α and 3β), allowing unphosphorylated glycogen synthase to catalyse the production of glycogen from glucose (Hajduch *et al.*, 1998). The physiological outcome is that mice lacking IHK1 remain insulin-responsive with lower insulin levels, blood glucose and reduced body weight (Bhandari *et al.*, 2008; Chakraborty *et al.*, 2010). Unfortunately, IHK1 knockout males were sterile, as spermatogenesis was impaired (Bhandari *et al.*, 2008). In contrast, knockdown of IHK1 has been shown to inhibit exocytosis of insulin-containing granules from pancreatic β cells. Overexpression of IHK1, 2 and 3 was seen to increase exocytosis of these insulin-containing granules (Illies *et al.*, 2007).

Glycogen synthase kinase 3α and 3β also bind, and are regulated by, IHK1. Akt-mediated GSK3β phosphorylation, which may already be reduced by 5-PP-InsP_5 (see above) is further inhibited as IHK1 binds and masks the Akt phosphorylation site. As unphosphorylated GSK3β is active, this would enhance GSK3β signalling. The consequences observed when IHK1 was knocked out in mice extended beyond glucose metabolism and included reduced locomotor activity in a novel environment and reduced social motivation, making the authors consider whether 'modulation of brain IHK1 could represent a previously unexplored avenue for the management of mental disorders' (Chakraborty *et al.*, 2014). Glycogen synthase kinase 3β has been proposed as a target for lithium therapy of manic depression (Freland and Beaulieu, 2012). Earlier data showed long-term treatment of mice with lithium significantly raised expression of diphosphoinositol polyphosphate phosphohydrolase 2 (DIPP-2), an enzyme involved in breakdown of inositol diphosphates, specifically in the cortex (Hua *et al.*, 2001).These observations show that both direct protein-protein interactions and altered metabolism of PP-InsP_5 are important in Akt signalling.

PP-InsP_5 is also regulated by insulin-like growth factor-1, accumulating over a 60-minute period in mouse embryonic fibroblasts; this was also observed in hepatocellular carcinoma cells stimulated with insulin-like growth factor-1 and insulin (Chakraborty *et al.*, 2010). Chakraborty *et al.* (2010) proposed that such increases in PP-InsP_5 could lead to insulin resistance and obesity.

In neutrophils, chemotactic signalling was augmented in cells lacking IHK1, showing that 5-PP-InsP_5 was acting as a negative regulator in immune cells too (Prasad *et al.*, 2011). Knockout of IHK1 in neutrophils showed increased receptor-dependent, NADPH oxidase-mediated superoxide production. Equally, activation of normal cells with the chemoattractant N-formylmethionyl-leucyl-phenylalanine caused a rapid decline in PP-InsP_5 levels, leading the authors to suggest that PP-InsP_5 is important in preventing resting neutrophils from inadvertent activation, while reduction of PP-InsP_5 allows sustained Akt signalling (Prasad *et al.*, 2011).

Inositol hexakisphosphate kinase 1 has also been implicated in bone development. Regulation of protein kinase C δ (PKCδ) by IHK1, but not IHK2 would suggest that cytoplasmic 5-PP-InsP_5 is required by PKCδ to translocate to the nucleus following stimulation of osteoblasts with fibroblast growth factor-2. Protein kinase C δ translocation leads to the expression of several osteogenic (bone producing) genes (Niger *et al.*, 2013). Furthermore, it is proposed that 5-PP-InsP_5 may be involved in intercellular signalling, diffusing through gap junctions and activating PKCδ in neighbouring cells, as activated PKCδ associates with connexin43 prior to delivery to the nucleus (Niger *et al.*, 2010). Could this pathway be delivering PP-InsP_5 to neighbouring cells, as suggested by Niger *et al.* (2013)? We are unaware of any publications assessing whether fibroblast growth factor-2 can raise inositol diphosphate levels and feel this would be required for such intercellular events to occur.

Effects on DNA regulation have been uncovered by York *et al.* (2005) who proposed that PP-InsP_4 was important in regulation of telomere length. Yeast lacking all inositol diphosphates had elongated telomeres, whereas mutants capable of producing only PP-InsP_4 showed telomeres shorter than wild type yeast. We believe this is an oversimplification of the situation and await studies with Kcs1p mutants that can selectively phosphorylate either InsP_5 or InsP_6 to validate such a claim.

A correlation between 5-PP-InsP_5 levels and the age of cells has been observed in bone marrow-derived mesenchymal stem cells (MSCs). Older cells had higher 5-PP-InsP_5 levels; they also showed decreased phosphorylation of Akt and a higher level of cells undergoing apoptosis (Zhang *et al.*, 2014b). As MSCs are an ideal therapy for treatment of myocardial infarction, inhibition of IHK was targeted and found to improve MSC survival and function in infarcted mouse hearts, suggesting that IHK inhibition after myocardial infarction could be of therapeutic benefit (Zhang *et al.*, 2014a). Zhang *et al.* (2014a) utilised N^2-(m-Trifluorobenzyl), Nω-(p-nitrobenzyl)purine (TNP) to inhibit IHK and lower PP-InsP_5 levels. This cell permeable small molecule was first shown to be a potent IHK inhibitor by Padmanabhan *et al.* (2009). TNP was originally described as an Ins(1,4,5)P_3 3-kinase inhibitor, yet was found to be approximately 1000-fold selective towards IHKs *in vivo*. As an ATP-competitive inhibitor, TNP's selectivity is aided by the low affinity IHKs have for ATP (1.35 mM) (Voglmaier *et al.*, 1996). It was also found not to interact with PP-InsP_5Ks. TNP also inhibited the *S. cerevisiae* form of IHK, Kcs1p. Inhibition of the kinase mimicked knockout phenotypes previously observed (Padmanabhan *et al.*, 2009).

IHK1 has been reported to be responsible for homologous recombination (HR) repair in mouse embryonic fibroblasts (MEFs). Treatment of MEFs with DNA-damaging agents showed that cells were unable to effect successful HR (Jadav *et al.*, 2013). Our findings showed that *S. cerevisiae* lacking inositol diphosphates were as sensitive to bleomycin, a DNA damaging agent, as the 'gold standard' mutants, Rad51 and Rad52 (Lettier *et al.*, 2006) (Figure 5.2) further indicating that inositol diphosphates may be involved in DNA repair through phosphorylation by IHK1 (Kilari, 2014).

This sensitivity was not observed when yeast were exposed to γ-irradiation as HR was not inhibited in these yeast strains (Padmanabhan *et al.*, 2005), thereby determining that the effects of such drugs is much oversimplified when used in biological systems (Dedon and Goldberg, 1992; Povirk, 1996). *S. cerevisiae* PP-InsP_5K deletion mutants (*vip1*Δ) were not

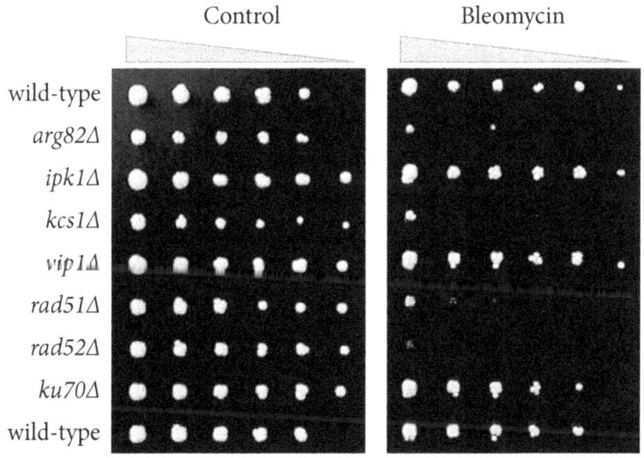

Figure 5.2. *Effect of bleomycin on mutant* S. cerevisiae *strains. Yeast (BY4741 from Euroscarf – all deletions were confirmed by PCR analysis) were grown to log-phase in yeast extract peptone dextrose (YPD) medium at 30 °C. The strains were serially diluted at 1/3 with YPD medium and spotted onto YPD agar plates (left to right; starting at 10,000 cells/spot) containing bleomycin at 10 mU/ml. A control, with yeast dilutions spotted onto an untreated YPD plate was also included. Plates were incubated at 30 °C for 48 h and photographed.*

sensitive to bleomycin, suggesting that 5-PP-InsP_5 and not 1-PP-insP_5 is important in this pathway (Figure 5.2).

Understanding the pathways and the roles of inositol diphosphates has left some of us feeling very battered and bruised. Our current understanding is that *S. cerevisiae* cells lacking inositol diphosphates are unable to process DNA containing 3' adducts, due to the mislocalisation of a key enzyme: while supernatants from homogenates of wild type *S. cerevisiae* are able to process U:G mismatches in DNA, using the protocol and reagents described by Owen *et al.* (2007), *kcs1Δ* supernatants were ineffective. More aggressive extraction procedures, requiring ethylenediamine tetraacetic acid (EDTA) or Tween 20 were able to rescue enzyme activity (Figure 5.3). The enzyme activity described resembled that of a uracil DNA glycosylase (UDG) (Owen *et al.*, 2007). We continue to study the mechanism(s) involved.

Knockout of a second, genetically unique inositol hexakisphosphate kinase, IHK2, renders mice predisposed to oral and oesophageal cancer when exposed chronically to low levels of 4-nitroquinolone, suggesting that 5-PP-InsP_5 acts as a pro-apoptotic in cells damaged by such agents (Morrison *et al.*, 2009). Although knockout mice showed no abnormality during development (Morrison *et al.*, 2009), zebrafish lacking IHK2 showed embryonic defects in craniofacial, slow muscle, somite and neural crest cells (Sarmah and Wente, 2010).

There is increasing evidence showing that 5-PP-InsP_5 is involved in apoptosis and autophagy. Interferon β (INF-β) drives apoptosis in ovarian cancer cells and upregulates IHK2 (Morrison *et al.*, 2001). Cells can be sensitised to INF-β by overexpression of IHK2 or made resistant by antisense approaches (Morrison *et al.*, 2001). Half of the enhancement

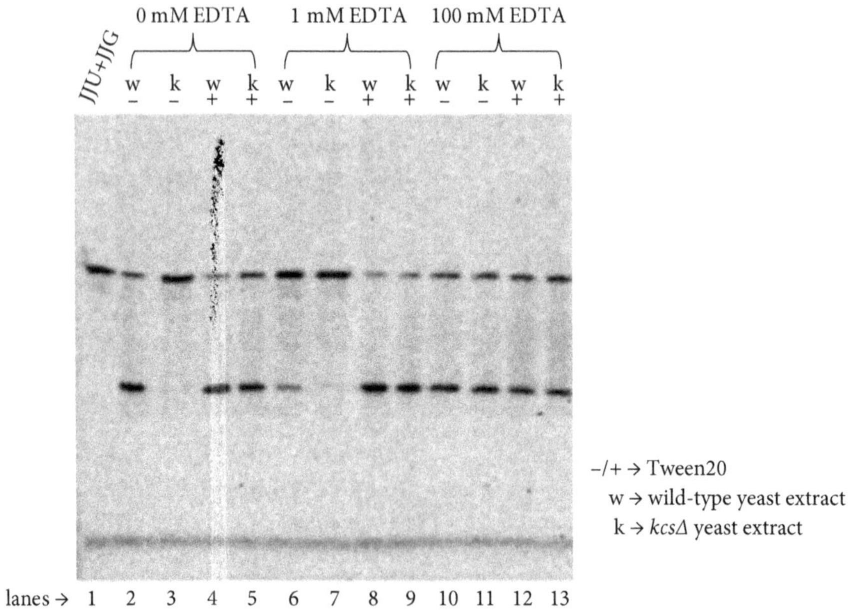

Figure 5.3. S. cerevisiae *yeast extracts and their resemblance to UDG-like enzyme. Extracts of wild type and* kcs1Δ *mutants were prepared in the presence or absence of 2% Tween-20 and in the presence of 0, 1 and 100 mM EDTA. The double stranded DNA (50 nM) containing a U:G mismatch was incubated with relevant yeast extracts for 30 min at 30 °C and heat treated at 96 °C for 5 min. Samples were loaded onto 7 M urea gel and run at 60 °C for 1 hour. Gel was scanned using phosphorimager. Lane 1: untreated substrate (50 nM). Lane 2-5: substrates treated with wild-type and* kcs1Δ *extracts prepared in the presence or absence of Tween-20 (no EDTA). Lane 6-9: substrates treated with yeast extracts prepared in the presence or absence of Tween-20, with 1 mM EDTA. Lane 10-13: substrates treated with yeast extracts prepared in presence or absence of 2% Tween-20, with 100 mM EDTA.*

by IHK2 was observed in a kinase-dead mutant, indicating either a second activity of IHK2 or direct protein interactions were involved (Morrison *et al.*, 2001). One such target is the tumour necrosis factor receptor associated factor-2 (TRAF-2) (Morrison *et al.*, 2007). Autophagy, the main pathways intracellular proteins and organelles are degraded, also appears to be regulated by inositol diphosphates. Autophagy is inhibited by the mechanistic/mammalian target of rapamycin (mTOR), which requires phosphorylation for activation. Overexpression of all mammalian IHKs reduced mTOR phosphorylation and hence activation, thus preventing the negative effects of mTOR on autophagy. S_iRNA did the reverse, increasing mTOR phosphorylation and decreasing autophagy, leading the authors to suggest IHKs and mTOR act as part of one pathway in the regulation of autophagy (Nagata *et al.*, 2010). Using knockout strategies in *S. cerevisiae*, Taylor *et al.* (2012) were able to show *kcs1*Δ, which are unable to synthesise 5-PP-InsP_5, but not *vip1*Δ mutants, which cannot synthesise 1-PP-InsP_5, are unable to activate autophagy following nitrogen starvation-induced autophagy. This phenotype was not merely due to the vacuole phenotype previously observed, as Taylor *et al.* (2012) showed that the autophagosome was mislocalised. With these studies, they were able to determine that

5-PP-$InsP_5$ is required for effective autophagy (Taylor *et al.*, 2012). Several neurological conditions exist in which autophagy is hindered, including Huntington's, Alzheimer's and Parkinson's diseases (Martin *et al.*, 2015). Huntington's disease is caused by elongation of the Huntingtin protein, with the length of the polyQ repeat being linked to the severity of the condition. This polyQ tract causes accumulation of toxic oligomers and aggregates. The toxicity of this protein can be reversed by phosphorylation by Akt (Humbert *et al.*, 2002). Increased levels of autophagosomes have been observed in cells from Huntington's patients, corresponding with cytoplasmic localisation of IHK2 and raised 5-PP-$InsP_5$ levels (Nagata *et al.*, 2011). These cytoplasmic actions of IHK2 and were proposed to drive cell death through the autophagic pathway (Nagata *et al.*, 2011).

Our own work with *Drosophila melanogaster* CG10082, the *D. melanogaster* IHK, and Aps, the DIPP homologue in *D. melanogaster,* (Winward *et al.*, 2010) would suggest inositol diphosphates are important in early development. We were able to acquire mutants overexpressing and lacking the *D. melanogaster* IHK. During routine stock maintenance, the EP(2)2319 flies, which would normally overexpress IHK when crossed with GAL4 flies, were found to be less viable than wild type flies, with 20% of the embryos dying. In comparison, only 1% of wild type embryos did not hatch. Even without the GAL4 cross, flies had increased PP-$InsP_5$ levels, possibly due to increased expression of IHK. It was also noted that they have a longer generation time and required higher numbers of parental flies to maintain a healthy stock. Similarly, BL13151, a mutant into which a P element has been inserted into the IHK gene, showed 25% of embryos dying. Measurement of inositol diphosphates suggested that BL13151 were defective in PP-$InsP_5$ production, steady-state levels being decreased by over 90% (Winward, 2005).

Overexpression of body-wide CG10082 using a heat-shock promotor led to eye phenotypes, only observable using scanning electron microscopy. In contrast to wild type flies, which showed rounded facets and orderly, cleaned bristles (Figure 5.4A), CG10082 overexpression led to fragile facet faces which collapsed under the electron beam. Bristles also showed evidence of damage, frequently irregular in their presentation (Figure 5.4B) (Winward, 2005).

The first details of IHK3 knockout mice appeared last year. These mice showed motor learning and coordination defects (Fu *et al.*, 2015), associated with abnormalities in Purkinje cell structure and synapse number. Time will tell whether these effects are mediated by kinase activity or purely protein-protein interaction.

The high energy pyrophosphate bond in PP-$InsP_5$, which has a free energy comparable to that of ADP (Stephens *et al.*, 1993) led many researchers to hunt for protein targets being directly phosphorylated by PP-$InsP_5$ and $[PP]_2$-$InsP_4$. PP-$InsP_5$ was shown to phosphorylate proteins in yeast extracts by a non-enzymatic mechanism (Saiardi *et al.*, 2004). Three targets were identified: NSR1 (a protein binding both topoisomerase I and DNA telomeres, allowing the former to act on the latter (Edwards *et al.*, 2000) – potentially explaining why inositol diphosphates regulate telomere length (York *et al.*, 2005)), YGR130c (a cytoplasmic protein whose function remains unknown) and NOPP140 (a chaperone for small nucleolar ribonucleoprotein particles involved in the modification of

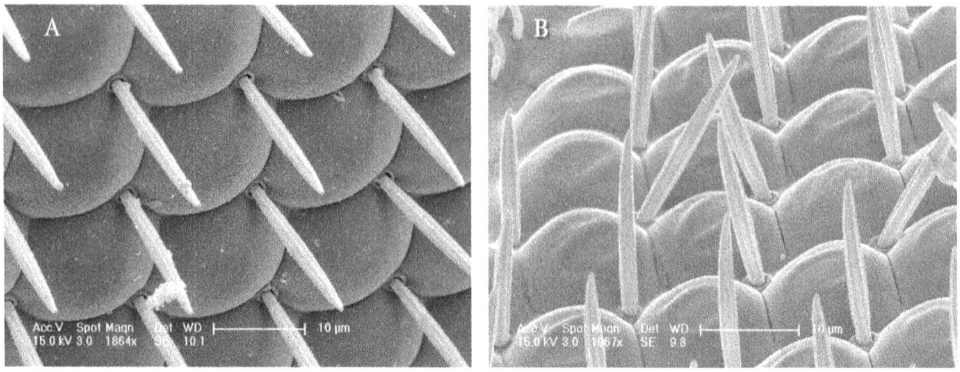

Figure 5.4. *Overexpression of IHK in* D. melanogaster *leads to an eye phenotype. The SEM images show the surface of the compound eye from (A) wild type and (B) EP(2) 2319 × HS GAL4* D. melanogaster. *Flies were 10 days old, heat shocked daily from 3° larval stage.*

ribosomal RNA (Yang *et al.*, 2000)). The context of phosphorylation was that the protein required 'several short serine stretches surrounded by acidic amino acids'. Truncated mouse NOPP140 was also shown to be phosphorylated, suggesting such phosphorylation is also possible in mammals (Saiardi *et al.*, 2004). A subsequent publication declared that this event was in fact pyrophosphorylation and it requires priming phosphorylation of the substrate protein by a kinase using ATP or GTP as a phosphate donor. As very few kinases can substitute ATP with GTP as their high-energy phosphate source, the authors stated their first choice of kinase was casein kinase 2. The β-phosphate from PP-$InsP_5$ then moves to this already existent phosphate, resulting in pyrophosphorylation of the substrate. These effects were also seen using $[PP]_2$-$InsP_4$ (Bhandari *et al.*, 2007). The pyrophosphorylated proteins were not substrates for the DIPPs which selectively remove the β-phosphate from inositol diphosphates. Indeed, they appear relatively resistant to phosphatases in general; however, a recent report suggests they are susceptible to alkaline phosphatases, showing this could be a physiological signalling event (Yates and Fiedler, 2015). The only other publication we are aware of is the pyrophosphorylation of AP-3 binding protein 1 (see above for PP-$InsP_5$ binding to AP-3). This pyrophosphorylation has been linked to the release of HIV-1 Gag protein (Azevedo *et al.*, 2009), suggesting that PP-$InsP_5$ and protein pyrophosphorylation could be a target in the treatment of HIV.

Clearly, there are many more interesting phenotypes awaiting to be discovered. Several studies using *S. pombe* lacking PP-$InsP_5$ kinase (PP-$InsP_5$K) show several phenotypes (Pohlmann and Fleig, 2010; Pohlmann *et al.*, 2014). No data on mammalian PP-$InsP_5$K knockouts have yet been published. Then we may start understanding the roles of 1-PP-$InsP_5$. A recent publication shows that in MEFs alone knockout of IHK1 alters the regulation of over 1,250 genes (Jadav *et al.*, 2016). The delineation of direct and indirect effects may take a while. One such gene has already been identified: IHK1, bound to phosphatidic acid can enter the nucleus, associate with chromatin and produce 5-PP-$InsP_5$. Raised localised PP-$InsP_5$ levels promote DNA methylation and silencing of the inositol 3-phosphate synthase gene (Yu *et al.*, 2016).

To summarise this section, inositol diphosphates appear to play a major role in regulating PH domain-containing proteins. By competing with inositol lipids, they prevent inappropriate over-activation of several key pathways involved in growth and proliferation, and insulin-mediated glycogen synthesis. As well as this negative regulation of proliferative signals, inositol diphosphates appear to be positive signals for autophagy and apoptosis, both physiological as well as pathological. DNA maintenance and repair can be affected, with inositol diphosphates being required for repair following specific insults. Finally, the story of protein pyrophosphorylation creating a long-lived species awaits further developments.

5.3 Biochemical regulation

5.3.1 Inositol hexakisphosphate kinases

Before starting this section, the reader should be discouraged from thinking that IHKs phosphorylate only $InsP_6$. It is now clear that IHKs as well as 'PP-$InsP_5$Ks' readily phosphorylate $InsP_6$. The crucial outcome is that IHKs add a β-phosphate to the 5-position and PP-$InsP_5$Ks add a β-phosphate to the 1 phosphate. Equally IHKs and PP-$InsP_5$Ks can both readily phosphorylate PP-$InsP_5$. Both are required to produce 1,5-$[PP]_2$-$InsP_4$. Other '$InsP_8$'s can be made in the absence of either; these are likely to be triphosphates (Draskovic *et al.*, 2008).

The initial purification of IHK showed it to be reversible, also allowing production of $InsP_6$ and ATP from PP-$InsP_5$ and ADP. A low affinity for PP-$InsP_5$ (1.97 μM) would suggest that, *in vivo*, the equilibrium would greatly favour PP-$InsP_5$ production although the reverse reaction had a significantly higher V_{max} (Voglmaier *et al.*, 1996). As stated by Voglmaier *et al.* (1996), this would suggest the energy of hydrolysis of the β-phosphate in PP-$InsP_5$ is considerable.

Subsequent cloning and characterisation of the three human IHKs suggests that these proteins have substantial overlap in characteristics and properties (Saiardi *et al.*, 2000; Saiardi *et al.*, 1999; Saiardi *et al.*, 2001; Schell *et al.*, 1999). Localisation studies with overexpressed GFP-tagged IHKs suggest that IHK1 is found in the cytoplasm and nucleus, showing a similar expression profile to GFP while IHK2, containing a nuclear localisation sequence, is predominantly nuclear and IHK3 is predominantly cytoplasmic (Saiardi *et al.*, 2001), although IHK2 has also been shown to reside in the cytoplasm, translocating to the nucleus following treatment of cells with INF-β (Morrison *et al.*, 2005).

Northern blots indicate IHK1 shows highest levels in brain and testes whereas IHK2 can be found in brain, lung, liver, kidney and testes (Saiardi *et al.*, 1999). IHK3 can be found in all tissues, but to varying degrees (Saiardi *et al.*, 2001). Direct regulation of the activity of IHKs has not been studied. However, several studies show consequences of IHK regulation: γ-irradiation of mammalian cells caused a post-translational increase in IHK2, an event leading to caspase-8-mediated apoptosis (Morrison *et al.*, 2002), and cisplatin- and staurosporine-induced cell death has been shown to be mediated by IHK2 driving a pulse of PP-$InsP_5$ to induce cell death (Nagata *et al.*, 2005). During this time, the localisation of IHK2 changes significantly with most moving from the nucleus to the mitochondria. No changes in localisation of IHK1 or IHK3 were observed (Nagata *et al.*, 2005). Only IHK2 knock-down lowered staurosporine-induced apoptosis. However, overexpression of IHK1, 2 and 3 all showed an increase in cell death (Nagata *et al.*, 2005) suggesting these effects are driven by PP-InsP$_5$ rather than a second function of IHK2. Studies with heat shock protein 90 (HSP90) have shown similar results from a different perspective: HSP90 has antiapoptotic properties as it binds and inhibits IHK2 (Chakraborty *et al.*, 2008). Overexpression of mutated IHK2 which does not interact with HSP90 released the cells allowing apoptosis to occur. The dodecameric HSP90 recognition site present in IHK2 (amino acids 132-143) is absent in both IHK1 and IHK3, explaining why such regulation is not observed with these. Some have suggested that small molecule inhibitors of this interaction could be a novel treatment as chemotherapeutic agents (Shames and Minna, 2008).

Inositol hexakisphosphate kinase 2 has been shown to be a target for phosphorylation by casein kinase II within a PEST region leading to ubiquitination and subsequent degradation (Chakraborty *et al.*, 2011). IHK1 also possesses a PEST region which can be targeted by casein kinase II, the phosphorylation of which has been reported to not affect activity; its effect on protein stability was not reported (Barker *et al.*, 2009). IHK3 lacks this PEST region, so is unlikely to be regulated in any way.

5.3.2 PP-InsP$_5$ kinases

Kinases favouring inositol diphosphates have also been purified, cloned and characterised. Whilst the original purification occurred at a similar time to that of IHKs (Huang *et al.*, 1998), the identification and cloning of this enzyme only surfaced several years later (Mulugu *et al.*, 2007). Vip1p from *S. cerevisiae* was show to be capable of phosphorylating InsP$_6$ as well as PP-InsP$_5$, although the product of InsP$_6$ phosphorylation was different to that when IHKs were used. Characterisation of two human PP-InsP$_5$Ks subsequently followed (Choi *et al.*, 2007; Fridy *et al.*, 2007), along with the structural identities of mammalian PP-InsP$_5$ and [PP]$_2$-InsP$_4$ (Lin *et al.*, 2009). These PP-InsP$_5$Ks phosphorylate at the 1-phosphate position and the proteins not only contain the kinase domain required for PP-InsP$_5$ phosphorylation, but both possess a histidine acid phosphatase (HAP) motif. In unstimulated cells, PP-InsP$_5$K1 appear to show cytoplasmic localisation, excluded from the nucleus (Fridy *et al.*, 2007). Whilst apparently inactive as a phosphatase (when tested with several standard substrates including InsP$_6$ and pNPP) (Choi *et al.*, 2007), this motif is used by PP-InsP$_5$K1 to regulate its cellular location (Gokhale *et al.*, 2011). Following platelet-derived growth factor activation of PI 3-K and production of PtdInsP$_3$, PP-InsP$_5$K1 translocates from the cytoplasm to the plasma membrane, as the PtdInsP$_3$ binds the HAP motif. It is proposed that this would, therefore, drive [PP]$_2$-InsP$_4$ production at the plasma membrane. Similarly, hyperosmotic shock, which also increases PtdInsP$_3$ levels, also caused translocation of PP-InsP$_5$K1 to the plasma membrane (Gokhale *et al.*, 2013). In contrast, PP-InsP$_5$K2 possesses a nuclear localisation signal. Phosphorylation of a key serine (S1006) determines its location and nuclear localisation was high when S1006 was unphosphorylated (or mutated to non-phosphorylatable alanine) and lowest when phosphorylated or replaced with a phosphomimic aspartate (Yong *et al.*, 2015).

Unfortunately, the authors were unable to determine which, if any, kinase was involved in such phosphorylation and, hence, whether such phosphorylation could be considered a signalling event (Yong *et al.*, 2015). A detailed characterisation of PP-$InsP_5$K2 has been performed and the ATP-synthase properties down-played. The binding affinity of ATP was also shown as much higher than that for IHKs (Weaver *et al.*, 2013).

5.3.3 Diphosphoinositol polyphosphate phosphohydrolases

The original family of enzymes involved in the dephosphorylation of inositol diphosphates belonged to the Nudix hydrolase family of enzymes. DIPP-1 was able to remove the β-phosphates from PP-$InsP_5$ and $[PP]_2$-$InsP_4$ with no activity towards any α-phosphate group (Safrany *et al.*, 1998). Cloning of similar proteins allowed the characterisation of DIPP-2α and 2β, which are only one amino acid different yet show greatly different activities towards inositol diphosphates and diadenosine polyphosphates, and DIPP-3α and 3β, also known as Aps2 and Aps1 (Caffrey *et al.*, 2000; Hidaka *et al.*, 2002; Kilari *et al.*, 2013; Leslie *et al.*, 2002; Safrany *et al.*, 1998). The DIPPs all show an absolute requirement for divalent cation cofactor and a broad range of substrates (Fisher *et al.*, 2002; Leslie *et al.*, 2002; Safrany *et al.*, 1999). Traditional substrates for Nudix hydrolases (also known as mutT and NudT proteins) contain a nucleoside diphosphate linked to X, including nucleoside tri-, tetra- and pentaphosphates, dinucleoside polyphosphates, nucleotide sugars and RNA caps (McLennan, 2006). Human DIPPs readily metabolised diadenosine hexaphosphate (Ap_6A), creating as the first products AMP and adenosine pentaphosphate (which was further hydrolysed), with smaller levels of ADP and adenosine tetraphosphate (p_4A) (which also further hydrolysed) being made. Diadenosine pentaphosphate was similarly hydrolysed, with the favoured cleavage creating AMP and p_4A (Leslie *et al.*, 2002; Safrany *et al.*, 1999). This is in contrast with the *S. cerevisiae* (Cartwright and McLennan, 1999) and *S. pombe* (Ingram *et al.*, 1999) homologues, which favoured the production of ADP. Other substrates include 5-phosphoribosyl 1-pyrophosphate (Fisher *et al.*, 2002) and, potentially, inorganic polyphosphate, although activity towards the final substrate appears to be limited to the yeast homologue, Ddp1p (Lonetti *et al.*, 2011). The pH optimum for even limited activity of the human enzymes occurred at 9.0, raising suspicion about whether this is physiologically relevant. Likewise, our earlier work comparing rates of hydrolysis of PP-$InsP_5$, $[PP]_2$ $InsP_4$ and dATP for DIPP-1 clearly showed it to be an inositol diphosphate-metabolising enzyme (Safrany *et al.*, 1998). The cellular localisation for all human DIPP proteins has been established in two different cell types (HEK293 and PC12 cells) using an N-terminal and a C-terminal GFP tag. All proteins showed no particular localisation, resembling the pattern presented by GFP alone (Leslie *et al.*, 2002). Tissue distribution of the DIPPs varies considerably: DIPP-1 is most highly expressed in brain, pancreas, heart and liver (Safrany *et al.*, 1998); DIPP-2α and β in heart, brain skeletal muscle, kidney and pancreas (Caffrey *et al.*, 2000); DIPP-3α in brain and liver and DIPP-3β in brain, pancreas and testis (Leslie *et al.*, 2002).

Regulation of the DIPPs has received little attention. The requirement for divalent cations as co-factors has received discussion in the case of manganese (II) regulating diadenosine polyphosphate hydrolysis by DIPP-3s (Leslie *et al.*, 2002). Of the divalent cations tested, cobalt (II) gave the highest activity against PP-$InsP_5$, whereas it was relatively ineffective when testing Ap_6A as substrate. It has not been possible to assess whether several cations would regulate inositol diphosphate metabolism, as they cause precipitation of the substrate. For this reason, we routinely perform similar experiments in the presence of EDTA, which chelates out low concentrations of contaminating metals, and then titrate in the metal of interest (Leslie *et al.*, 2002; Winward *et al.*, 2010). While magnesium (II) is an effective cofactor for inositol diphosphate hydrolysis, it is ineffective as cofactor for nucleotide metabolism in the case of DIPP-3α and 3β (Leslie *et al.*, 2002). Clearly, we are a long way from understanding the context in which these enzymes work in a cellular environment, where a complex mixture of metals, both free and protein-bound may play interacting roles.

DIPPs appear, for the most part, to be inhibited by fluoride. The only detailed analysis appears in the original purification and cloning paper (Safrany *et al.*, 1998) which showed DIPP-1 being non-competitively inhibited (IC_{50} unaffected by substrate concentration) when PP-$InsP_5$ was used as a substrate; conversely, $[PP]_2$-$InsP_4$ was inhibited in an uncompetitive manner (IC_{50} *decreasing* as substrate concentration *increases*), suggesting these two substrates were bound to the enzyme in different orientations. This subtle difference may explain the earlier conclusion that $[PP]_2$-$InsP_4$ and PP-$InsP_5$ were hydrolysed by different phosphatases: Shears *et al.* (1995) found that the IC_{50} for fluoride was substantially lower for PP-$InsP_5$ than $[PP]_2$-$InsP_4$ when using trace amounts. We revisited the nature of this fluoride-mediated inhibition of Nudix hydrolases when characterising a closely related *D. melanogaster* protein Apf, which does not metabolise inositol diphosphates, but favours instead diadenosine tetraphosphate (Ap_4A). Zinc (II) and magnesium (II) showed similar levels of activation, yet fluoride only inhibited the enzyme in the presence of magnesium (II). We therefore proposed that Nudix hydrolases were affected by MgF_3^-. This has also been observed with other Nudix hydrolases (see Winward *et al.*, 2007 and references therein).

While only the *D. melanogaster* DIPP homologue, Aps has been shown to form dimers under oxidising conditions, all human DIPPs contain the N-terminal CxC motif which allows Aps to form covalent bonds with neighbours. This motif is a considerable distance from the active site and kinetic parameters of Aps are unaffected by a range of redox potentials. However, the stability of the enzyme is greatly affected. For example, in an oxidising environment (absence of dithiothreitol, a potent reducing agent) and no oxidising agent added, Aps is temperature and acid-labile. A five minute incubation at 36 °C caused loss of activity by over 65%. In contrast, the addition of 1 mM dithiothreitol altered the conditions so that the protein lost no activity following treatment at 76 °C. Equally, perchloric acid precipitation did not affect activity once the enzyme was reconstituted and pH adjusted (Winward *et al.*, 2010). Other forms of regulation considered include free calcium, cAMP-dependent protein kinase and protein kinase C, none of which appeared to have any effect (Safrany *et al.*, 1998).

Following a proposal that DIPPs were unable to metabolise 5-PP-$InsP_5$ (Lonetti *et al.*, 2011), we revisited our earlier work using pure, chemically-defined substrates that were unavailable when our earlier studies were performed. We were able to show all DIPPs metabolised 5-PP-$InsP_5$, 1-PP-$InsP_5$ and 1,5-$[PP]_2$-$InsP_4$ with similar affinities. The favoured target of attack in 1,5-$[PP]_2$-$InsP_4$ appeared to be the 5-phosphate, giving a first on-first off profile (Kilari *et al.*, 2013).

A later entrant into the group of proteins involved in removal of inositol diphosphates has now emerged. Siw14p from *S. cerevisiae* resembles a protein phosphatase and has no human homologue (Steidle *et al.*, 2016). Siw14p contains an active site cysteine and hydrolyses pNPP, but was shown to metabolise only 5-PP-$InsP_5$. Controls showed that 1-PP-$InsP_5$ was readily metabolised by DIPP-1, suggesting the selectivity of this protein is real. Whilst the k_{cat} values were comparable, the affinity (34 μM) is significantly lower for 5-PP-$InsP_5$ than we found for the DIPP proteins 35-146 nM, but as we have previously declared (Kilari *et al.*, 2013), this probably reflects the quality of substrate used and the analysis methods.

Studies of these proteins allow us to speculate that inositol diphosphate metabolism can be regulated in a myriad of ways. The kinases involved can be trafficked around cells, allowing local synthesis in microenvironments. In addition to the ability of these enzymes to regulate inositol diphosphates, the proteins themselves show a range of direct effects, which, at times, is difficult to dissect from the indirect, inositol diphosphate-mediated, consequences of regulation. Of course, while the biology of these proteins is of interest, *we* are more interested in the turnover, levels, and biological activity of the molecules they regulate – the inositol diphosphates. We now turn to briefly consider the pharmacological regulation of these molecules which can lead us to propose further effects on physiology and pathology.

5.4 Pharmacological regulation of the levels of inositol diphosphates

The first evidence that inositol diphosphates were present in mammalian cells and that they were readily regulated was presented in 1993 (Menniti *et al.*, 1993). Cell homogenates were able to phosphorylate $InsP_5$ and $InsP_6$ to 'pyrophosphates', an effect enhanced by the addition of fluoride, a wide-spectrum phosphatase inhibitor. Fluoride treatment of a wide range of cells also showed the accumulation of pyrophosphates with similar elution times, showing that the belief held at the time, that $InsP_5$ and $InsP_6$ were metabolically lethargic and not involved in signalling, was incorrect (Menniti *et al.*, 1993). The high turnover of $InsP_5$ (20%/hour) and $InsP_6$ (50%/hour) through these pyrophosphates led the authors to conclude it was for a 'profound and, presumably worthwhile, investment of cellular energy'. Inositol diphosphates were also present in primary hepatocytes, PP-$InsP_4$ and PP-$InsP_5$ synthesis was decreased following treatment with the sarcoplasmic/endoplasmic reticulum calcium ATPase inhibitor thapsigargin (Glennon and Shears, 1993). These effects were not mimicked by activation of phospholipase C, making Glennon and Shears suggest that a prolonged increase in cytoplasmic calcium was required to inhibit the kinase(s) involved. The introduction of EDTA into assay buffers (Shears *et al.*, 1995) greatly enhanced the recovery of $[PP]_2$-$InsP_4$, allowing it to be readily observed, albeit at very low levels. The ability of fluoride to raise its levels in a biphasic matter (raise at 0.8 mM, lower at 10 mM while not affecting kinase activity) was considered 'intriguing' (Shears *et al.*, 1995).

Understanding this observation, even partially, required a further three years: $[PP]_2$-InsP_4 were found to be lowered by agents raising cyclic AMP and cyclic GMP (Safrany and Shears, 1998). High concentrations of fluoride can stimulate heterotrimeric G-proteins, in this case G_s. This had been previously considered, but the recovery of $[PP]_2$-InsP_4 was too low for it to be observed (Menniti *et al.*, 1993). Activation of β_2-adrenergic receptors caused a dose-dependent decrease in $[PP]_2$-InsP_4 which was not mediated by cyclic AMP- or cyclic GMP-dependent kinases. The effect was mimicked by cell permeable analogues of cyclic AMP and cyclic GMP as well as phosphodiesterase inhibitors, which prevent the hydrolysis of the cyclic moieties (Safrany and Shears, 1998). This would also exclude a role for AMP-dependent kinases, a point further reinforced several years later (Choi *et al.*, 2008). That $[PP]_2$-InsP_4 levels decrease in the presence of high concentrations of inhibitor (fluoride) continues to suggest that the phosphate is not merely removed by DIPP or a similar enzyme. Protein pyrophosphorylation (Bhandari *et al.*, 2007; Saiardi *et al.*, 2004) is the strongest candidate yet put forward. In contrast, cyclic AMP has been shown to increase intracellular PP-InsP_5 and $[PP]_2$-InsP_4 levels in *D. discoideum*. Cyclic AMP acts on cAR1, a G-protein-coupled cell surface receptor, acting as a chemoattractant to elicit aggregation during times of nutritional stress. While starvation (removal of bacteria, transfer of cultures to phosphate-buffered agar) increased inositol diphosphate levels 25-fold (Laussmann *et al.*, 2000), the increase seen in response to cyclic AMP was a modest 4-fold (Luo *et al.*, 2003). Starvation causes aggregation of *D. discoideum* and production of a multicellular stage with fruiting body. Separation of stalks and spores showed the inositol diphosphates were found exclusively in the spores, with levels of 5,6-$[PP]_2$-InsP_4 in the region of 1.5 – 2 mM (Laussmann *et al.*, 2000). Germination at low cell density saw a rapid decline of inositol diphosphates. Again, the mechanism by which this occurs, and the consequences remain unknown.

Effects on cellular inositol diphosphates have also been shown to occur during infection with African swine fever virus (ASFV). ASFV contains g5R (also known as D250), a Nudix hydrolase which resembles mammalian Nudix hydrolases favouring Ap_4A. *In vitro* biochemical characterisation showed, however, that the favoured substrate was PP-InsP_5. Infection of African green monkey kidney epithelial Vero cells with ASFV showed a decline in PP-InsP_5, ATP and GTP (Cartwright *et al.*, 2002). Whether this decline was required for virus invasion was not investigated. Previous studies with *Salmonella* species showed that SigD/SopB was important for bacterial invasion. SigD, an enzyme that metabolises InsP_5 *in vitro* and *in vivo*, is required for efficient invasion. The focus of this mechanism has fallen upon the product, inositol 1,4,5,6-tetrakisphosphate, which activates calcium-activated chloride channels (Norris *et al.*, 1998). Future analysis should take into account the loss of InsP_5 and, therefore, its phosphorylation products InsP_6 (Zhou *et al.*, 2001) and the inositol diphosphates.

Continued evidence that inositol diphosphates could act in a signalling capacity was highlighted by two concomitant papers showing that $[PP]_2$-InsP_4 levels were increased in mammalian cells during hyperosmotic stress following treatment with sucrose (Safrany, 2004) or sorbitol (Pesesse *et al.*, 2004; Safrany, 2004). These effects were not downstream of extracellular signal-related kinases (ERKs) and p38 MAPα/β kinases, a conclusion determined only later when the tools used by Pesesse *et al.* (2004) were found to be less selective than originally thought (Choi *et al.*, 2008). Indeed, epidermal growth factor receptor activation of ERKs and p38 MAPα/β kinases was unable to replicate the effect of sorbitol (Pesesse *et al.*, 2004), suggesting other pathways were also involved. It was subsequently shown that 5-aminoimidazole-4-carboxamide ribonucleoside (AICAR) was able to reverse the effects of hyperosmotic shock by stimulating AMP-dependent protein kinase and also lowered basal $[PP]_2$-$InsP_4$ levels. These effects remained when AMP-dependent protein kinase levels were reduced using SiRNA protocols. The authors considered the effect on $[PP]_2$-$InsP_4$ to be mediated through 'cellular adenosine nucleotide balance' (Choi *et al.*, 2008), an idea further developed by Wundenberg *et al.* (2014) who showed IHKs can also use $InsP_6$ to add a phosphate to ADP when ATP:ADP ratios fall.

Genistein, a broad-range tyrosine kinase inhibitor lowered $[PP]_2$-$InsP_4$ in Syrian hamster vas deferens DDT_1MF-2 cells whereas herbimycin A, an unrelated tyrosine kinase inhibitor did not. This would suggest PP-$InsP_5K$ may be directly inhibited by genistein, rather than being mediated through tyrosine kinases (Safrany, 2004). Herbimycin A is also described as an inhibitor of HSP90, so may be expected to regulate IHK2 (Chakraborty *et al.*, 2008). It is unknown whether a homologue of IHK2 is present in DDT_1MF-2 cells. One could speculate that as genistein and cyclic AMP both lower $[PP]_2$-$InsP_4$ (Safrany, 2004; Safrany and Shears, 1998) and enhance vesicle transport in cells (Hamm-Alvarez *et al.*, 1993) that these two events are linked – lowering $[PP]_2$-$InsP_4$ removes the break discussed above. The calmodulin inhibitors W-7 and chlorpromazine were also found to lower inositol diphosphate levels, but non-selectively, also affecting $InsP_5$, $InsP_6$, PtdIns(4,5)P_2, PtdIns(4)P and ATP (Safrany, 2004), so while used to consider the roles of inositol diphosphates in cell biology (Byrum *et al.*, 2004) they are far from ideal. PP-$InsP_4$ levels are also sensitive to thermal shock. Raising or lowering the temperature raised levels in a MAP kinase kinase-dependent manner. Conversely, oxidative or mechanical stress did not affect levels (Choi *et al.*, 2005).

From these observations, we can determine that one inositol diphosphate – $[PP]_2$-$InsP_4$ – is physiologically regulated by adenylyl cyclase. Both bacterial and viral infection reduces inositol diphosphates, which may be required in the release of some viruses. Finally, several widely used drugs and treatments affect levels of inositol diphosphates. These include ATP-competitive kinase inhibitors genistein and TNP. The consequences of such regulation remain to be determined.

5.5 Conclusions

Inositol diphosphates are high energy molecules able to pyrophosphorylate proteins. Many roles have been proposed; many more will be proposed. At this time we propose areas of interest can include:

1. Cancer biology and DNA repair.
2. Insulin signalling, diabetes and obesity.
3. Mental health, depression and bipolar disorder.

References

Albert, C., Safrany, S.T., Bembenek, M.E., Reddy, K.M., Reddy, K., Falck, J., Brocker, M., Shears, S.B. and Mayr, G.W., 1997. Biological variability in the structures of diphosphoinositol polyphosphates in *Dictyostelium discoideum* and mammalian cells. Biochemical Journal 327:553-560.

Ali, N., Duden, R., Bembenek, M.E. and Shears, S.B., 1995. The interaction of coatomer with inositol polyphosphates is conserved in *Saccharomyces cerevisiae*. Biochemical Journal 310:279-284.

Azevedo, C., Burton, A., Ruiz-Mateos, E., Marsh, M. and Saiardi, A., 2009. Inositol pyrophosphate mediated pyrophosphorylation of AP3B1 regulates HIV-1 Gag release. Proceedings of the National Academy of Sciences of the United States of America 106:21161-21166.

Barker, C.J., Illies, C., Gaboardi, G.C. and Berggren, P.O., 2009. Inositol pyrophosphates: structure, enzymology and function. Cellular and molecular life sciences 66:3851-3871.

Bhandari, R., Juluri, K.R., Resnick, A.C. and Snyder, S.H., 2008. Gene deletion of inositol hexakisphosphate kinase 1 reveals inositol pyrophosphate regulation of insulin secretion, growth, and spermiogenesis. Proceedings of the National Academy of Sciences of the United States of America 105:2349-2353.

Bhandari, R., Saiardi, A., Ahmadibeni, Y., Snowman, A.M., Resnick, A.C., Kristiansen, T.Z., Molina, H., Pandey, A., Werner, J.K., Jr., Juluri, K.R., Xu, Y., Prestwich, G.D., Parang, K. and Snyder, S.H., 2007. Protein pyrophosphorylation by inositol pyrophosphates is a posttranslational event. Proceedings of the National Academy of Sciences of the United States of America 104:15305-15310.

Byrum, J., Jordan, S., Safrany, S.T. and Rodgers, W., 2004. Visualization of inositol phosphate-dependent mobility of Ku: depletion of the DNA-PK cofactor $InsP_6$ inhibits Ku mobility. Nucleic Acids Research 32:2776-2784.

Caffrey, J.J., Safrany, S.T., Yang, X. and Shears, S.B., 2000. Discovery of molecular and catalytic diversity among human diphosphoinositol-polyphosphate phosphohydrolases. An expanding Nudt family. Journal of Biological Chemistry 275:12730-12736.

Cartwright, J.L. and McLennan, A.G., 1999. The *Saccharomyces cerevisiae* YOR163w gene encodes a diadenosine 5', 5'''- P^1,P^6-hexaphosphate (Ap_6A) hydrolase member of the MutT motif (Nudix hydrolase) family. Journal of Biological Chemistry 274:8604-8610.

Cartwright, J.L., Safrany, S.T., Dixon, L.K., Darzynkiewicz, E., Stepinski, J., Burke, R. and McLennan, A.G., 2002. The g5R (D250) gene of African swine fever virus encodes a nudix hydrolase that preferentially degrades diphosphoinositol polyphosphates. Journal of Virology 76:1415-1421.

Chakraborty, A., Koldobskiy, M.A., Bello, N.T., Maxwell, M., Potter, J.J., Juluri, K.R., Maag, D., Kim, S., Huang, A.S., Dailey, M.J., Saleh, M., Snowman, A.M., Moran, T.H., Mezey, E. and Snyder, S.H., 2010. Inositol pyrophosphates inhibit Akt signaling, thereby regulating insulin sensitivity and weight gain. Cell 143:897-910.

Chakraborty, A., Koldobskiy, M.A., Sixt, K.M., Juluri, K.R., Mustafa, A.K., Snowman, A.M., Van Rossum, D.B., Patterson, R.L. and Snyder, S.H., 2008. HSP90 regulates cell survival via inositol hexakisphosphate kinase-2. Proceedings of the National Academy of Sciences of the United States of America 105:1134-1139.

Chakraborty, A., Latapy, C., Xu, J., Snyder, S.H. and Beaulieu, J.M., 2014. Inositol hexakisphosphate kinase-1 regulates behavioral responses via GSK3 signaling pathways. Molecular Psychiatry 19:284-293.

Chakraborty, A., Werner, J.K., Jr., Koldobskiy, M.A., Mustafa, A.K., Juluri, K.R., Pietropaoli, J., Snowman, A.M. and Snyder, S.H., 2011. Casein kinase-2 mediates cell survival through phosphorylation and degradation of inositol hexakisphosphate kinase-2. Proceedings of the National Academy of Sciences of the United States of America 108:2205-2209.

Chaudhary, A., Gu, Q.M., Thum, O., Profit, A.A., Qi, Y., Jeyakumar, L., Fleischer, S. and Prestwich, G.D., 1998. Specific interaction of Golgi coatomer protein alpha-COP with phosphatidylinositol 3,4,5-trisphosphate. Journal of Biological Chemistry 273:8344-8350.

Choi, J.H., Williams, J., Cho, J., Falck, J.R. and Shears, S.B., 2007. Purification, sequencing, and molecular identification of a mammalian PP-InsP(5) kinase that is activated when cells are exposed to hyperosmotic stress. Journal of Biological Chemistry 282:30763-30775.

Choi, K., Mollapour, E., Choi, J.H. and Shears, S.B., 2008. Cellular energetic status supervises the synthesis of bis-diphosphoinositol tetrakisphosphate independently of AMP-activated protein kinase. Molecular Pharmacology 74:527-536.

Choi, K., Mollapour, E. and Shears, S.B., 2005. Signal transduction during environmental stress: $InsP_8$ operates within highly restricted contexts. Cellular Signalling 17:1533-1541.

Dedon, P.C. and Goldberg, I.H., 1992. Free-radical mechanisms involved in the formation of sequence-dependent bistranded DNA lesions by the antitumor antibiotics bleomycin, neocarzinostatin, and calicheamicin. Chemical Research in Toxicology 5:311-332.

Draskovic, P., Saiardi, A., Bhandari, R., Burton, A., Ilc, G., Kovacevic, M., Snyder, S.H. and Podobnik, M., 2008. Inositol hexakisphosphate kinase products contain diphosphate and triphosphate groups. Chemistry and Biology 15:274-286.

Edwards, T.K., Saleem, A., Shaman, J.A., Dennis, T., Gerigk, C., Oliveros, E., Gartenberg, M.R. and Rubin, E.H., 2000. Role for nucleolin/Nsr1 in the cellular localization of topoisomerase I. Journal of Biological Chemistry 275:36181-36188.

Falck, J.R., Reddy, K.K., Ye, J.H., Saady, M., Mioskowski, C., Shears, S.B., Tan, Z. and Safrany, S., 1995. Synthesis and structure of cellular mediators – inositol polyphosphate diphosphates. Journal of the American Chemical Society 117:12172-12175.

Fisher, D.I., Safrany, S.T., Strike, P., McLennan, A.G. and Cartwright, J.L., 2002. Nudix hydrolases that degrade dinucleoside and diphosphoinositol polyphosphates also have 5-phosphoribosyl 1-pyrophosphate (PRPP) pyrophosphatase activity that generates the glycolytic activator ribose 1,5-bisphosphate. Journal of Biological Chemistry 277:47313-47317.

Fleischer, B., Xie, J., Mayrleitner, M., Shears, S.B., Palmer, D.J. and Fleischer, S., 1994. Golgi coatomer binds, and forms K^+-selective channels gated by, inositol polyphosphates. Journal of Biological Chemistry 269:17826-17832.

Freland, L. and Beaulieu, J.M., 2012. Inhibition of GSK3 by lithium, from single molecules to signaling networks. Frontiers in Molecular Neuroscience 5:14.

Fridy, P.C., Otto, J.C., Dollins, D.E. and York, J.D., 2007. Cloning and characterization of two human VIP_1-like inositol hexakisphosphate and diphosphoinositol pentakisphosphate kinases. Journal of Biological Chemistry 282:30754-30762.

Fu, C., Xu, J., Li, R.J., Crawford, J.A., Khan, A.B., Ma, T.M., Cha, J.Y., Snowman, A.M., Pletnikov, M.V. and Snyder, S.H., 2015. Inositol hexakisphosphate kinase-3 regulates the morphology and synapse formation of cerebellar purkinje cells via spectrin/adducin. Journal of Neuroscience 35:11056-11067.

Gaidarov, I., Chen, Q., Falck, J.R., Reddy, K.K. and Keen, J.H., 1996. A functional phosphatidylinositol 3,4,5-trisphosphate/phosphoinositide binding domain in the clathrin adaptor AP-2 alpha subunit. Implications for the endocytic pathway. Journal of Biological Chemistry 271:20922-20929.

Gaidarov, I. and Keen, J.H., 1999. Phosphoinositide-AP-2 interactions required for targeting to plasma membrane clathrin-coated pits. Journal of Cell Biology 146:755-764.

Glennon, M.C. and Shears, S.B., 1993. Turnover of inositol pentakisphosphates, inositol hexakisphosphate and diphosphoinositol polyphosphates in primary cultured hepatocytes. Biochemical Journal 293:583-590.

Gokhale, N.A., Zaremba, A., Janoshazi, A.K., Weaver, J.D. and Shears, S.B., 2013. PPIP5K1 modulates ligand competition between diphosphoinositol polyphosphates and PtdIns(3,4,5)P_3 for polyphosphoinositide-binding domains. Biochemical Journal 453:413-426.

Gokhale, N.A., Zaremba, A. and Shears, S.B., 2011. Receptor-dependent compartmentalization of PPIP5K1, a kinase with a cryptic polyphosphoinositide binding domain. Biochemical Journal 434:415-426.

Hajduch, E., Alessi, D.R., Hemmings, B.A. and Hundal, H.S., 1998. Constitutive activation of protein kinase B alpha by membrane targeting promotes glucose and system A amino acid transport, protein synthesis, and inactivation of glycogen synthase kinase 3 in L6 muscle cells. Diabetes 47:1006-1013.

Hamm-Alvarez, S.F., Kim, P.Y. and Sheetz, M.P., 1993. Regulation of vesicle transport in CV-1 cells and extracts. Journal of Cell Science 106:955-966.

Hao, W., Tan, Z., Prasad, K., Reddy, K.K., Chen, J., Prestwich, G.D., Falck, J.R., Shears, S.B. and Lafer, E.M., 1997. Regulation of AP-3 function by inositides. Identification of phosphatidylinositol 3,4,5-trisphosphate as a potent ligand. Journal of Biological Chemistry 272:6393-6398.

Hidaka, K., Caffrey, J.J., Hua, L., Zhang, T., Falck, J.R., Nickel, G.C., Carrel, L., Barnes, L.D. and Shears, S.B., 2002. An adjacent pair of human NUDT genes on chromosome X are preferentially expressed in testis and encode two new isoforms of diphosphoinositol polyphosphate phosphohydrolase. Journal of Biological Chemistry 277:32730-32738.

Hua, L.V., Green, M., Warsh, J.J. and Li, P.P., 2001. Molecular cloning of a novel isoform of diphosphoinositol polyphosphate phosphohydrolase: a potential target of lithium therapy. Neuropsychopharmacology 24:640-651.

Huang, C.F., Voglmaier, S.M., Bembenek, M.E., Saiardi, A. and Snyder, S.H., 1998. Identification and purification of diphosphoinositol pentakisphosphate kinase, which synthesizes the inositol pyrophosphate bis(diphospho)inositol tetrakisphosphate. Biochemistry 37:14998-15004.

Huang, K.N. and Symington, L.S., 1995. Suppressors of a *Saccharomyces cerevisiae* pkc1 mutation identify alleles of the phosphatase gene PTC1 and of a novel gene encoding a putative basic leucine zipper protein. Genetics 141:1275-1285.

Humbert, S., Bryson, E.A., Cordelieres, F.P., Connors, N.C., Datta, S.R., Finkbeiner, S., Greenberg, M.E. and Saudou, F., 2002. The IGF-1/Akt pathway is neuroprotective in Huntington's disease and involves Huntingtin phosphorylation by Akt. Developmental Cell 2:831-837.

Illies, C., Gromada, J., Fiume, R., Leibiger, B., Yu, J., Juhl, K., Yang, S.N., Barma, D.K., Falck, J.R., Saiardi, A., Barker, C.J. and Berggren, P.O., 2007. Requirement of inositol pyrophosphates for full exocytotic capacity in pancreatic beta cells. Science 318:1299-1302.

Ingram, S.W., Stratemann, S.A. and Barnes, L.D., 1999. *Schizosaccharomyces pombe* Aps1, a diadenosine 5',5'''-P^1, P^6- hexaphosphate hydrolase that is a member of the nudix (MutT) family of hydrolases: cloning of the gene and characterization of the purified enzyme. Biochemistry 38:3649-3655.

Jadav, R.S., Chanduri, M.V., Sengupta, S. and Bhandari, R., 2013. Inositol pyrophosphate synthesis by inositol hexakisphosphate kinase 1 is required for homologous recombination repair. Journal of Biological Chemistry 288:3312-3321.

Jadav, R.S., Kumar, D., Buwa, N., Ganguli, S., Thampatty, S.R., Balasubramanian, N. and Bhandari, R., 2016. Deletion of inositol hexakisphosphate kinase 1 (IP6K1) reduces cell migration and invasion, conferring protection from aerodigestive tract carcinoma in mice. Cellular Signalling 28:1124-1136.

Kijima, Y., Mayrleitner, M., Timerman, A.P., Saito, A., Schindler, H. and Fleischer, S., 1993. A cardiac clathrin assembly protein forms a potassium channel in planar lipid bilayers. Journal of Biological Chemistry 268:16253-16258.

Kilari, R.S., 2014. Roles of inositol diphosphates in DNA repair and effects of aspirin analogues on oesophageal cancer. Doctoral Thesis, University of Wolverhampton, UK.

Kilari, R.S., Weaver, J.D., Shears, S.B. and Safrany, S.T., 2013. Understanding inositol pyrophosphate metabolism and function: kinetic characterization of the DIPPs. FEBS Letters 587:3464-3470.

Komander, D., Fairservice, A., Deak, M., Kular, G.S., Prescott, A.R., Peter Downes, C., Safrany, S.T., Alessi, D.R. and van Aalten, D.M., 2004. Structural insights into the regulation of PDK1 by phosphoinositides and inositol phosphates. Embo Journal 23:3918-3928.

Laussmann, T., Hansen, A., Reddy, K.M., Reddy, K.K., Falck, J.R. and Vogel, G., 1998. Diphospho-*myo*-inositol phosphates in *Dictyostelium* and *Polysphondylium*: identification of a new bisdiphospho-*myo*-inositol tetrakisphosphate. FEBS Letters 426:145-150.

Laussmann, T., Pikzack, C., Thiel, U., Mayr, G.W. and Vogel, G., 2000. Diphospho-*myo*-inositol phosphates during the life cycle of *Dictyostelium* and *Polysphondylium*. European Journal of Biochemistry 267:2447-2451.

Laussmann, T., Reddy, K.M., Reddy, K.K., Falck, J.R. and Vogel, G., 1997. Diphospho-*myo*-inositol phosphates from *Dictyostelium* identified as D-6- diphospho-*myo*-inositol pentakisphosphate and D-5,6-bisdiphospho-*myo*- inositol tetrakisphosphate. Biochemical Journal 322:31-33.

Lemmon, M.A., Ferguson, K.M., O'Brien, R., Sigler, P.B. and Schlessinger, J., 1995. Specific and high-affinity binding of inositol phosphates to an isolated pleckstrin homology domain. Proceedings of the National Academy of Sciences of the United States of America 92:10472-10476.

Leslie, N.R., McLennan, A.G. and Safrany, S.T., 2002. Cloning and characterisation of hAps1 and hAps2, human diadenosine polyphosphate-metabolising Nudix hydrolases. BMC Biochemistry 3:20.

Lettier, G., Feng, Q., de Mayolo, A.A., Erdeniz, N., Reid, R.J., Lisby, M., Mortensen, U.H. and Rothstein, R., 2006. The role of DNA double-strand breaks in spontaneous homologous recombination in *S. cerevisiae*. PLOS Genetics 2:e194.

Lin, H., Fridy, P.C., Ribeiro, A.A., Choi, J.H., Barma, D.K., Vogel, G., Falck, J.R., Shears, S.B., York, J.D. and Mayr, G.W., 2009. Structural analysis and detection of biological inositol pyrophosphates reveal that the family of VIP/diphosphoinositol pentakisphosphate kinases are 1/3-kinases. Journal of Biological Chemistry 284:1863-1872.

Lonetti, A., Szijgyarto, Z., Bosch, D., Loss, O., Azevedo, C. and Saiardi, A., 2011. Identification of an evolutionarily conserved family of inorganic polyphosphate endopolyphosphatases. Journal of Biological Chemistry 286:31966-31974.

Luo, H.R., Huang, Y.E., Chen, J.C., Saiardi, A., Iijima, M., Ye, K., Huang, Y., Nagata, E., Devreotes, P. and Snyder, S.H., 2003. Inositol pyrophosphates mediate chemotaxis in *Dictyostelium* via pleckstrin homology domain-PtdIns(3,4,5)P_3 interactions. Cell 114:559-572.

Martin, D.D., Ladha, S., Ehrnhoefer, D.E. and Hayden, M.R., 2015. Autophagy in Huntington disease and Huntington in autophagy. Trends in Neurosciences 38:26-35.

Martin, J.B., Bakker-Grunwald, T. and Klein, G., 1993. ^{31}P-NMR analysis of *Entamoeba histolytica*. Occurrence of high amounts of two inositol phosphates. European Journal of Biochemistry 214:711-718.

Martin, J.B., Laussmann, T., Bakker-Grunwald, T., Vogel, G. and Klein, G., 2000. neo-inositol polyphosphates in the amoeba *Entamoeba histolytica*. Journal of Biological Chemistry 275:10134-10140.

McLennan, A.G., 2006. The Nudix hydrolase superfamily. Cellular and Molecular Life Sciences 63:123-143.

Menniti, F.S., Miller, R.N., Putney, J.W., Jr. and Shears, S.B., 1993. Turnover of inositol polyphosphate pyrophosphates in pancreatoma cells. Journal of Biological Chemistry 268:3850-3856.

Morrison, B.H., Bauer, J.A., Hu, J., Grane, R.W., Ozdemir, A.M., Chawla-Sarkar, M., Gong, B., Almasan, A., Kalvakolanu, D.V. and Lindner, D.J., 2002. Inositol hexakisphosphate kinase 2 sensitizes ovarian carcinoma cells to multiple cancer therapeutics. Oncogene 21:1882-1889.

Morrison, B.H., Bauer, J.A., Kalvakolanu, D.V. and Lindner, D.J., 2001. Inositol hexakisphosphate kinase 2 mediates growth suppressive and apoptotic effects of interferon-beta in ovarian carcinoma cells. Journal of Biological Chemistry 276:24965-24970.

Morrison, B.H., Bauer, J.A., Lupica, J.A., Tang, Z., Schmidt, H., DiDonato, J.A. and Lindner, D.J., 2007. Effect of inositol hexakisphosphate kinase 2 on transforming growth factor beta-activated kinase 1 and NF-kappaB activation. Journal of Biological Chemistry 282:15349-15356.

Morrison, B.H., Haney, R., Lamarre, E., Drazba, J., Prestwich, G.D. and Lindner, D.J., 2009. Gene deletion of inositol hexakisphosphate kinase 2 predisposes to aerodigestive tract carcinoma. Oncogene 28:2383-2392.

Morrison, B.H., Tang, Z., Jacobs, B.S., Bauer, J.A. and Lindner, D.J., 2005. Apo2L/TRAIL induction and nuclear translocation of inositol hexakisphosphate kinase 2 during IFN-beta-induced apoptosis in ovarian carcinoma. Biochemical Journal 385:595-603.

Mulugu, S., Bai, W.L., Fridy, P.C., Bastidas, R.J., Otto, J.C., Dollins, D.E., Haystead, T.A., Ribeiro, A.A. and York, J.D., 2007. A conserved family of enzymes that phosphorylate inositol hexakisphosphate. Science 316:106-109.

Nagata, E., Luo, H.R., Saiardi, A., Bae, B.I., Suzuki, N. and Snyder, S.H., 2005. Inositol hexakisphosphate kinase-2, a physiologic mediator of cell death. Journal of Biological Chemistry 280:1634-1640.

Nagata, E., Saiardi, A., Tsukamoto, H., Okada, Y., Itoh, Y., Satoh, T., Itoh, J., Margolis, R.L., Takizawa, S., Sawa, A. and Takagi, S., 2011. Inositol hexakisphosphate kinases induce cell death in Huntington disease. Journal of Biological Chemistry 286:26680-26686.

Nagata, E., Saiardi, A., Tsukamoto, H., Satoh, T., Itoh, Y., Itoh, J., Shibata, M., Takizawa, S. and Takagi, S., 2010. Inositol hexakisphosphate kinases promote autophagy. The International Journal of Biochemistry and Cell Biology 42:2065-2071.

Niger, C., Hebert, C. and Stains, J.P., 2010. Interaction of connexin43 and protein kinase C-delta during FGF2 signaling. BMC Biochemistry 11:14.

Niger, C., Luciotti, M.A., Buo, A.M., Hebert, C., Ma, V. and Stains, J.P., 2013. The regulation of runt-related transcription factor 2 by fibroblast growth factor-2 and connexin43 requires the inositol polyphosphate/ protein kinase Cdelta cascade. Journal of Bone and Mineral Research 28:1468-1477.

Norris, F.A., Wilson, M.P., Wallis, T.S., Galyov, E.E. and Majerus, P.W., 1998. SopB, a protein required for virulence of *Salmonella dublin*, is an inositol phosphate phosphatase. Proceedings of the National Academy of Sciences of the United States of America 95:14057-14059.

Owen, R.M., Baker, R.D., Bader, S., Dunlop, M.G. and Nicholl, I.D., 2007. The identification of a novel alternatively spliced form of the MBD4 DNA glycosylase. Oncology Reports 17:111-116.

Padmanabhan, U., Cosgrove, A.J., Downes, C.P., Donaldson, A.D. and Safrany, S.T., 2005. Role of inositol diphosphates in repair of bleomycin-induced DNA damage, Biochemical Society Annual Symposium – The Cell Biology of Inositol Lipids and Phosphates, Birmingham, UK.

Padmanabhan, U., Dollins, D.E., Fridy, P.C., York, J.D. and Downes, C.P., 2009. Characterization of a selective inhibitor of inositol hexakisphosphate kinases: use in defining biological roles and metabolic relationships of inositol pyrophosphates. Journal of Biological Chemistry 284:10571-10582.

Pesesse, X., Choi, K., Zhang, T. and Shears, S.B., 2004. Signaling by higher inositol polyphosphates. Synthesis of bisdiphosphoinositol tetrakisphosphate ('$InsP_8$') is selectively activated by hyperosmotic stress. Journal of Biological Chemistry 279:43378-43381.

Pohlmann, J. and Fleig, U., 2010. Asp1, a conserved 1/3 inositol polyphosphate kinase, regulates the dimorphic switch in Schizosaccharomyces pombe. Molecular and Cellular Biology 30:4535-4547.

Pohlmann, J., Risse, C., Seidel, C., Pohlmann, T., Jakopec, V., Walla, E., Ramrath, P., Takeshita, N., Baumann, S., Feldbrugge, M., Fischer, R. and Fleig, U., 2014. The Vip1 inositol polyphosphate kinase family regulates polarized growth and modulates the microtubule cytoskeleton in fungi. PLOS Genetics 10:e1004586.

Povirk, L.F., 1996. DNA damage and mutagenesis by radiomimetic DNA-cleaving agents: bleomycin, neocarzinostatin and other enediynes. Mutation Research 355:71-89.

Prasad, A., Jia, Y., Chakraborty, A., Li, Y., Jain, S.K., Zhong, J., Roy, S.G., Loison, F., Mondal, S., Sakai, J., Blanchard, C., Snyder, S.H. and Luo, H.R., 2011. Inositol hexakisphosphate kinase 1 regulates neutrophil function in innate immunity by inhibiting phosphatidylinositol-(3,4,5)-trisphosphate signaling. Nature Immunology 12:752-760.

Safrany, S.T., 2004. Protocols for regulation and study of diphosphoinositol polyphosphates. Molecular Pharmacology 66:1585-1591.

Safrany, S.T. and Shears, S.B., 1998. Turnover of bis-diphosphoinositol tetrakisphosphate in a smooth muscle cell line is regulated by beta2-adrenergic receptors through a cAMP- mediated, A-kinase-independent mechanism. Embo Journal 17:1710-1716.

Safrany, S.T., Caffrey, J.J., Yang, X., Bembenek, M.E., Moyer, M.B., Burkhart, W.A. and Shears, S.B., 1998. A novel context for the 'MutT' module, a guardian of cell integrity, in a diphosphoinositol polyphosphate phosphohydrolase. Embo Journal 17:6599-6607.

Safrany, S.T., Ingram, S.W., Cartwright, J.L., Falck, J.R., McLennan, A.G., Barnes, L.D. and Shears, S.B., 1999. The diadenosine hexaphosphate hydrolases from *Schizosaccharomyces pombe* and *Saccharomyces cerevisiae* are homologues of the human diphosphoinositol polyphosphate phosphohydrolase. Overlapping substrate specificities in a MutT-type protein. Journal of Biological Chemistry 274:21735-21740.

Saiardi, A., Bhandari, R., Resnick, A.C., Snowman, A.M. and Snyder, S.H., 2004. Phosphorylation of proteins by inositol pyrophosphates. Science 306:2101-2105.

Saiardi, A., Caffrey, J.J., Snyder, S.H. and Shears, S.B., 2000. The inositol hexakisphosphate kinase family. Catalytic flexibility and function in yeast vacuole biogenesis. Journal of Biological Chemistry 275:24686-24692.

Saiardi, A., Erdjument-Bromage, H., Snowman, A.M., Tempst, P. and Snyder, S.H., 1999. Synthesis of diphosphoinositol pentakisphosphate by a newly identified family of higher inositol polyphosphate kinases. Current Biology 9:1323-1326.

Saiardi, A., Nagata, E., Luo, H.R., Snowman, A.M. and Snyder, S.H., 2001. Identification and characterization of a novel inositol hexakisphosphate kinase. Journal of Biological Chemistry 276:39179-39185.

Saiardi, A., Sciambi, C., McCaffery, J.M., Wendland, B. and Snyder, S.H., 2002. Inositol pyrophosphates regulate endocytic trafficking. Proceedings of the National Academy of Sciences of the United States of America 99:14206-14211.

Sarmah, B. and Wente, S.R., 2010. Inositol hexakisphosphate kinase-2 acts as an effector of the vertebrate Hedgehog pathway. Proceedings of the National Academy of Sciences of the United States of America 107:19921-19926.

Schell, M.J., Letcher, A.J., Brearley, C.A., Biber, J., Murer, H. and Irvine, R.F., 1999. PiUS (Pi uptake stimulator) is an inositol hexakisphosphate kinase. Febs Letters 461:169-172.

Shames, D.S. and Minna, J.D., 2008. IP6K2 is a client for HSP90 and a target for cancer therapeutics development. Proceedings of the National Academy of Sciences of the United States of America 105:1389-1390.

Shears, S.B., Ali, N., Craxton, A. and Bembenek, M.E., 1995. Synthesis and metabolism of bis-diphosphoinositol tetrakisphosphate *in vitro* and *in vivo*. Journal of Biological Chemistry 270:10489-10497.

Stauffer, T.P., Ahn, S. and Meyer, T., 1998. Receptor-induced transient reduction in plasma membrane PtdIns(4,5)P_2 concentration monitored in living cells. Current Biology 8:343-346.

Steidle, E.A., Chong, L.S., Wu, M., Crooke, E., Fiedler, D., Resnick, A.C. and Rolfes, R.J., 2016. A novel inositol pyrophosphate phosphatase in *Saccharomyces cerevisiae*: Siw14 protein selectively cleaves the beta-phosphate from 5-diphosphoinositol pentakisphosphate (5PP-IP_5). Journal of Biological Chemistry 291:6772-6783.

Stephens, L., Radenberg, T., Thiel, U., Vogel, G., Khoo, K.H., Dell, A., Jackson, T.R., Hawkins, P.T. and Mayr, G.W., 1993. The detection, purification, structural characterization, and metabolism of diphosphoinositol pentakisphosphate(s) and bisdiphosphoinositol tetrakisphosphate(s). Journal of Biological Chemistry 268:4009-4015.

Taylor, R., Jr., Chen, P.H., Chou, C.C., Patel, J. and Jin, S.V., 2012. KCS1 deletion in *Saccharomyces cerevisiae* leads to a defect in translocation of autophagic proteins and reduces autophagosome formation. Autophagy 8:1300-1311.

Timerman, A.P., Mayrleitner, M.M., Lukas, T.J., Chadwick, C.C., Saito, A., Watterson, D.M., Schindler, H. and Fleischer, S., 1992. Inositol polyphosphate receptor and clathrin assembly protein AP-2 are related proteins that form potassium-selective ion channels in planar lipid bilayers. Proceedings of the National Academy of Sciences of the United States of America 89:8976-8980.

Voglmaier, S.M., Bembenek, M.E., Kaplin, A.I., Dorman, G., Olszewski, J.D., Prestwich, G.D. and Snyder, S.H., 1996. Purified inositol hexakisphosphate kinase is an ATP synthase: diphosphoinositol pentakisphosphate as a high-energy phosphate donor. Proceedings of the National Academy of Sciences of the United States of America 93:4305-4310.

Weaver, J.D., Wang, H. and Shears, S.B., 2013. The kinetic properties of a human PPIP5K reveal that its kinase activities are protected against the consequences of a deteriorating cellular bioenergetic environment. Bioscience Reports 33:e00022.

Winward, L., Whitfield, W.G., McLennan, A.G. and Safrany, S.T., 2010. Oxidation of the diphosphoinositol polyphosphate phosphohydrolase-like Nudix hydrolase Aps from Drosophila melanogaster induces thermolability--A possible regulatory switch? The International Journal of Biochemistry and Cell Biology 42:1174-1181.

Winward, L., Whitfield, W.G., Woodman, T.J., McLennan, A.G. and Safrany, S.T., 2007. Characterisation of a bis(5'-nucleosyl)-tetraphosphatase (asymmetrical) from Drosophila melanogaster. The International Journal of Biochemistry and Cell Biology 39:943-954.

Winward, L.J., 2005. Roles of inositol diphosphates and diadenosine polyphosphates in *Drosophila melanogaster* development. Doctoral Thesis, University of Dundee.

Wundenberg, T., Grabinski, N., Lin, H. and Mayr, G.W., 2014. Discovery of $InsP_6$-kinases as $InsP_6$-dephosphorylating enzymes provides a new mechanism of cytosolic $InsP_6$ degradation driven by the cellular ATP/ADP ratio. Biochemical Journal 462:173-184.

Yang, Y., Isaac, C., Wang, C., Dragon, F., Pogacic, V. and Meier, U.T., 2000. Conserved composition of mammalian box H/ACA and box C/D small nucleolar ribonucleoprotein particles and their interaction with the common factor Nopp140. Molecular Biology of the Cell 11:567-577.

Yates, L.M. and Fiedler, D., 2015. Establishing the stability and reversibility of protein pyrophosphorylation with synthetic peptides. Chembiochem 16:415-423.

Ye, W., Ali, N., Bembenek, M.E., Shears, S.B. and Lafer, E.M., 1995. Inhibition of clathrin assembly by high affinity binding of specific inositol polyphosphates to the synapse-specific clathrin assembly protein AP-3. Journal of Biological Chemistry 270:1564-1568.

Yong, S.T., Nguyen, H.N., Choi, J.H., Bortner, C.D., Williams, J., Pulloor, N.K., Krishnan, M.N. and Shears, S.B., 2015. Identification of a functional nuclear translocation sequence in hPPIP5K2. BMC Cell Biology 16:17.

York, S.J., Armbruster, B.N., Greenwell, P., Petes, T.D. and York, J.D., 2005. Inositol diphosphate signaling regulates telomere length. Journal of Biological Chemistry 280:4264-4269.

Yu, W., Ye, C. and Greenberg, M.L., 2016. Inositol hexakisphosphate kinase 1 (IP6K1) regulates inositol synthesis in mammalian cells. Journal of Biological Chemistry 291:10437-10444.

Zhang, Z., Liang, D., Gao, X., Zhao, C., Qin, X., Xu, Y., Su, T., Sun, D., Li, W., Wang, H., Liu, B. and Cao, F., 2014a. Selective inhibition of inositol hexakisphosphate kinases (IP6Ks) enhances mesenchymal stem cell engraftment and improves therapeutic efficacy for myocardial infarction. Basic Research in Cardiology 109:417.

Zhang, Z., Zhao, C., Liu, B., Liang, D., Qin, X., Li, X., Zhang, R., Li, C., Wang, H., Sun, D. and Cao, F., 2014b. Inositol pyrophosphates mediate the effects of aging on bone marrow mesenchymal stem cells by inhibiting Akt signaling. Stem Cell Research and Therapy 5:33.

Zhou, D., Chen, L.M., Hernandez, L., Shears, S.B. and Galán, J.E., 2001. A *Salmonella* inositol polyphosphatase acts in conjunction with other bacterial effectors to promote host cell actin cytoskeleton rearrangements and bacterial internalization. Molecular Microbiology 39:248-259.

Chapter 6
Low phytate nutrition – what is the pig and poultry industry doing to counter dietary phytate as an anti-nutrient and how is it being applied?

P. Wilcock and C.L. Walk*
AB Vista, Woodstock Court, Blenheim Road, Marlborough Business Park, Marlborough, Wiltshire SN8 4AN, United Kingdom; pete.wilcock@abvista.com

Abstract

Understanding both dietary phytate levels and phytase characteristics is important in optimising phytase use in monogastric feeds. This can be associated with maximising P release and feed cost savings or the near complete phytate degradation to improve animal performance and improve production efficiencies. This paper reviews pig and poultry research conducted with low phytate nutrition through superdosing phytase and how these results are being applied commercially today. This includes the impact that superdosing phytase has on mineral utilisation with a specific focus on zinc (Zn), copper (Cu) and iron (Fe) utilisation in nursery pigs as well as the performance or cost saving benefits of improved nutrient utilisation in growing and finishing pigs.

Keywords: phytase, superdose, minerals

6.1 Introduction

This chapter is focused on how research targeting low phytate nutrition through the use of superdosing phytase is being applied commercially. Low phytate nutrition is achieved by the breakdown of dietary phytate (>85%) by superdose levels of phytase (>1,500 FTU/kg), into inositol-phosphates (IP) with a lower degree of phosphorylation mainly IP_5, IP_4, IP_3 and inositol. It is not the intention of this chapter to cover in detail the mechanistic effects that phytate and low phytate esters have on nutrient (particularly amino acid) utilisation as that is covered in Chapter 3 of this book (Bedford, 2016). This review will focus on the importance of understanding the dietary phytate levels to optimise phytase usage as well as superdosing as described for high phytase dosing by Santos *et al.* (2013) and its effects on animal performance.

6.2 Phytate phosphorus levels in raw materials

Diets for pigs and poultry are predominantly composed of vegetable-based raw materials as they are generally more cost efficient than animal proteins. Even piglet starter feeds, traditionally high in animal proteins (plasma protein, fishmeal, meat meal, skim milk,

DOI 10.3920/978-90-8686-836-0_6,

whey) to maximise nutrient digestibility, have seen a decline in these protein sources due to costs in favour of vegetable proteins. Phytate is the main phosphorus (P) form in seeds and grains and is a mixed cation salt of phytic acid (*myo*-inositol 1,2,3,4,5,6-hexakis dihydrogen phosphate). The molecule consists of a fully phosphorylated *myo*-inositol ring with a total of 12 proton dissociation sites. Of these, six dissociate at acidic pH, three at neutral pH and 3 at a basic pH, which means that the phytate is negatively charged at a wide range of pH conditions (Maenz *et al.*, 1999; Maenz, 2001). Phytate is therefore effective at binding minerals (e.g. Ca^{2+}, Zn^{2+}, Fe^{2+}, Cu^{2+}, Na^{2+} and Mg^{2+}) as well as having the ability to precipitate proteins and is now seen as a major anti-nutrient in animal feed.

The ability of pigs and poultry to hydrolyse P from phytate is limited (Bedford, 2000) and this has resulted in >90% of global commercial broiler and pig feed now containing a phytase. The release of P from the phytate by phytases allows for the reduction of inorganic phosphates in the diet, saving feed costs, while phytate degradation is also linked to an increased amino acid and energy utilisation by the animal producing additional savings.

Phytate levels in raw materials can vary depending on the plant type, the growing conditions, the breeding program and how the final feed ingredient is processed. The site of phytate accumulation differs depending on the seed type. In many dicotyledonous plants, such as rapeseed, soybeans and peas, the phytate is concentrated within globoids found in protein structures present in the cotyledons. However, many cereal grains such as wheat, barley and oats have the phytate present within globoids located predominantly in the aleurone layer. Hence milling to produce wheat bran, for example, results in a high concentration of phytate. Corn differs from other cereals in that phytate is mainly found in the germ (O'Dell *et al.*, 1972). With corn distillers dried grains with solubles (DDGS) produced from the ethanol industry, the protein and fibre fractions of corn are expected to be concentrated, by approximately three times the level of corn, and the expectation would be the same for phytate. However, the fermentation process results in lower and more variable levels of phytate in DDGS (Stein and Shurson, 2009). In addition, phytase is often added to the ethanol production process as it reduces phytate concentration and thus increases calcium (Ca) availability, which is important in stabilising amylase activity and increasing production yields (Khullar *et al.*, 2011). These factors increase the variability of phytate concentration in DDGS that enter the feed sector and analyses of three sources of DDGS being used by one swine production system highlight those differences in phytate levels (Table 6.1), depending on source. Furthermore, by-products from the wet milling industry, corn germ meal, corn gluten meal and corn gluten feed are low in phytate due to the steeping process hydrolysing the phytate within the by-product (Rojas *et al.*, 2013).

Hence, the ability to measure phytate in raw materials and feeds is important to maximise phytase use and cost benefits. With recent advances in near infrared technology (NIR) it is now possible to quickly and at a low cost measure phytate within raw materials and feed. This provides a fast, cost-effective tool for the nutritionist to measure phytate. Understanding the variation in raw material phytate levels (Table 6.2) can be important; for example, a simple corn (60%), soya meal (20%) based ration may vary by up to 0.06% (based on ±1 SD) in dietary phytate-P. Therefore monitoring the phytate levels in raw materials and feeds is important in optimising phytase use.

Table 6.1. *Phytate phosphorus (P) concentration (%) in corn distillers grains from three ethanol plants.*

	n	Phytate-P[1]	
		Mean (%)	Std. dev.
Plant 1	10	0.21	0.05
Plant 2	10	0.11	0.01
Plant 3	10	0.31	0.03

[1] Determined using assay procedure K-Phyt 08/14 (Megazyme, Wicklow, Ireland).

Table 6.2. *Phytate phosphorus (P) content of different raw materials as determined by NIR.*

Raw materials[1]	n	Phytate-P (%, as fed)			
		Mean	Std. dev.	Min	Max
Corn	68,229	0.187	0.032	0.109	0.270
Sorghum	7,336	0.178	0.026	0.105	0.250
Wheat	1,421	0.208	0.025	0.139	0.281
Barley	91	0.199	0.023	0.134	0.253
Millet	25	0.238	0.045	0.134	0.363
Oats	14	0.214	0.030	0.138	0.256
Soya bean – meal	2,561	0.441	0.051	0.296	0.579
Soya bean – expeller	529	0.405	0.034	0.320	0.495
Soya bean – full fat	342	0.372	0.066	0.234	0.533
Canola meal	229	0.634	0.115	0.393	0.805
DDGS (corn)[2]	291	0.252	0.042	0.167	0.347
DDGS (wheat)	102	0.243	0.038	0.167	0.334
Wheat bran	285	0.829	0.040	0.711	0.939
Corn gluten meal	144	0.425	0.076	0.221	0.535
Bakery meal	67	0.370	0.194	0.155	0.854
Corn germ	29	0.192	0.004	0.181	0.200

[1] Only raw materials with >10 samples are included from the database that covers >80,000 samples.
[2] DDGS = distillers dried grains with solubles.

6.3 Optimising phytase use for phosphorus release

Traditionally, phytases have been used in feed for pigs and poultry to target a moderate release of 0.10% to 0.12% available phosphorus (avP) and this use, in part, has been limited due to the efficacy of the phytase as well as the inability to analyse dietary phytate levels. For example, the original *Aspergillus niger* phytase, if included at 500 FTU/kg in a pig diet containing 0.25% phytate-P, would be expected to give an avP release of 0.10% which means a phytate hydrolysis of 40%. In comparison, the new phytases, such as those from

Buttiauxella sp. and the modified *Escherichia coli*, would be expected to give an avP release of 0.14 to 0.15% respectively, in pigs meaning phytate hydrolysis of 55-60%.

These next generation phytases have better characteristics for phytate breakdown under conditions of the gastro-intestinal tract, with the modified *E. coli* phytase having the ability to rapidly degrade intact phytate and IP_x esters (Menezes-Blackburn *et al.*, 2015). This becomes important when looking for a consistent high P release, as the phytase has to have the ability to hydrolyse phytate and IP_x esters when their gastric levels are increasingly reduced. For example, a typical corn soy-based diet contains approximately 0.90% phytate, and each phytate molecule contains six phosphates. If only one phosphate is removed from each phytate molecule, that would only yield 0.04% P (or 16% of phytate hydrolysed). Therefore, to achieve 0.175% P released from phytate hydrolysis of both phytate and IP_x esters are required.

To ensure sufficient dietary phytate-P is present, it is essential to have the ability to measure phytate. It is this ability to measure dietary phytate levels, coupled with the better understanding of phytase characteristics and the IP_x degradation that has led to a move in the industry to target a higher avP release from phytases. More broiler and pig producers are now targeting a 0.15% avP release from phytase as a standard, while a higher avP release between 0.17 and 0.20% is gaining traction in markets such as Latin America (T. Santos, personal communication) and Asia (U. Aftab, personal communication).

However, it must be noted that the ability to target a higher P release from phytase is not only affected by the dietary phytate content and the phytase source used, but also the composition of the diet. For example, if the P release from phytase and the native P in the diet is sufficient to reach the required nutrient specification, whereby there is no inorganic P source or potentially meat and bone meal to remove from the diet, then it is generally not cost effective to go any higher in phytase dose using just a mineral (P and, Ca) matrix, as the cost of the diet could potentially be negatively affected. Table 6.3 highlights the effect of increasing phytase dose in feed formulations and potential cost savings in a typical UK pig grower diet. In this example only the digestible P and Ca release is accounted for and any expected amino acid or energy benefit is ignored. The diet is high in phytate-P (approximately 0.27%) which can be attributed to the inclusion of alternative protein sources, such as rapeseed expellers and sunflower meal, that both have high phytate-P levels relative to soybean meal. The formulation exercise shows that the level of phytase in this example could be increased beyond the typical dose of 500 FTU/kg to 750 FTU/kg as there is sufficient dietary phytate to achieve the expected digestible P release from phytase. In fact, there is sufficient dietary phytate to increase the phytase dose up to 1000 FTU/kg but as all inorganic P has already been removed by the inclusion of 750 FTU/kg, there is no further cost benefit to the dietary digestible P reduction. In contrast, if this diet had contained 0.20% phytate-P, then phytase inclusion would have been limited to 500 FTU/kg to ensure that the expected digestible P was released. In this case any further increase in phytase would increase the digestible P release but at a level less than expected due to limited substrate i.e. phytate-P.

Table 6.3. *Formulation exercise showing the effect of increasing phytase levels on pig grower feed formulations and final costs.*

	Diet 1	Diet 2	Diet 3
Raw materials (%)			
Barley	15.0	15.0	15.0
Wheat	55.6	56.1	56.4
Soya bean meal	14.3	14.2	14.1
Rapeseed expellers	5.00	5.00	5.00
Lysine	0.75	0.75	0.75
Methionine	0.078	0.077	0.076
Threonine	0.17	0.17	0.17
MCP[1]	0.35	0.11	0
Salt	0.46	0.46	0.46
Fat	2.06	1.90	1.83
Premix	0.50	0.50	0.50
Limestone	0.76	0.72	0.70
Modified *Escherichia coli* phytase	0.005	0.01	0.015
Sunflower extract	3.00	3.00	3.00
Molasses	Bal	Bal	Bal
Feed cost difference to Diet 1	-	-£1.50	-£2.00
Dig lysine (%)	1.08	1.08	1.08
DE (MJ/kg)	14.0	14.0	14.0
Digestible P (%)	0.28	0.28	0.28
Calcium (%)	0.58	0.58	0.58
Phytase[2] (FTU/kg)	250	500	750

[1] Monocalcium phosphate was fixed at 83% digestible P.

[2] Modified *E. coli* phytase (5,000 FTU/g) was included at the following levels and assumed reductions in available P (avP) and Ca (250 FTU/kg with 0.10% avP and 0.11% Ca, 500 FTU/kg with 0.15% avP and 0.165% Ca, and 750 FTU/kg with 0.175% and 0.192% Ca) and adjusted to digestible P by the relative digestibility of P in MCP.

This exercise highlights that understanding the level of dietary phytate and the phytase used are important in determining if there is an opportunity to increase phytase dose in respect to further P release to maximise feed cost savings.

6.4 Phytate as an anti-nutrient

More recently the anti-nutritional effect of phytate has been studied in relation to its effect on animal performance and how this can be mitigated by phytase. In broilers, it has been demonstrated that increasing dietary phytate levels has a negative effect on growth performance (Cabahug *et al.*, 1999; Linares *et al.*, 2007; Liu *et al.*, 2008, 2009; Onyango and

Adeola, 2009). Liu *et al.* (2008) reported that in 28 day old broilers, increasing phytate-P from 0.22% to 0.44% reduced bodyweight by 73 g/bird (P<0.001) and resulted in a poorer feed conversion (1.40 vs 1.50 g/g; P<0.001). Studies in weaned pigs (Woyengo *et al.*, 2012) demonstrated that at 21 days, both ADG (316 vs 198 g/day; P<0.05) and FCR (1.82 vs 1.32 g/g P<0.05) were reduced due to an increase in dietary phytate-P from 0 to 0.56%. This was supported by a recent piglet study (Walk *et al.*, 2014b) that compared growth performance of pigs weaned at 21 days and fed an iso-nutrient diet with low (0.18%) or high (0.22%) phytate-P for 21 days. The increased phytate-P was achieved by the addition of 2.5% rice bran which is high in phytate-P and had a minimal effect on dietary formulation. Results indicated that even a marginal increase in phytate-P of 0.04% (i.e. equivalent to approximately a 10% soybean meal inclusion) resulted in a reduced ADG (P<0.08) from wean to 21 days (Figure 6.1) with a numerical reduction in FCR.

This negative impact that phytate can have on poultry and pig performance has led to a greater focus on superdosing phytase (>1,500 FTU/kg feed) to target near-complete breakdown of dietary phytate. This phytate degradation can result in a reduced endogenous amino acid loss as well as a lower level of energy needed for maintenance, resulting in an overall improvement in amino acid and energy utilisation by the animal (Cowieson *et al.*, 2011; Onyango *et al.*, 2009). Furthermore, superdosing phytase leads to the breakdown of phytate and IP_x esters, resulting in an increased inositol release (Holloway *et al.*, 2016a; Walk *et al.*, 2014a; Wilcock *et al.*, 2014). Inositol release from phytate degradation has been correlated to better performance in both broilers (Walk *et al.*, 2014a) and pigs (Wilcock *et al.*, 2014). Dietary inositol supplementation has been directly linked to better performance with both Zyla *et al.* (2004) and Bedford and Rama Rao (2013) reporting that inositol added to the diet improves broiler performance. Results of the latter study indicated that 30% of the superdosing response may be associated with inositol release. The hydrolysis of phytate by phytase produces smaller phytate esters such as IP_5, IP_4 and IP_3 (Greiner *et al.*, 1993), and it was initially thought that these did not negatively impact nutrient utilisation. This seems to not be the case, with data showing that not only does IP_6 have a negative

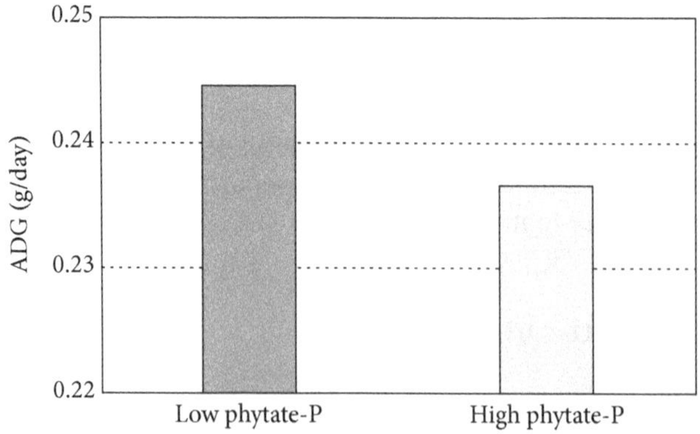

Figure 6.1. *The effect of increasing dietary phytate on average daily gain in piglets for 21 d post-weaning (P<0.08; Walk* et al., *2014b).*

effect on pepsin activity (Liu and Cowieson, 2010) but that IP_4 and IP_3 can also have an effect (Yu *et al.*, 2012) which may in part explain the reduced amino acid digestibility associated with phytate. Furthermore, smaller phytate esters have also been shown to negatively impact energy utilisation (Beeson *et al.*, 2014a,b; Lui *et al.*, 2008) and mineral utilisation (Beeson *et al.*, 2014a,b; Persson *et al.*, 1998; Xu *et al.*, 1992). This may in part explain the benefit of using superdose levels of phytase on growth performance. An initial dose (250-500 FTU/kg) of an *E. coli* phytase may be sufficient to break down phytate to IP_4 by the removal of two phosphates, but insufficient to target full IP_4 destruction. With IP_4 and IP_3 accumulation, there is still the potential to bind to proteins, pepsin and minerals, similarly to phytate potentially impairing nutrient utilisation and animal performance. Recent work in broilers (Beeson *et al.*, 2014a,b) indicates that there may be a correlation between low phosphorylated IP_x esters, generated through phytase supplementation, and poor mineral, energy and protein digestibility (Table 6.4) although more *in vivo* work in this area should be conducted to determine the relative effect of the negative effects of phytate and IP_x esters on nutrient digestibility.

With phytate having the ability to bind cations (e.g. Ca^{2+}, Zn^{2+}, Fe^{2+} and Cu^{2+}), the effect of superdosing phytase on mineral utilisation has interest in relation to mineral excretion and environmental concerns as well as animal performance. When related to iron (Fe), zinc (Zn) and copper (Cu) and the application in nursery pigs, for example, superdosing may play an important role by improving the availability of these minerals, improving performance or achieve similar performance with a lower dietary mineral level resulting in feed cost saving and reduced mineral excretion. This area of low phytate nutrition through superdosing phytase has gained traction in commercial production systems. This has led to a greater interest from nutritionists in understanding the key phytase characteristics that allow for the rapid breakdown of IP_6 and less phosphorylated IP esters and how these may differ between the different commercial phytases.

The following section will focus on some areas of application of the low phytate feeding concept used in commercial swine production.

Table 6.4. *Correlation between IP_4 and IP_3 and ileal nutrient digestibility in broilers (n=40 samples; Beeson* et al., *2014a,b).*

Phytate ester	Nutrient	Correlation of ileal IP_3/IP_4 level and ileal nutrient digestibility	Significance
Ileal IP_3	AME (kcal)	-0.85	*P*<0.0001
	DM	-0.85	*P*<0.0001
	N	-0.69	*P*<0.0001
	Na	-0.68	*P*<0.0001
Ileal IP_4	Mg	-0.69	*P*<0.0001
	Fe	-0.61	*P*<0.0001

6.5 Low phytate nutrition – swine application

6.5.1 Nursery pigs: enhanced growth performance

Use of a low phytate nutrition program through superdosing phytase in the nursery results in an enhanced growth performance, with trials using a modified *E. coli* phytase demonstrating an average improvement in gain by 8% and in FCR by 5 points (unpublished data). Recently, in a commercial trial in the USA this type of program was evaluated in nursery pigs. Weaned pigs were either fed following a standard commercial three-phase feeding program formulated to meet the requirements for pigs up to 42 days post-weaning or fed the same diets supplemented with 2,000 FTU/kg of a modified *E. coli* phytase (low phytate nutrition program). From weaning to 21 days post-weaning, the phytase was added on top of the control diet, after which up to day 42 a mineral reduction as recommended for the use of 500 FTU/kg (-0.15% avP and -0.16% Ca) was applied. The results were consistent with previous data and demonstrated that the low phytate nutrition program resulted in an enhanced gain of 1.6 kg ($P<0.01$) and a 9 point improvement in FCR (1.40 vs 1.31; $P<0.05$; Table 6.5). More importantly this extra gain was achieved at approximately $0.05/kg gain lower cost, or approximately $1.00 per pig saving for a 20 kg gain in the nursery by using the lower phytate nutrition program.

Table 6.5. *The effect of a low phytate nutrition program achieved through the use of superdosing phytase on nursery pig performance (42 days post-weaning).*

	Treatments[1]	
	Control[2]	Low phytate nutrition program[2]
Phytase (FTU/kg)[3]	0/500[4]	2,000[5]
Start weight (kg)	7.00	6.96
Day 42 weight (kg)	24.29[a]	25.88[b]
Average daily gain (g/d)	414[a]	451[b]
Average daily feed intake (g/d)	582	588
Feed conversion ratio (g/g)	1.40[a]	1.31[b]

[1] [ab] Means in a row not sharing a common superscript differ ($P<0.05$).

[2] Diets were based on corn, soya, whey, plasma, fishmeal, fat from weaning to day 21 post-weaning and based on corn, soya, and fat from day 21 to 42 post-weaning.

[3] Phytase used was Quantum Blue 5G (AB Vista, UK).

[4] In the control diet, phytase was included from weaning to 21 day with no nutrient matrix associated with the phytase and with a 500 FTU/kg mineral matrix (0.15% avP and 0.16% Ca) from 21 to 42 day.

[5] In the low phytate nutrition program, phytase was as the control diet with an additional 1,500 FTU/kg of phytase (no matrix associated with extra phytase) added to give 2,000 FTU/kg in the final feed.

6.5.2 Nursery pigs: alternative raw materials

In piglet starter diets (fed for 21 days after weaning) there has been a focus on reducing further the level of highly digestible animal proteins in the diet to reduce costs. This has resulted in the higher inclusion of vegetable-based proteins such as fermented soybean meal, enzyme treated soybean meal, and other enhanced soybean products as well as increased usage of conventional soybean meal and an increased use of yeast-based products as a protein source. As an estimate, a 10% increase in soybean meal at the expense of animal proteins will increase phytate-P in the diet by approximately 0.03-0.04%. As previously described, this small increase in dietary phytate may have a detrimental effect on performance. Hence, using new strategies such as superdosing phytase to negate this adverse effect of higher soybean meal, and therefore higher phytate-P, on early pig growth performance is important. This effect is shown in two recent studies when the combination of high soybean meal and superdose levels of a modified *E.* coli phytase was used to replace poultry meal (Zier-Rush *et al.*, 2012) and plasma protein (Bradley *et al.*, 2015) without negatively affecting performance (Table 6.6). Furthermore the cost savings on a per pig basis for the starter feed period alone was estimated at $0.44 per pig for replacement of poultry meal and $0.46 per pig for replacement of plasma. These trials support earlier work

Table 6.6. *The effect of superdosing phytase in piglet starter feeds devoid of plasma protein from weaning to 21 days (Bradley* et al., *2015).*

Feeding phase (P)	Treatments[1]		
	Control	NC	NC + phytase
	P1/P2	P1/P2	P1/P2
Raw materials			
Plasma (%)	6/3	0/0	0/0
Soya bean meal (%)	15/20	25/30	25/30
Phytase (FTU/kg)[2]	0/0	0/0	2,000/2,000
Nutrients			
Dig lysine	1.56/1.51	1.56/1.51	1.56/1.51
ME (Kcal/kg)	3,415/3,360	3,415/3,360	3,415/3,360
Calcium (%)	0.85/0.80	0.85/0.80	0.85/0.80
Available phosphorus (%)	0.50/0.45	0.50/0.45	0.50/0.45
Phytate phosphrous (%)	0.18/0.21	0.21/0.24	0.21/0.24
Performance (weaning – d21)			
ADG (g/d)	270[a]	243[b]	274[a]
ADFI (g/d)	308	292	308
FCR (g/g)	1.14[a]	1.20[b]	1.13[a]

[1] [ab] Means in a row not sharing a common superscript differ (P<0.05).

[2] Phytase used was Quantum Blue 5G (AB Vista, UK) and included at the expense of corn from weaning to 21 days post-weaning with no nutrient matrix associated with the phytase.

(Wilcock *et al.*, 2010) that indicated pigs fed a simple diet (high in soybean meal and low in lactose) could perform similarly to pigs fed a complex diet (low in soybean meal and high in lactose) when a modified *E. coli* phytase was added at 2,500 FTU/kg to the simple diet.

6.5.3 Nursery pigs: minerals – iron

With increasing litter size and reduced weaning weights, there has been an increased interest in the effect that reducing phytate within the gastro intestinal tract can have on Fe utilisation in the weaned pig. Newly born pigs typically have low Fe reserves (estimated 50 mg) with the sows' milk only contributing an estimated 1 mg of Fe per day (Brady *et al.*, 1978). The pig will, within 3 to 6 days after birth, be Fe deficient based on the Fe requirement of 8-16 mg per day (dependent on growth rate). Typically an Fe injection is used at 3-4 days of age to provide sufficient Fe to maintain growth rate through to weaning. Even with the practice of Fe injection, the larger, faster growing piglets may be more susceptible to anaemia (Bhattarai, 2015), and Fe may become limiting in these pigs by 17 days of age (Jolliff and Mahan, 2011). These fast growing pigs may be marginally Fe deficient at weaning when they are transferred onto starter feed.

With the lower use of animal proteins in starter feeds, the background dietary Fe levels have been reduced, but this commonly is not compensated by a change in supplemental Fe, with typical levels still being at 100 to 200 mg/kg feed. Higher levels than this are not often considered due to the concern of increased *E. coli* growth in the lower gastrointestinal tract (Knight *et al.*, 1983). Furthermore, the lowering of animal protein will result in an increase in dietary phytate, due to the higher inclusion of vegetable proteins, with typical starter feeds having a dietary phytate level of 0.8 to 1.0%. Phytate is a potent chelator of Fe (Yu *et al.*, 2012) with not only phytate binding Fe but also less phosphorylated IP esters (IP_4 and IP_3) having a similar ability to do so.

Hurrell (2003) has demonstrated in human adults that decreasing phytate levels and the phytate:Fe ratio in bread increased the level of Fe absorption. Based on this human model in standard pig starter feeds the typical dietary phytate (0.80 to 0.90%) level could therefore potentially limit Fe absorption to 20-30%. This may produce a situation where weaned pigs are marginally Fe deficient and then moved to a starter diet that limits dietary Fe utilisation. Hence, reducing phytate through superdosing phytase within the gasto-intestinal tract in this post-weaning period may be beneficial in terms of increasing Fe status and early pig performance.

This hypothesis was confirmed in a recent study in weaned pigs when a greater phytate breakdown (i.e. lower ileal phytate levels) was achieved through the use of a modified *E. coli* phytase (>1,500 FTU/kg). This resulted in an improved Fe status, as measured by haemoglobin (Hb, g/dl; Figure 6.2) and red blood cell distribution width, both sensitive indicators of Fe status (unpublished data).

Gastro-intestinal phytate degradation in the post-weaning period for pigs through the use of superdosing phytase may be beneficial in eliciting a response to supplemental Fe. When comparing two commercial trials in weaned pigs; trial 1 without superdosing phytase and

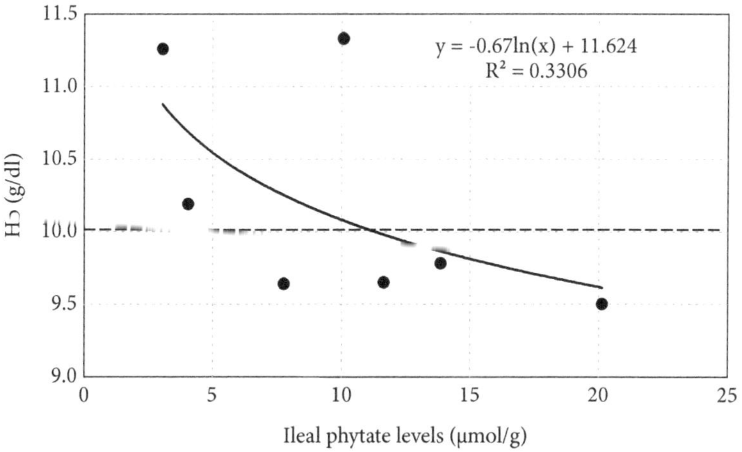

Figure 6.2. *The effect of ileal phytate concentration on blood haemoglobin in piglets at 21 d post-weaning (S. Mansbridge, A. Stewart, I. Wellock, and C. L. Walk, personal communication). Each data point is a mean from 30 pigs. --- minimum level of Hb (g/dl) assumed to be adequate for growth (Zimmerman, 1980).*

trial 2 with 2,000 FTU/kg phytase, supplemental Fe improved performance (gain, +10%; FCR, -7%) only in those diets containing superdosing phytase (Table 6.7).

This has led to an increased interest in the area of Fe and phytate interaction, with a small number of companies in the swine sector increasing the level of Fe in the presence of superdosing phytase to improve early weaned pig performance. With a minimal cost of <$0.05 per pig to increase Fe supplementation by 200 mg/kg (iron sulphate) in the initial 14 days post-weaning, this is potentially a cost effective opportunity to improve early pig performance post-weaning.

Table 6.7. *Effects of supplemental iron on 20 day post-wean pig performance with or without superdosing phytase.*

Weaning to day 20 post-weaning	Phytase (FTU/kg)[1]	Iron (ppm)	ADG (g/day)[2]	FCR (g/g)[3]
Trial 1[4]	0	200	243	1.30
	0	350	247	1.33
Trial 2[4]	2,000	200	256[B]	1.30[a]
	2,000	350	282[A]	1.22[b]

[1] Phytase used was Quantum Blue 5G, included from weaning to 20 days post-weaning with no nutrient matrix associated with the phytase.

[2] [AB] Means in a column not sharing a common superscript are different ($P<0.05$).

[3] [ab] Means in a column not sharing a common superscript are different ($P<0.10$).

[4] Trial 1 and trial 2 were run in the same research facility and feed type.

6.5.4 Nursery pigs: minerals – copper

The Cu requirements for swine are low and vary between 5 and 10 mg/kg, depending on phase of feeding. However, Cu is often fed at higher levels (125 to 250 mg/kg depending on dietary phase) due to its benefits in terms of improved growth rate and purported antimicrobial properties. Similarly to other divalent cations such as Fe and Zn, Cu is readily bound by phytate. The ability to form insoluble Cu-phytate complexes can depend on the form of Cu supplied as well as the pH at the time of interaction (Pang and Applegate, 2006). In an *in vitro* system at pH 5.5, copper sulphate most readily bound to phytate, which resulted in the highest rate of insoluble Cu-phytate (41.4%) complex formation (Table 6.8). However, copper citrate was less likely to bind to phytate and this resulted in the lowest rate of insoluble (9.6%) Cu-phytate complex formation. At pH 6.5, copper sulphate formed the highest amount of insoluble Cu-phytate (48.2%) complexes, with copper citrate, copper-lysine and tri-basic copper chloride all forming the least insoluble complexes. Similar to Fe and Zn, Cu is also bound by less highly phosphorylated inositol phosphates, with a higher proportion of Cu bound at a more neutral pH (Persson *et al.*, 1998). Thus, the gastrointestinal reduction of phytate through superdosing phytase may be beneficial in terms of improving Cu digestibility and performance (Adeola *et al.*, 1995; Jongbloed and Bikker, 2010), thereby allowing lower levels of dietary Cu to be used without a detrimental effect on performance.

With copper sulphate still being the main Cu source used in the swine industry, a recent nursery pig trial from weaning to 42 days examined the effect of dietary supplemental Cu level (0, 62.5, 125 and 250 mg/kg from copper sulphate) on growth performance with and without the use of a modified *E. coli* phytase (0 or 2,000 FTU/kg feed). There was no interaction between Cu level and phytase, with phytase superdosing giving significant ADG (453 vs 415 g/d; $P<0.05$) and FCR (1.30 vs 1.41; $P<0.05$) benefits. The 250 mg/kg supplemental Cu was the only Cu treatment significantly better than the control for both ADG (458 vs 408 g/d; $P<0.05$) and FCR (1.39 vs 1.32; $P<0.05$). The data indicate that

Table 6.8. *Insoluble copper-phytate complexes as influenced by copper source and pH (modified from Pang and Applegate, 2006).*

Copper source	Insoluble copper (%)[1]	
	pH 5.5	pH 6.5
Copper citrate	9.6[e]	4.2[c]
Copper chloride	25.4[c]	39.1[b]
Copper lysine	17.6[d]	5.7[c]
Copper sulphate	41.4[a]	48.2[a]
Tribasic copper chloride	33.0[b]	6.1[c]
SEM	0.69	0.62

[1] [ae] Means in a column not sharing a common superscript differ ($P<0.05$).

superdosing phytase significantly improves growth at all levels of Cu and that supplemental Cu may potentially be reduced from 250 to 125 mg/kg with the addition of superdosing phytase while maintaining animal growth performance (Figure 6.3).

Further work in this area needs to be conducted although in the current swine segment the savings of reducing copper sulphate in diets is minimal (100 mg/kg reduction equates to <$0.04 per pig saving through the nursery). But the environmental concern, especially in Europe where reductions in allowable Cu concentrations in diets is under discussion, may lead to greater interest in this mineral-phytate interaction.

6.5.5 Nursery pigs: minerals – zinc

Zinc oxide is traditionally used in starter feeds at pharmacological levels (>2,500 mg/kg) as a way to improve growth and mitigate post-weaning diarrhoea (Poulsen, 1995; Sales, 2013). Although the exact mechanism is unknown, it may involve a number of mechanisms including bacterial modulation, electrolyte balance, and tight junction function or regulating gene expression (Roselli *et al.*, 2003; Zhang and Guo, 2009). Known to be relatively insoluble and not available for absorption, increasing the inclusion of zinc oxide in pig diets increased serum and tissue Zn concentrations (Walk *et al.*, 2013), highlighting that Zn from zinc oxide is, to some extent, absorbed and stored in tissues. Hence, improving Zn solubility within the intestine may be a factor in improving the efficacy of zinc oxide and allow for a reduction in starter diets. Phytate, including IP_x esters, may bind Zn, especially between a pH of 4 and 6 (Lönnerdal *et al.*, 1989; Xu *et al.*, 1992). The impact of this can be two-fold; it may limit the efficacy of pharmacological

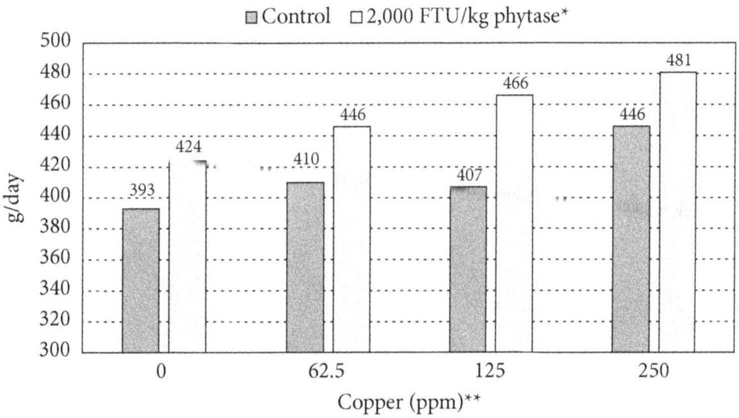

Figure 6.3. *The effect of copper on nursery pig daily gain (42d post-weaning) with and without superdosing phytase (unpublished). * Phytase used was Quantum Blue 5G (AB Vista, Marlborough, UK) included from weaning to 21 days post-weaning with no nutrient matrix associated with the phytase and from 21 to 42 days post-weaning with a 500 FTU/kg mineral matrix (0.15% available P and 0.16% calcium). ** 15 mg/kg copper was included in trace mineral mixture in all diets, and copper supplemented through the addition of copper sulphate.*

levels of zinc oxide, and may potentially reduce pancreatic protease activities that are Zn dependent (Mills *et al.*, 1967).

Low phytate nutrition through the use of superdosing phytase is now a common application in piglet starter feeds and the interaction with zinc oxide levels has been well studied. Walk *et al.* (2013) reported that increasing pharmacological levels of zinc oxide increased serum Zn concentration while the addition of phytase at 2,500 FTU/kg had a similar effect. Jongbloed and Bikker (2010) also reported an increase in Zn serum levels as phytase dose increased in the diet. This was attributed to the hydrolysis of phytate by the phytase, thereby reducing the Zn-phytate interaction, which resulted in increased Zn absorption. Based on this mechanism, a series of five studies (Walk *et al.*, 2013, 2014b, 2014c, 2015; unpublished data) was conducted to determine the effect of phytase at 2,500 FTU/kg on pig performance at differing supplementary dietary Zn levels (0, 1,750-2,500 and 2,500-3,500 mg/kg) from weaning to 21 d post-weaning. The trials were also conducted to ascertain if lower levels of Zn could be used in the presence of phytase superdosing and achieve the same level of performance as pigs fed a higher zinc oxide level without phytase. A summary of the trial results (Table 6.9) shows that increasing Zn supplementation improved ADG and FCR, and the addition of phytase at 2,500 FTU/kg diet improved growth performance above that of Zn supplementation for each dietary Zn level when supplemented without phytase. Furthermore, the use of phytase at 2,500 FTU/kg feed allowed for the reduction of Zn while maintaining or even improving growth performance, for example, reducing Zn supplementation from 2,500 mg/kg to 1,750 mg/kg with added phytase at 2,500 FTU/kg feed would improve growth performance (116 vs 130%) while allowing for a reduction in dietary Zn levels. The concern of reducing pharmacological Zn levels is that there may be

Table 6.9. *The effect of supplemental zinc on 21 day post-wean pig performance with or without superdosing phytase (extracted from Walk* et al., *2013, 2014b,c, 2015; unpublished).*

Supplementary zinc (mg/kg)[1]	0 FTU/kg phytase	2,500 FTU/kg phytase[2]
Average daily gain		
(relative improvement compared to pigs fed diets without zinc and phytase supplementation)		
0	100%	110.1%
1,750-2,500	112.4%	130.0%
2,500-3,500	116.6%	120.7%
Feed conversion ratio		
(relative improvement compared to pigs fed diet without zinc and phytase supplementation)		
0	100%	96.3%
1,750-2,500	94.9%	88.7%
2,500-3,500	92.4%	90.2%

[1] Supplementary zinc levels relate to the addition of zinc oxide above the base level already included in all diets (100-150 mg/kg).

[2] Phytase used was Quantum Blue 5G (AB Vista, Marlborough, UK) and included from weaning to day 21 post-weaning with no nutrient matrix associated with the phytase.

an associated increase in post-wean scouring, but these studies did not reflect this with pigs fed the superdosing diets having a 6% reduction in post-weaning scouring when compared to a control diet with no phytase (Walk *et al.*, 2014c). This may in part be due to increased Zn availability associated with superdosing phytase.

This research opens up the opportunity of maintaining or improving post-weaning pig performance while lowering pharmacological levels of dietary zinc oxide. This concept already has been adapted by a number of swine companies that are looking at the prudent use of zinc oxide in starter diets by balancing the key role that a pharmacological level of zinc oxide has in terms of growth performance and health with the ability to lower Zn levels with phytase superdosing to mitigate environmental concerns.

6.5.6 Finisher pigs: enhanced performance

In finishing pigs the application of low phytate nutrition through superdosing phytase has been used to enhance growth performance or maintain growth performance at a lower cost. This may be attributed to the increased amino acid and energy utilisation associated with dietary phytate degradation as well as with an increase in the inositol absorption, as described previously.

Superdosing phytase to improve finisher performance is becoming more established, as demonstrated in recently published data (Holloway *et al.*, 2016b; Jang *et al.*, 2016; Santos *et al.*, 2014). Wilcock *et al.* (2014) reported that in pigs from 20 to 60 kg, for every addition of 500 FTU/kg diet of a modified *E. coli* phytase above a base level of 500 FTU/kg, ADG (+21 g/day) and FCR (-2.3 points) were linearly ($P<0.05$) improved up to a maximum inclusion of 2,000 FTU/kg diet. In addition, stomach inositol concentrations were analysed and there was a positive correlation between inositol levels and pig growth performance, which supported previous work in broilers (Walk *et al.*, 2014a). Seven commercial comparison trials (unpublished data) have demonstrated that in finisher pigs (22 to 102 kg), feeding a modified *E. coli* phytase at 2,000 FTU/kg diet while reducing Ca and avP as recommended when 500 FTU/kg diet are applied (-0.15% avP and -0.16% Ca) resulted in 1.6 kg extra gain and a 5 point improvement in FCR compared to pigs fed a base level of phytase (400 to 500 FTU/kg). This application of improving nutrient utilisation and performance through superdosing is gaining traction, especially in hotter climates where feed intake can be lower, thereby limiting nutrient intake.

6.5.7 Finisher pigs: cost savings

Using phytase to reduce feed cost through taking account of the nutrient release is a more common approach in the swine market, with feed producers often taking a Ca, avP, amino acid and energy matrix into consideration for a standard phytase inclusion (340 to 750 FTU/kg) level. However, the introduction of superdosing phytase has resulted in higher amino acids supply assumed for phytases, resulting in further cost saving opportunities. This has led to research looking at the use of higher phytase doses (>2,500 FTU/kg) to determine the impact on amino acid nutrition and if getting closer to total dietary phytate destruction is of benefit.

A recent trial (unpublished data) conducted in a commercial US research barn in pigs from 35 to 120 kg investigated the effect of reducing dietary digestible lysine by 10% (NC) compared with a control diet (PC), and the effect that higher doses of phytase may have on NC performance. In all diets, the other essential amino acids were balanced to lysine using the same ideal amino acid profile. The PC was P and Ca adequate with no phytase supplementation, whereas the phytase supplemented diets were formulated based on the mineral release associated with 250 FTU/kg (0.10% avP and 0.11% Ca). The results (Table 6.10) indicated that there was a reduction ($P<0.05$) in ADG with a reduction in dietary amino acids and that this was restored back to the level of the PC by addition of 3,000 FTU/kg of a modified *E. coli* phytase. For FCR, there was no significant effect of treatment, but numerically, it followed a similar trend as for ADG, with pigs fed 3,000 FTU/kg diet achieving the same performance as the PC. The highest dose of phytase reduced feed cost by $1.85 per pig without negatively affecting growth performance, and this resulted in a lower cost/kg gain compared to the PC.

This is an area of research that needs further work and greater understanding of what cost advantages high phytase doses may have. It does, however, help support the more aggressive amino acid matrices, linked to higher phytase doses (>1,500 FTU/kg), that are being used successfully in large scale production systems today.

Table 6.10. *The effect of phytase dose on finisher (35 to 120 kg LW) pig performance in diets formulated to be 10% lower in amino acid than requirement.*

	Treatment[1]				**Probability**
	PC	**NC**	**NC**	**NC**	
Relative digestible lysine	100%	90%	90%	90%	
Phytase (FTU/kg)[2]	0	0	1,500	3,000	
Performance					
End weight (kg)	124.9	120.9	122.3	123.8	$P<0.07$
ADG (g/d)	858[a]	812[b]	825[b]	848[a]	$P<0.01$
FCR (g/g)	2.60	2.67	2.66	2.62	$P<0.08$
Feed cost per pig ($) (relative to PC)	0[a]	-3.26[b]	-2.08[b]	-1.85[b]	$P<0.007$
Feed cost/kg gain ($)	0[a]	-0.02[b]	-0.01[ab]	-0.02[b]	$P<0.05$

[1] [ab] Means in a row not sharing a common superscript differ ($P<0.05$).

[2] Phytase used was Quantum Blue 5G (AB Vista, Marlborough, UK), formulated in using a 250 FTU/kg mineral matrix (0.10% avP and 0.11% Ca).

6.6 Conclusions

The ability to commonly measure dietary phytate and a greater understanding of phytase characteristics enables phytase dose to be optimised to release the maximum P and the cost benefits associated with this. Furthermore, it is concluded that phytate may have a negative effect on both pig and broiler growth performance through reduced nutrient digestibility by phytate binding to digestive enzymes, proteins and minerals. This results in an increased level of nutrient losses including amino acids, minerals as well as an increased energy maintenance requirement. Through the use of superdosing phytase it is possible to target the near complete destruction of phytate which not only improves amino acid and energy utilisation but also inositol release. Inositol absorption has been linked to improved growth performance in broilers and pigs and is thought to explain approximately 30% of the superdosing phytase response.

1. The increased nutrient utilisation associated with superdosing phytase can be used to improve cost effective broiler or pig performance or as in the case of finisher pigs be used to allow for the use of a lower nutrient dense diets without negatively impacting performance.
2. The adoption of superdosing phytase continues to gain traction in commercial broiler and pig production systems and more research focused on improving amino acid utilisation with higher phytase doses should be conducted to understand the maximal opportunity for phytase application.
3. Other areas of development may focus on expanding the research on phytate and trace minerals to determine if the effect of superdosing phytase changes relative to the mineral source used (i.e. inorganic vs chelated minerals).

References

Adeola, O., Lawrence, B.V., Sutton, A.L. and Cline, T.R., 1995. Phytase induced changes in mineral utilization in zinc supplemented diets for pigs. Journal of Animal Science 73:3384-3391.

Bedford, M.R., 2000. Exogenous enzymes in monogastric nutrition – their current value and future benefits. Animal Feed Science and Technology 86:1-13.

Bedford, M.R. and Walk, C.L., 2016. Reduction of phytate to tetrakisphosphate (IP_4) to trisphosphate (IP_3), or perhaps even lower does not remove its antinutritive properties. In: Walk, C.L., Kühn, I., Stein, H.H., Kidd, M.T. and Rodehutscord, M. (eds.) Phytate destruction – consequences for precision animal nutrition. Wageningen Academic Publishers, Wageningen, the Netherlands, pp. 45-51.

Bedford, M.R. and Rama Rao, S.V., 2013. Effects of inositol or phytase alone or in combination in diets deficient or in excess of dietary phosphate. Poultry Science Association Meeting, San Diego, CA, USA, Abstr 110.

Beeson, L., Walk, C.L. and Olukosi, O., 2014a. Growth performance and nutrient digestibility responses of broilers to high supplemental levels of phytase in diets adequate or deficient in non-phytate phosphorus and calcium. World's Poultry Science Association UK Spring meeting (abstract).

Beeson, L., Walk, C.L. and Olukosi, O., 2014b. The flow of inositol phosphate esters and phytate phosphorus in the proximal and distal parts of the digestive tract of broilers receiving diets adequate in available phosphorus and supplemented with high levels of phytase. Journal of Animal Science. 92:Supplement 2 223 (Abstr).

Bhattarai, S., 2015. Early indicators of iron deficiency in large piglets at weaning. Journal of Swine Health and Production 23:10-17.

Bradley, C.L., Walk, C.L., Walker, N.D. and Wilcock, P., 2015. The effect of superdosing phytase with or without the addition of live yeast in diets void of spray dried plasma in pigs from weaning to 21 days post-weaning. American Society of Animal Science - Midwest Section, Des Moines, IA, USA, Abstr. 125.

Brady, P.S., Ku. P.K., Ullrey, D.E. and Miller, E.R., 1978. Evaluation of an amino acid-iron chelate hematinic for the baby pig. Journal of Animal Science 47:1135-1140.

Cabahug, S., Ravindran, V., Selle, P.H. and Bryden, W.L., 1999. Response of broiler chickens to microbial phytase supplementation as influenced by dietary phytic acid and non-phytate-phosphorus contents. 1. Effects of bird performance and toe ash. British Poultry Science 40:660-666.

Cowieson, A.J., Bedford, M.R., Ravindran, V. and Selle, P.H., 2011. Increased dietary sodium chloride reduces endogenous amino acid flow and influences the physiological response to the digestion of phytic acid by broiler chickens. British Poultry Science 52:613-624.

Greiner, R., Konietzny, U. and Jany, K.D., 1993. Purification and characterization of two phytases from *Escherichia* coli. Archives of Biochemistry and Biophysics 303:107-113.

Holloway, C.L., Boyd, R.D., Walk, C. L. and Patience, J.F., 2016a. Impact of super-dosing phytase in diets fed to 40 kg, 60 kg and 80 kg pigs on phytate catabolism. American Society of Animal Science - Midwest Section, Des Moines, IA, USA, Abstr 238.

Holloway, C.L., Boyd, R.D., Zier-Rush, C.E., Walk, C.L. and Patience, J.F., 2016b. Impact on growth performance and carcass characteristics of super-dosing phytase in growing pig diets. American Society of Animal Science - Midwest Section, Des Moines, IA, USA, Abstr 239.

Hurrell, R.F., 2003. Influence of vegetable protein sources on trace element and mineral bioavailability. Journal of Nutrition 133:2973-2977.

Jang, Y.D., Wilcock, P., Boyd, R.D. and Lindemann, M.D., 2016. Effect of combined xylanase and phytase supplementation on growth performance, carcass characteristics, and apparent total tract digestibility in pigs fed corn-based diets containing multiple by-products. American Society of Animal Science - Midwest Section, Des Moines, IA, USA, Abstr 237.

Jolliff, J.S. and Mahan. D.C., 2011. Effect of injected and dietary iron in young pigs on blood haematology and postnatal pig growth performance. Journal of Animal Science 89:4068-4080.

Jongbloed, A.W. and Bikker, P., 2010. Meta-analysis to quantify the effect if microbial phytase on the digestibility and bioavailability of several macro and micro minerals in growing pigs and broilers. Proceedings of the 1st International Phytase Summit, Washington pp 107-130.

Khullar, E., Shetty, J.K., Rausch, K.D., Tumbleson, M.E. and Singh, V., 2011. Use of phytase in ethanol production from E Mill corn processing. Cereal Chemistry 88:223-227.

Knight, C.D., Klasing, K.C. and Forsyth, D.M., 1983. *Escherichia* coli growth in serum of iron dextran supplemented pigs. Journal of Animal Science 57:387-395.

Linares, L.B., Broomhead, J.N., Guaiume, E.A., Ledoux, D.R., Veum, T.L. and Rabov, V., 2007. Effect of low phytate barley (hordeum vulgare L.) on zinc utilization in young broiler chicks. Poultry Science 86:299-308.

Liu, N. and Cowieson, A., 2010. The effect of phytic acid on the activity of chicken pepsinogen. Proceedings of the 1st International Phytase Summit, Washington pp 142-14.

Liu, N., Ru, Y.J., Cowieson, A.J., Li, D. and Cheng, X.C., 2008. Effects of phytate and phytase on the performance and immune function of broilers fed nutritionally marginal diets. Poultry Science 87:1105-1111.

Liu, N., Ru, Y.J., Li, D., Wang, J. and Lei, X., 2009. Effect of dietary phytate and phytase on the proteolytic digestion and growth regulation of broilers. Archives of Animal Nutrition 63:292-303.

Lönnerdal, B., Sandberg, A.S., Sandstrom, B. and Kunz, C., 1989. Inhibitory effects of phytic acid and other inositol phosphates on zinc and calcium absorption in suckling rats. Journal of Nutrition 119:211-213.

Maenz, D.D., 2001. Enzymatic and other characteristics of phytases as they relate to their use in animal feeds. In Bedford, M.R. and Partridge, G.G. (eds.) Enzymes in farm animal nutrition. CABI Publishing, Wallingford, UK.

Maenz, D.D., Engele-Schaan, C.M., Newkirk, R.W. and Classen, H.L. 1999. The effects of mineral chelators on the formation of phytase resistant and phytase susceptible forms of phytic acid in solution and in a slurry of canola meal. Animal Feed Science and Technology 81:177-192

Menezes-Blackburn, D., Gabler, S. and Greiner, R., 2015. Performance of seven commercial phytases in an *in vitro* simulation of poultry digestive tract. Journal of Agricultural Food Chemistry 63:6142-6149.

Mills, C.F., Quarterman, J., Williams, R.B. and Dalgarno, A.C., 1967. The effects of zinc deficiency on pancreatic carboxypeptidase activity and protein digestion and absorption in the rat. Biochemistry Journal 102:712-718.

O'Dell, B.L., de Boland, A.R. and Koirtyohann, S.R., 1972. Distribution of phytate and nutritionally important elements among the morphological components of cereal grains. Journal of Agricultural Food Chemistry 20:718-723.

Onyango, E.M. and Adeola, O., 2009. Dietary phytate (inositol hexaphosphate) regulates the activity of intestinal mucosa phytase. Journal of Animal Physiology and Animal Nutrition 93:639-646.

Onyango, E.M., Asem, E.K. and Adeola, O., 2009. Phytic acid increases mucin and endogenous amino acid losses from gastrointestinal tract of chickens. British Journal of Nutrition 101:836-842.

Pang, Y. and Applegate, T.J., 2006. Effects of copper source and concentration on *in vitro* phytate-phosphorus hydrolysis by phytase. Journal of Agricultural Food Chemistry 54:1792-1796.

Persson, H., Turk, M., Nyman, M. and Sandberg, A.S., 1998. Binding of Cu^{2+}, Zn^{2+}, and $Cd^{2+}$to inositol tri-, tetra-, penta-, and hexaphosphates. Journal of Agricultural Food Chemistry 46:3194-3200.

Poulsen, H.D., 1995. Zinc oxide for weaning piglets. Acta Agriculturae Scandinavica Animal Science 45:159-167.

Rojas, O.J., Liu, Y. and Stein, H.H., 2013. Phosphorus digestibility and concentration of digestible and metabolizable energy in corn, corn coproducts, and bakery meal fed to growing pigs. Journal of Animal Science 91:5326-5335.

Roselli, M., Finamore, A., Garaguso, I., Britti, M.S. and Mengheri, E., 2003. Zinc oxide protects cultured enterocytes from the damage induced by *Escherichia* coli. Journal of Nutrition 133:4077-4082.

Sales, J., 2013. Effects of pharmacological concentrations of dietary zinc oxide on growth of post-weaning pigs: a meta-analysis. Biological Trace Element Research 152:343-349.

Santos, T.T., Srinongkote, S., Bedford, M.R. and Walk, C.L., 2013. Effect of high phytase inclusion rates on performance of broilers fed diets not severely limited in available phosphorus. Asian-Australian Journal of Animal Science 26:227-232.

Santos, T.T., Walk, C.L., Wilcock, P., Cordero, G and Chewning, J.J., 2014. Performance and bone characteristics of growing pigs fed diets marginally deficient in available phosphorus and a novel microbial phytase. Canadian Journal of Animal Science 94:93-497.

Stein, H.H. and Shurson, G.C., 2009. Board-Invited Review: The use and application of distillers dried grains with solubles in swine diets. Journal of Animal Science 87:1292-1303.

Walk, C.L, Srinongkote, S. and Wilcock, P., 2013. Influence of a microbial phytase and zinc oxide on young pig growth performance and serum minerals. Journal of Animal Science 91:286-291.

Walk, C.L, Wellock, I.J., Toplis, P., Chewning, J.J. and Wilcock P., 2014c. Influence of increasing pharmacological ZnO dose to 3500 ppm and superdoses of phytase on piglet growth performance and fecal scores from d 0 to 21 post-weaning. American Society of Animal Science – Midwest Section, Des Moines, IA, USA, Abstr 351.

Walk, C.L., Chewning, J. and Wilcock, P., 2014b. Influence of the zinc to phytate ratio and superdoses of phytase on piglet growth performance. Journal of Animal Science 92: Supplement 2:70 (Abstr).

Walk, C.L., Santos. T.T. and Bedford, M.R., 2014a. Influence of superdoses of a novel microbial phytase on growth performance, tibia ash, and gizzard phytate and inositol in young broilers. Poultry Science 90:543-549.

Walk, C.L., Wilcock, P. and Magowan, E., 2015. Evaluation of the effects of pharmacological zinc oxide and phosphorus source on weaned piglet growth performance, plasma minerals and mineral digestibility. Animal 7:1145-1152.

Wilcock, P., Bedford. M.R., Walk, C.L., Bass. B. and Maxwell, C.V., 2010. The effect of a novel phytase on the growth performance of nursery pigs. American Society of Animal Science – Midwest Section, Des Moines, IA, USA, Abstr 181.

Wilcock, P., Bradley, C.L., Chewning, J.J. and Walk, C.L., 2014. The effect of superdosing on inositol and phytate concentration in the gastrointestinal tract and its effect on pig performance. American Society of Animal Science – Joint Animal Meeting; Kansas City, KS, USA, Abstr 761.

Woyengo, T.A., Weihrauch, D. and Nyachoti, C.M., 2012. Effect of dietary phytic acid on performance nutrient uptake in the small intestine of piglets. Journal of Animal Science 90:543-549.

Xu, P., Price, J. Wise, A. and Aggett, P.J., 1992. Interaction of inositol phosphates with calcium, zinc, and histadine. Journal of Inorganic Biochemistry 47:119-130.

Yu, S., Cowieson, A., Gilbert, C., Plumstead, P. and Dalsgaard, S., 2012. Interactions of phytate and *myo*-inositol phosphate esters (IP_{1-5}) including IP_5 isomers with dietary protein and iron and inhibition of protein. Journal of Animal Science 90:1824-1832.

Zhang, B. and Guo, Y., 2009. Supplemental zinc reduced intestinal permeability by enhancing occludin and zonula occludens protein 1 (ZO-1) expression in weaning pig. British Journal of Nutrition 102:687-693.

Zier-Rush, C., Smith, S., Palan, R., Steckel, J., Wilcock, P. and Boyd, R.D., 2012. The effect of a supra-phytase level on post-weaning growth and stool firmness of pigs fed low or high soybean meal diets. XII International Symposium on Digestive Physiology of Pigs. Abstr. 1086.

Zimmerman, D. R., 1980. Iron in swine nutrition. In: National Feed Ingredients Association Literature Review on Iron in Animal and Poultry Nutrition. National Feed Ingredients Association, Des Moines, IA, USA.

Zyła, K., Mika, M., Stodolak, B., Wikiera, A., Koreleski, J. and Swiatkiewicz, S., 2004. Towards complete dephosphorylation and total conversion of phytates in poultry feeds. Poultry Science 83:1175-1186.

Chapter 7
Current and future amino acid formulation trends – phytate and phytase history and implications

M.T. Kidd[1], T.S. Nelson[1], R.D. Brister[2] and M. Donohue[3]*
[1]Center of Excellence for Poultry Science, University of Arkansas System Division of Agriculture, POSC O106, Fayetteville, AR 72701, USA; [2]Tyson Foods, Inc., 2200 W Don Tyson Pkwy, Springdale, AR 72762, USA; [3]Agri Stats, Inc., 6510 Mutual Dr., Fort Wayne, IN 46825, USA; mkidd@uark.edu

Abstract

The dietary inclusion of the ingredients phytase and feed grade amino acids differ. Both have a history of over a half a century of research, but each limiting feed grade amino acid entered dietary formulation cautiously after a decade of research whereas phytase took almost four decades of research, and now it is the most accepted and widely used enzyme in poultry diets. The clear acceptance of phytase in poultry feed was hinged on the heightened availability of phytate phosphorus (P) and reduced inorganic P diet addition, and thus reduced bird P excretion. Clarity in other nutrient availability from dietary phytate destruction, i.e. amino acids, remains subject to debate. Clearly, phytase increases amino acid availability by acting on protein molecules bound to phytate in binary and ternary forms, but diet type and bird health contribute to the consistency of amino acid release. The implication of phytase on intestinal mucin warrants attention and has significant implications in broilers fed on various diet types. As phytase is poised to play a significant role in amino acid digestibility, future research with varying experimental designs is warranted to better quantify protein-phytate interactions and resultant phytase effects on amino acid digestibility.

Keywords: enzyme, protein digestibility, mucin

7.1 Introduction

Protein-phytate interrelationships regarding solubility and nutrient availability have been the subject of research investigations for over half a century (Fontaine *et al.*, 1946; Hill and Tyler, 1954). Research in the early 1990's with phytase demonstrated improved utilization of phytate phosphorus (P) in non-ruminants (Simons *et al.*, 1990) and although subsequent use followed P pollution legislation, commercialisation and dietary least cost optimisation of phytase in non-ruminant diets occurred throughout the 1990's and early 2000's. But commercial acceptance and dietary addition throughout the 2000's was not only tied to increased dietary P and divalent cation availability, but improved dietary protein, carbohydrate, and fat availability as well (Bryden *et al.*, 2007).

As commercial primary breeder companies have provided a broiler capable to attaining a given weight while consuming less feed each year (National Chicken Council, 2016), dietary pressure exists to assure adequate amino acid ingestion and digestion. Phytate, phytase, and protein interactions play an important role in amino acid utilisation in broilers. This review focuses on dietary ingredients, diet amino acids levels, and phytate-phytase implications.

7.2 Early phytase research

Prior to phytase commercialisation in the 1990's, phytase research among laboratories was sparse, except for early phytate solubility research (Fontaine *et al.*, 1946; Hill and Tyler, 1954) and the focused effort at the International Minerals and Chemicals research laboratory (as reviewed by Wodzinski and Ullah, 1996) in Skokie, Illinois, USA. With specific reference to the effort in the International Minerals and Chemicals research laboratory, the effort began in 1962 and was led by Rudy Wodzinski, Tsuong Shieh, and Tal Nelson. Dr. Shieh was responsible for screening over 2,000 organisms for phytase activity. Per Dr. Shieh's laboratory phytase elucidation protocols, the organism *Aspergillus ficuum* was identified as the best high yield phytase producer. However, subsequent *Aspergillus* nuclear DNA studies showed that *A. Ficuum* was not a valid species and *Aspergillus niger* nomenclature should be conserved (Frisvad *et al.*, 1990).

Successful phytase isolation in the International Minerals and Chemicals research laboratory led to studies that demonstrated its growth in moistened soybean and cotton seed meals, which led to *in vivo* work demonstrating that phytate P was available to chicks if subsequently hydrolysed by phytase (Nelson *et al.*, 1968). In addition, research at the International Minerals and Chemicals research laboratory (Table 7.1) in the 1960's with Single Comb White Leghorn male chicks fed diets with phytase treated soybean meal pointed to beneficial impacts of phytase above that related to the release of phytate P (i.e. protein bioavailability). The authors in the former cited work were the first to pretreat a corn and soybean meal based diet with phytase from *A. niger* and provide it in a diet for day old chicks. However, to hydrolyse phytate-P in the soybean meal by the phytase culture filtrate it was necessary to add moisture for enzyme distribution. Because the heat energy used to dry the meal cost more than the phytate P release, this phytase application approach was abandoned (Wodzinski and Ullah, 1996).

As more phytase became available through enzyme production and isolation, larger *in vivo* feeding studies were carried out and Nelson *et al.* (1971) demonstrated that chicks fed diets based in corn and soybean meal with phytase (solvent precipitated *A. niger*) improved bone deposition of P most probably based on improved phytate-P availability. The former pioneering work from the team at International Minerals and Chemicals research laboratory was affirmed by work in the Netherlands demonstrating improved utilization of phytate P in non-ruminants (Simons *et al.*, 1990); sparking commercial use in the early 1990's in Europe (The Netherlands) and the USA (Maryland) primarily for P

Table 7.1. *Evaluation of chick protein efficiency ratio from phytase treated soybean meal as partial replacement of soybean meal in chicks.*[1]

Protein source[2]	Diet CP% (%)	7 to 21 d performance[3]		
		BW gain (g)	FCR (g/g)	PER (g/g)
Soybean meal	17	279	1.93	3.05
Soybean meal	20	306	1.70	2.94
Soybean meal	23	352	1.54	2.83
Soybean meal + phytase	17	268	1.87[a]	3.15[a]
Soybean meal + phytase	20	331[a]	1.66[b]	3.01[b]
Soybean meal + phytase	23	354	1.51	2.88

[1] T. S. Nelson, unpublished work conducted in 1967.

[2] The phytase treated soybean meal provided 3%CP at each dietary CP level and was added to the diet based on glucose and soybean meal.

[3] FCR represents feed conversion ratio and PER represents protein efficiency ratio.

[a] Differs from soybean meal without phytase at the same CP level ($P<0.01$).

[b] Differs from soybean meal without phytase at the same CP level ($P<0.05$).

pollution mitigation. However, it appears that the team from the International Minerals and Chemicals research laboratory in the 1960's was the first to suggested protein efficiency improvements with dietary phytase.

Throughout the 1990's, phytase commercialisation resulted in numerous basic and applied research studies delineating the efficacy of exogenous phytase applied in poultry diets. Phytase and phytate studies led to a dramatic decrease in dietary inorganic P supplementation as well as P content of manure. Concurrently, research elucidated the strong chemical chelation of phytate for divalent cations, carbohydrates, and amino acids (Kornegay, 1996). Hence, high levels of dietary phytate were shown to hinder amino acid digestibility (Selle *et al.*, 2000). The following section addresses phytate-protein complexes and future research to better quantify the phytase effect on amino acid digestibility improvements.

Phytase in the 1960's

There were extra benefits of phytase, in addition to available phosphorus (P), in chicks during our research work at International Minerals and Chemical Corporation in the mid 1960's. A major problem was that the organism that produced phytase at a low level was not conducive to commercial development. The comments here are delved from summaries of each experiment and unpublished results obtained 50 years ago. At first, emphasis was on the effect of phytase on phytate P. Later, after we had several experiments on P utilisation, we had started to study the phytase effect but were thwarted in our efforts as described below. One effect was on calcium. Diets containing phytase treated soybean meal required less supplemental calcium than chicks fed diets containing untreated soybean meal. Another observation: the results of several experiments (unpublished data) was that phytase treated ingredients had a higher metabolisable energy content than untreated ingredients. This fact was confirmed by two experiments conducted independent of International Minerals and Chemical Corporation. The last International Minerals and Chemical Corporation study consisted of three experiments (unpublished data) to study the effect of phytase on the protein efficiency ratio. Diets containing untreated and phytase treated soybean meal were fed to chicks. The results were not clear cut but indicated an increase in the protein efficiency ratio in the phytase treated diets. As indicated above additional work was needed to confirm the findings but this was not to happen.

The first chicken experiment was in 1963 and further tests continued until 1967 and shortly thereafter the phytase project was terminated. The reason given was that the low phytase production by the organism (*Aspergillus niger*) was too low to be profitable for commercial production.

University Professor Tal Nelson, July 2016

7.3 Phytate-protein interactions

Up to twelve anionic protons surrounding the phytate molecule make it a powerful binder of minerals, carbohydrates, and proteins with exposed cations. The original phytate-protein interaction was that of positively charged protein possessing a pH less than its isoelectric point binding with one or more of the negative phytate protons forming binary protein-phytate complexes (Cosgrove, 1966). It is the basic amino acid residues of lysine, arginine, and histidine that form salt complexes to yield the original binary protein-phytate complexes (Cosgrove, 1966) rendering the protein complex with basic amino acids as insoluble.

The uniqueness of the amino acid lysine being the reference standard for the ideal protein concept has resulted in an ease of quantifying limiting amino acid needs for broilers fed on corn, soybean meal, or a corn–soybean meal blended type diet (Baker, 2009). Hence, as the lysine need for broilers of differing ages and strains is known, sequential amino acid needs can be predicted without need of experimentation per the ideal protein concept. And because dietary lysine is the second limiting amino acid in most poultry diets, feed grade lysine is added to virtually every broiler and turkey diet. It has been shown that dietary lysine in the monohydrochloride form will bind to phytate (Rutherfurd *et al.*, 1997). It has been established that the availability of positively charged-basic amino acid residues in digesta dictates the propensity of the binary protein-phytate interaction (Anderson, 1985). The steady increase of dietary feed grade threonine from 2000 to present has concomitantly increased the quantity of feed grade lysine being added to poultry feeds. Indeed, the degree of free dietary lysine forming phytate complexes would decrease the amount of binary-phytate protein complexes. Moreover, the presence of high phytate feeds can reduce lysine availability rendering all downstream amino acids as being in excess to the point of lysine adequacy. The opposite would also apply rendering dietary phytase as the conduit for optimal lysine digestibility and reduce the need for safety margins on downstream amino acids.

Ternary protein-phytate interactions in poultry digesta are more complex as they involve macro and micro minerals that are in the divalent cation form (e.g. calcium (Ca), iron (Fe), zinc (Zn)). Presumably phytate-Ca-protein interactions are prevalent as Ca levels in poultry diets are fed up to 1% in meat producing birds and 3.5% in egg producing birds. Much of the research on ternary protein-phytate interactions in poultry is based on micro minerals as they are added to diets in mg/kg amounts, and phytate complexes can have profound effects on mineral availability, as well as protein digestion. For example, Jongeneel *et al.* (1989) demonstrated that phytate-Zn-protein interactions not only reduced Zn availability, but also protein digestion per a reduced activity of Zn-dependent proteases. Although ternary protein-phytate interactions may constitute amino acids or small peptides, their ability to reduce overall amino acid digestibility should be subject to further investigation due to variation in the literature (Selle *et al.*, 2000).

Baldwin (1996) reviewed the mechanistic approach of how Hofmeister ions interact to impact protein stability. This cationic and anionic binding and destabilization of protein differs widely from the previously discussed binary and ternary protein–phytate interactions, and has the potential to vary based on feed ingredient contents of carbonates, phosphates, sulfates and chlorides. The former effect was assessed by Cowieson and Cowieson (2011) with a phytate *in vitro* assay that demonstrated the anionic ability of phytate to bind protein concomitantly with increasing protein concentrations. Future phytate-protein interrelationship studies should consider the cationic and anionic binding and destabilisation of protein in the dietary designs. Moreover, the protein-phytate interactions may involve the myriad of phytate-anti-nutritional effects. For example, the divalent cation Zn is important for Zn mediated DNA translation to protein, rendering both mineral and amino acid status, as unimpeded by phytate, impacting protein synthesis.

7.4 Phytate, mucin, and amino acid nutrition

Research with phytate-amino acid digestibility interactions continued to show the deleterious impact of phytate on amino acid digestibility (Rutherfurd *et al.*, 2002), but diet type and bird age produced differences in amino acid digestibility (Selle *et al.*,2000; Rutherfurd *et al.*, 2002). Diet formulations examples solved for phytate P content are presented in Table 7.2. Sorghum based diets had lower phytate P, follow by corn and wheat based diets. Additional formulation results to phytate P were poultry meal (7% maximum) and canola meal (7% maximum), which lowered and increased phytate P, respectively, on a consistent basis (Table 7.2). The use of meat meals in the U.S.A has been reduced due to some broiler integrator complexes feeding all vegetable based diets. Broiler complexes using meat products has been reduced from 93% of complexes in 2000 to 80% of complexes in 2016, but the average inclusion of 4.0 to 4.5% has remained relatively constant (M. Donohue, personal communication). Although the use of distillers grains with solubles has had little effect on the dietary phytate P level, its use has been decreased from over 90% of broiler complexes using it in 2010 compared to 50% in 2016 (M. Donohue, personal communication). In addition to varying phytate content, seed phytate origination may also contribute to the variability of phytate-amino acid digestibility variation (Weremko *et al.*, 2001; Zimmermann *et al.*, 2002).

The feed grade amino acids in chemical forms to provide available methionine, lysine, threonine, and valine were first included in U.S.A. broiler diets in the 1950's, 1970's, 1990's, and 2010's, respectively (Kidd *et al.*, 2013). As each feed grade amino acid has entered formulation, dietary CP has the flexibility to fall to the level necessary to satisfy the next limiting amino acid. From 1950 to present, numerous research studies revealed feed grade amino acid optimisation after various amino acid needs throughout the broilers' growth cycle were assessed. Indeed, each limiting feed grade amino acid added to the bird's diet is more difficult to assess consistent efficacy the lesser the degree of limitation order. For example, threonine needs of broilers were investigated for nearly three decades before commercial use in the U.S.A. in 1997 (as reviewed by Kidd and Hackenhaar, 2005). Much of the hesitation in the use of feed grade threonine was attributable to variation in requirement estimates and digestibility coefficients.

Threonine was the third limiting amino acid in all diets formulated in Table 7.2, but phytate P varied from 0.18 to 0.27% of diet. In recent threonine research, a factor

Table 7.2. *Predicted formulation results of phytate phosphorus and the fourth limiting amino acid in diets varying in cereals, oilseeds, and meat meals.*[1]

Diet type[2]	Formulation results		
	CP (% of diet)	4th limiting amino acid	Phytate P (% of diet)
Corn/soybean meal	20.6	Val	0.25
Corn/soybean meal/DDGS	21.2	Arg	0.25
Corn/soybean meal/DDGS/poultry meal	22.2	Arg	0.22
Corn/soybean meal/DDGS/meat meal	21.5	Arg	0.24
Corn/soybean meal/poultry meal	21.7	Trp	0.21
Corn/soybean meal & canola meal	20.9	Val	0.27
Corn/soybean meal & canola meal/poultry meal	21.9	Arg	0.24
Sorghum/soybean meal	20.9	Arg	0.22
Sorghum/soybean meal/DDGS	21.6	Arg	0.23
Sorghum/soybean meal/DDGS/poultry meal	22.8	Arg	0.20
Sorghum/soybean meal/DDGS/meat meal	21.9	Arg	0.21
Sorghum/soybean meal/poultry meal	22.1	Arg	0.18
Sorghum/soybean meal & canola meal	21.3	Arg	0.24
Sorghum/soybean meal & canola meal/poultry meal	22.4	Arg	0.21
Wheat/soybean meal	20.7	Val	0.24
Wheat/soybean meal/DDGS	21.0	Arg	0.24
Wheat/soybean meal/DDGS/poultry meal	22.1	Arg	0.21
Wheat/soybean meal/DDGS/meat meal	22.0	Arg	0.23
Wheat/soybean meal/poultry meal	21.3	Val	0.20
Wheat/soybean meal & canola meal	20.9	Val	0.26
Wheat/soybean meal & canola meal/poultry meal	21.7	Arg	0.23

[1] Diets were formulated for a broiler grower bird and amino acid specifications were: Lys 1.15 digestible; and ratios to Lys for TSAA, Thr, Val, Ile, Arg, and Trp were 75, 66, 76, 66, 105 and 17. The available P minimum was set at 0.40% with the remaining P, non-phytate, originating from phytase and inorganic feed grade P.

[2] DDGS represents distillers grains with solubles. DDGS, canola meal, and poultry meal were allowed to enter formulation up to 7%, whereas meat and bone meal had a 3% maximum.

impacting threonine needs is its need for intestinal health. Research in pigs shed light on the variability of threonine needs as it was established that pigs fed via parenteral nutrition require 45% less threonine as pigs fed diets orally (Bertolo *et al.*, 1998). Stoll *et al.* (1998) found that 40% of the intestinal threonine reached the portal blood in piglets. Further, intestinal mucins are important for intestinal maintenance and function, and threonine accounts for more than 40% of the mucin amino acid residues (Bengmark and Jeppsson, 1995).

Cowieson *et al.* (2004) used a precision feeding assay to assess the response of phytate and phytase on endogenous amino acid and mucin secretions by feeding a combination of phytic acid, glucose, and phytase. They found that broiler chickens have higher secretions of mucin in the presence of phytic acid and this endogenous effect can be mitigated with dietary phytase (Cowieson *et al.* 2004). The increased mucin production in diets devoid of phytase could lead to first limiting amino acid (i.e. cysteine) and third limiting amino acid (i.e. threonine) deficiencies from its costly de novo synthesis. Regardless of the degree of deficiency in the former amino acids, the fourth limiting amino acid and beyond can only respond to the adequacy of the first, second, and third limiting amino acids. Moreover, housing environment, antibiotic program, and diet make-up can also increase intestinal issues that would increase mucin production concomitant with phytate. The resultant bird performance in the former scenarios makes the case for super dosing of phytase to remove as much dietary phytate as possible and helps explain the inconsistent outcomes of amino acid digestibility studies with phytase. For example, Walk *et al.* (2013) fed Cobb 500 male broilers from 0 to 1,500 U/kg of microbial phytase and found that the improved feed conversion may have been associated with the reduction of antinutrient factors of phytate rather than improved P utilisation per se.

7.5 Precision nutrition and phytase research considerations

Gaps in our knowledge of amino acid needs, as well as the phytate sparing degree on amino acid needs, exist in the literature. As it is difficult to realise a nutrient, e.g. amino acid, response near the asymptote of the biological response, assessment of amino acid 'requirements' has been conducted with diets reduced in a given amino acid approximately 50% of the estimated need (Lewis, 1992), resulting in a curvilinear trend line to estimate the amino acid need. Although all amino acids from deficiency to adequacy follow a curvilinear response, it is more difficult to assess less limiting amino acids, i.e. those beyond threonine (valine, isoleucine, arginine, and tryptophan), and because they are 'less limiting', diet make-up for experimental testing usually contains ingredients that are not representative of commercial conditions. The former conditions add to the variability of results of amino acid needs in the literature, as well as specific variables: strain, sex, age, diet make-up, and environment. Moreover, improved efficiencies of commercial poultry as achieved from trait selection by primary breeder companies results in modern commercial poultry consuming less feed yearly to attain equal or greater whole body protein composition as compared to the previous year (National Chicken Council, 2016).

Although P release from phytate as a function of phytase is more predictable than that of amino acid release from phytate, much variation exists in the Ca and P release from dietary phytate (Rosen, 2002). Factors attributable to the variation include: bird performance relative to the control; diet P level and diet type; experimental environment and year of test; bird age and test length; and phytase dose. In addition, recent work on using high doses of phytase, i.e. superdosing, will add to the heightened variability of P, Ca, and other nutrients released (Walk *et al.*, 2013).

7.6 Conclusions

Commercial assessment of moderate (500 U/kg) to high (1,500 U/kg) levels of phytase are warranted in industrial settings with commercial diets marginal in amino acids, except methionine and lysine, in factorial and dose type experimental designs. The difficulty with the former is two fold: a design issue due to the need of a large number of pens and a diet issue in interpreting the amino acid need that resulted in the improved response. However, dose response type studies could be used to assess a specific amino acid response (e.g. threonine from 85 to 110% of the predicted need) in diets with and without phytase. Clearly, studies should be conducted to assess threonine, serine, or glycine sparing effects of phytase in birds in environmental conditions producing different gut health status. In addition, evaluation of phytase implications on amino acid needs for intestinal health should also be assessed in valine, isoleucine, and leucine per their transamination to mucin via the alpha-ketoglutarate pathway. In the former assessment of amino acid needs with and without phytase and in birds with mild to moderate enteric stress, it is emphasised to utilise diets of commercial significance. The blending of a summit diet marginal in the amino acid in question will achieve the former.

1. Unlike the rapid commercialization of amino acids, half a century of phytase research was conducted prior to its commercialization in the 1990's.
2. Additional research is warranted to assess amino acid digestibility as improved by dietary phytase, and why variation in the literature exists in amino acid digestibility studies with dietary phytase.
3. Dietary phytate-protein complexes occur through binary, ternary, and Hofmeister ion interactions.
4. Increased bird mucin production and excretion from dietary phytate is metabolically costly and has the potential to increase key amino acid needs (e.g. cysteine, threonine, serine, glycine, valine, isoleucine, and leucine).
5. Amino acid needs and digestibility trials, with and without various phytase doses, should be assessed in commercial settings (i.e. diet type and environmental conditions) with key amino acids that are pressure points in formulation linear programing.

References

Anderson, P.A., 1985. Interactions between proteins and constituents that affect protein quality. In: Finley, J.W. and Hopkins, D.T. (eds.) Digestibility and amino acid availability in cereals and oilseeds. American Association of Cereal Chemists, St. Paul, MN, USA, pp. 31-45.

Baker, D.H., 2009. Advances in protein-amino acid nutrition of poultry. Amino Acids 37:29-41.

Baldwin, R.L., 1996. How Hofmeister ion interactions affect protein stability. Biophysical Journal 71:2056-2063.

Bengmark, S. and Jeppsson, B., 1995. Gastrointestinal surface protection and mucosa reconditioning. Journal of Parenteral and Enteral Nutrition 19:410-415.

Bertolo, R.F.P., Chen, C.Z.L., Law, G., Pencharz, P.B. and Ball, R.O., 1998. Threonine requirement of neonatal piglets receiving total parenteral nutrition is considerably lower than that of piglets receiving an identical diet intragastrically. Journal of Nutrition 128:1752-1759.

Bryden, W.L., Selle, P.H. and Ravindran, V., 2007. Phytate: an anti-nutritive factor in animal diets. In: Planter, K.E., Wieregna, T.L. and Pfister, J.A. (eds.) Poisonous plants: global research and solutions. CABI Publishing, Oxon, Wallingford, United Kingdom, pp. 279-284.

Cosgrove, D.J., 1966. The chemistry and biochemistry of inositol polyphosphates. Reviews of Pure and Applied Chemistry 16:209-224.

Cowieson, A.J. and Cowieson, N.P., 2011. Phytate and the thermodynamics of water. Proceedings of the Australian Poultry Science Symposium 22:22-25.

Cowieson A.J., Acamovic T. and Bedford M.R., 2004. The effects of phytase and phytic acid on the loss of endogenous amino acids and minerals from broiler chickens. British Poultry Science 45:101-108.

Fontaine, T.D., Pons, W.A. and Irving, G.W., 1946. Protein-phytic acid relationships in peanuts and cottonseed. The Journal of Biological Chemistry 164:487-507.

Frisvad, J.C., Hawksworth, D.L., Kozakiewicz, Z., Pitt, J.I., Samson, R.A. and Stolk, A.C., 1990. Proposals to conserve important species names in *Aspergillus* and *Penicillium*. In: Samson, R.A. and Pitt, J.I. (eds.) Modern concepts in *Penicillium* and *Aspergillus* classification. Plenum Press, New York, NY, USA, pp. 83-88.

Hill, R. and Tyler, C., 1954. The reaction between protein and phytate. Journal of Agricultural Science 44:324-326.

Jongeneel, C.V., Bouvier, J. and Bairoch, A., 1989. A unique signature identifies a family of zinc-dependent metallopeptidases. Federation of European Biochemical Societies. 2:211-214.

Kidd, M.T. and Hackenhaar, L., 2005. Dietary threonine for broilers: Dietary interactions and feed additive supplement use. CAB Reviews: perspectives in agriculture, veterinary science, nutrition and natural resources. DOI: http://doi.org/10.1079/PAVSNNR20061005.

Kidd, M.T., Tillman, P.B., Waldroup, P.W. and Holder, W., 2013. Feed-grade amino acid use in the United States: The synergetic inclusion history with linear programming. Journal of Applied Poultry Research 22:583-590.

Kornegay, R.T., 1996. Nutritional, environmental and economic considerations for using phytase in pig and poultry diets. In: Kornegay, E.T. (eds.) Nutrient management of food animals to enhance and protect the environment. CRC Press, Boca Raton, FL, USA, pp. 277-302.

Lewis, A.J., 1992. Determination of the amino acid requirements of animals. In: Nissen, S. (ed.) Modern methods in protein nutrition and metabolism. Academic Press, Inc., New York, NY, USA, pp. 67-85.

National Chicken Council, 2016. U.S. broiler performance. National Chicken Council, Washington, D.C., USA. Available at: http://tinyurl.com/hr9hvpl.

Nelson, T.S., Shieh, T.R., Wodzinski, R.J. and Ware, J.H., 1968. The availability of phytate phosphorus in soybean meal before and after treatment with a mold phytase. Poultry Science 47:1842-1848.

Nelson, T.S., Shieh, T.R., Wodzinski, R.J. and Ware, J.H., 1971. Effect of supplemental phytase on the utilization of phytate phosphorus in chicks. Journal of Nutrition 101:1289-1294.

Rosen, G.D., 2002. Multifactorial analysis of the effects of microbial phytase in broiler nutrition. In: Proceedings of the 49th Maryland Nutrition Conference, March 27-28, 2002. College Park, Maryland, USA, pp. 88-101.

Rutherfurd S.M., Edwards A.C.E. and Selle P.H., 1997. Effect of phytase on lysine-rice pollard complexes. In: Cranwell, P.D. (ed.) Manipulating pig production, vol. VI. Australasian Pig Science Association, Werribee, VIC, Australia, p. 248

Rutherfurd, S.M., Chung, T.K. and Moughan, P.J., 2002. The effect of microbial phytase on ileal phosphorus and amino acid digestibility in the broiler chicken. British Poultry Science. 44:598-606.

Selle, P.H., Ravindran, V., Caldwell, R.A., and Bryden, W.L., 2000. Phytate and phytase: consequences for protein utilisation. Nutrition Research Reviews 13:255-278.

Simons, P.V.M., Versteegh, H.A.J. and Jongbloed, A.W., 1990. Improvement of phosphorus availability by microbial phytase in broiler and pigs. British Journal of Nutrition 64:525-540.

Stoll, B., Henry, J., Reeds, P.J., Yu, H., Jahoor, F. and Burrin, D.G., 1998. Catabolism dominates the first-pass intestinal metabolism of dietary essential amino acids in milk-fed piglets. Journal of Nutrition 128:606-614.

Walk, C.L., Bedford, M.R., Santos, T.T., Paiva, D., Bradley, J.R., Wladecki, H., Honaker, C. and McElroy, A.P., 2013. Extra-phosphoric effects of superdoses of a novel microbial phytase. Poultry Science 92:719-725.

Weremko, D., Fandrejewski, H., Raj, S.T. and Skiba, G., 2001. Enzymatic efficiency of plant and microbial phytase in cereal-rapeseed diets for growing pigs. Journal of Animal and Feed Sciences. 10:649-660

Wodzinski, R.J. and Ullah, A.H.J., 1996. Phytase. In: Neidleman S.L. and Laskin, A.I. (eds.) Applied microbiology. Academic Press, New York, NY, USA, pp. 263-302.

Zimmermann, B., Lantzsch, H.J., Mosenthin, R., Schöner, F.J., Biesalski, H.K. and Drochner, W., 2002. Comparative evaluation of the efficacy of cereal and microbial phytases in growing pigs fed diets with marginal phosphorus supply. Journal of the Science of Food and Agriculture. 82:1298-1304.

Chapter 8
Molecular pathways involved in amino acid and phosphorus utilization

S. Dridi[*] and M. Kidd
Center of Excellence for Poultry Science, Division of Agriculture, University of Arkansas, 1260 W. Maple Street, Fayetteville, AR 72701, USA; dridi@uark.edu

Abstract

Meeting the increase in food demand and food security, that is being driven by increased human population growth, severe drought conditions and environmental constraints (heat waves) due to climate changes, combined with limited natural resources (land, water, and energy), is challenging. Broiler (meat-type) chickens play a key role in meat production worldwide and poultry meat is highly regarded globally as one of the most efficient food sources without religious taboos. However poultry production is also facing several challenges including loss of productivity during hot weather and emerging increase of metabolic disorders such as woody breast, white striping, and lameness. It is likely that nutrition plays a major role in mitigating or ameliorating these incidences. Effective dosing of two key nutritional components amongst several others, amino acids and phosphorus (P) (inorganic phosphate and phytase), may provide both economic (low diet cost while improving growth performance) and environmental advantages (decrease of P and nitrogen excretion to manure). This review will summarize recent advances in our understanding of molecular pathways involved in amino acid and phosphorus utilization to further understand nutrient requirements of broiler chickens and to ensure efficient and sustainable poultry production in the future.

Keywords: BCAA, phytase, energy homeostasis, metabolism, mTOR, AMPK

8.1 Introduction

The poultry industry is facing challenges from a steep projected increase in global demand for high quality animal protein, the need to resolve the high incidence of metabolic disorders, and to adapt to environmental constraints driven by climate changes. Due to the economic growth in developing countries and a predicted increase in world human population to between 9 and 10 billion, United Nations Food and Agriculture Organization estimates that by 2050 there will be a 73% increase in meat and egg consumption over 2011 levels (Alexandratos, 2012). Broiler (meat-type) chickens play a key role in worldwide meat production with an estimated average annual production of 100 million tonnes. Poultry meat is highly regarded worldwide as one of the most efficient food sources without religious taboos; it contains high quality nutrients and is relatively inexpensive. However, it is facing several challenges including the increasing prevalence of metabolic disorders such as woody breast and white striping, lameness and fat accumulation which

C.L. Walk *et al.* (ed.) **Phytate destruction – consequences for precision animal nutrition**
DOI 10.3920/978-90-8686-836-0_8, © Wageningen Academic Publishers 2016

may have a negative impact on bird well-being, health and production. In addition, drought conditions and heat waves driven by climate change are impacting not only crop production, but also every aspect of animal lives and their very existence (Chen *et al.*, 2011). Since global demand for grains is continuously increasing and since feed costs make up about 70% of total cost of producing a live bird, mechanism-based formulation of a balanced diet is critical for the poultry industry to increase feed efficiency, which defines the bird's ability to convert feed into body weight, and which represents a vital economic and agricultural trait.

A large percentage of cost in a broiler chicken diet is attributed to ingredients that contribute to crude protein. Indispensable amino acids, including the branched-chain amino acids (BCAA), are not synthesized endogenously and need to be supplemented to the diet. They exert several key roles in metabolism homeostasis and cell functions including immunity (Calder, 2006; De Simone *et al.*, 2013; Thornton *et al.*, 2006), survival and growth (Farran and Thomas, 1992; Joshi *et al.*, 2006; Tavernari *et al.*, 2013), energy homeostasis (Wullschleger *et al.*, 2006), and regulation of protein and lipid metabolism (Arakawa *et al.*, 2011; Du *et al.*, 2012; Goichon *et al.*, 2013; Lansard *et al.*, 2011; Mao *et al.*, 2013). Recent evidence showed that BCAAs are not only substrates for various metabolic pathways, but can also serve as signalling molecules controlling signal transduction pathways and gene transcription (Avruch *et al.*, 2009; Efeyan *et al.*, 2012; Jousse *et al.*, 2004; Kimball and Jefferson, 2006). Thus, their dosage needs to be closely managed because insufficient or excessive levels of BCAA in the diet can be detrimental to cell metabolism and cell growth. To do so, mechanistic understanding of their roles in normal and stress conditions as well as their signalling pathways is needed.

Another important nutrient in poultry diet formulation is inorganic phosphate, which needs to be closely managed. The majority of feed phosphate is derived from rock phosphate that is a non-renewable resource and becoming increasingly scarce and expensive as poultry account for approximately 50% of animal feed consumption worldwide (Neset and Cordell, 2012). Excess of dietary phosphorus (P) concentrations increases the cost of poultry production and P excretion into manure, and deficiency in P is detrimental to the animal as P is an essential element for all form of life. Phosphorus is an important component of nucleic acids and phospholipids, and it is also a co-factor of many enzyme activities. It plays a vital role in energy homeostasis, osmotic and acid-base balance, amino acid metabolism, and protein synthesis.

This chapter integrates recently defined cellular and molecular mechanisms mediating BCAA effects on energy homeostasis and fat metabolism, and its role in stress response and in health and disease. The last section will discuss potential molecular pathways involved in P metabolism and utilization. The objective is that this chapter will facilitate discussion by asking new questions, advancing knowledge, and developing mechanism-based formulation and nutritional strategies to improve feed efficiency and poultry sustainability while preserving the environment.

8.2 Branched-chain amino acids, energy homeostasis and fat metabolism

Genetic selection for rapid growth rate, driven by economic demands, has resulted in hyperphagic, heavy and obese broiler chickens. They voraciously consume approximately 1.1 kg of feed to achieve a 40-fold increase in body weight arising from breast (pectoralis) muscle and abdominal fat during a period of 42 days (Hood, 1982; Scheuermann *et al.*, 2003). Research conducted in the 1990's reported that as dietary energy level increased, birds were able to satisfy their energy needs by decreasing feed intake which indicated that birds adjust their feed intake according to their metabolisable energy requirement (Leeson *et al.*, 1996; Plavnik *et al.*, 1997). However, this conclusion has been revised as broilers have been shown to decrease or increase their feed intake when dietary protein content in the diet was increased or reduced, respectively (Bouvarel *et al.*, 2008). Combined, these results indicate that the mechanisms controlling appetite can be altered by genetic selection, nutrient requirement of the bird and/or diet composition.

At molecular levels, a series of highly integrated regulatory mechanisms, yet not completely defined at least in avian species, exists for the control of energy intake and expenditure at both peripheral and central sites. Within the central nervous system (CNS), the hypothalamic satiety (arcuate nucleus, ARC and ventromedial nucleus, VMH) and hunger (paraventricular nucleus, PVH and lateral hypothalamus, LHA) centres play key roles in the regulation of feed intake and energy homeostasis (Elmquist *et al.*, 1999; Morgane, 1961). Two separate populations of neuronal cell types are located in the ARC, one synthesises the powerfully orexigenic peptides (neuropeptide Y, NPY and agouti-related peptide, AgRP), while the other produces the anorexigenic pro-opiomelanocortin (POMC) and cocaine and amphetamine regulated transcript (CART) (Friedman and Halaas, 1998). Cota *et al.* (2006) has showed that central administration of leucine decreased feed intake and body weight in rodents and this effect was mediated through an activation of hypothalamic mechanistic target of rapamycin (mTOR) and its downstream mediator ribosomal S6 kinase (S6K1). Immunofluorescence staining demonstrated that S6K1 was co-localized with AgRP in the ARC of the hypothalamus. mTOR is a highly conserved serine-threonine kinase and a master nutrient sensor that controls growth and metabolism. Early experiments, depriving rodents of a single amino acid, leucine, caused profound weight loss and muscle waste, followed by death (Said and Hegsted, 1970). Moreover, it was observed that amino acid withdrawal from cells and organisms triggered autophagy, a process of cellular self-digestion where pre-existing proteins and organelles are broken down into simpler metabolites via lysosomal degradation (Kroemer *et al.*, 2010). Following the discovery of mTOR, it was observed that withdrawal of amino acids from the culture media suppressed mTOR signaling in mammalian cells, and inhibition of mTOR by starvation or pharmacologically (rapamycin) induced autophagy. From these data, a feedback loop began to emerge connecting amino acids, mTOR, and autophagy in a mechanism that drives growth under nutrient abundance and mediates growth arrest under nutrient depletion, allowing amino acid stores to be replenished. Biochemical and genetic screening studies uncovered a group of mediators of amino acid signalling to mTOR such as Rag GTPases, Rheb, and lipid kinase Vps34. In chickens, Izumi *et al.* (2004) showed that intracerebroventricular injection of leucine, but not isoleucine or valine, increased feed intake and pecking rhythm. Results of recently conducted experiments indicate that peripheral administration of rapamycin or L-leucine reduced feed intake in a dose-dependent manner in quails (unpublished data). The discrepancy between the latter study and that of Izumi *et al.* (2004) may be that avian species (chicken vs quail), age, and/or experimental conditions (route of administration) were different between the two studies. We also found that the mTOR pathway is expressed in quail hypothalamus and it is likely to be involved in regulation of feed efficiency (Blankenship *et al.*, in press). However, further studies are needed to unravel the molecular mechanisms by which BCAAs regulate feed intake in avian species which, in turn, may help to determine the ideal BCAA ratio during the different phases of rearing.

Although the effects of BCAAs are well established in muscle, i.e. stimulation of protein synthesis (Everaert *et al.*, 2010) and reduction of proteasome-mediated proteolysis (Combaret *et al.*, 2005), its effects on hepatic fat metabolism have been poorly documented in mammals (Krause *et al.*, 2002) and amphibians (Lansard *et al.*, 2010) and are still unknown in chickens. Chickens liver is the primary site for lipid synthesis (Goodridge and Ball, 1967; Leveille *et al.*, 1968). Furthermore, chickens are usually hyperglycemic compared with mammals (Krzysik-Walker *et al.*, 2008), insulin resistant (Simon *et al.*, 1977), and they lack the intracellular glucose transporter GLUT4 (Seki *et al.*, 2003). To this effect, we have recently completed three valine dose-response trials in broilers from 14-28 days of age. Female Cobb 500 broilers were used and the digestible valine/lysine-isoleucine/lysine ratios were 77/69% (control group, which conforms to levels used in commercial feeding of birds), 72/63% (L group), and 72/63% with additional crystalline Val and Ile to reach 77/69% (L+S group). All groups exhibited similar body weight and feed conversion ratio (FCR) (Bai *et al.*, 2015). There was no significant difference in circulating BCAA levels, percentage of immune cells, and between weights of several organs indicating that optimal yields could be attained with low BCAA levels. To gain further insights at molecular levels and by using different molecular techniques, we observed that low BCAA levels down regulated the hepatic expression of lipogenic genes including acetyl-CoA carboxylase (ACCα) and stearoyl-CoA desaturase (SCD-1). We also observed that low BCAA levels induced the phosphorylation of adenosine monophosphate-activated protein kinase (AMPKα) at Thr^{172} site and decreased mTOR phosphorylation at Ser^{2448} and its downstream S6K at Thr^{389} sites compared with the control group. The serine threonine kinase AMPK is an evolutionary conserved signalling molecule that is emerging as one of the most important energy sensors and monitors for many organisms, from yeast to mammals (Carling *et al.*, 2012). The low levels of BCAA induced the hepatic expression of uncoupling protein (UCP) and nuclear respiratory factor (NRF-1), key genes involved in β-oxidation. However, low levels of BCAA did not affect autophagy-related genes. Furthermore, low BCAA levels induced both mRNA expression and phosphorylation of Forkhead box protein (FoxO1) on Thr^{24} but not on Ser^{26} compared with the control group. FoxO1 is a transcription factor that mediates several cellular processes from metabolic homeostasis to oxidative stress responses. Combined, our data indicated that low BCAA levels inhibited hepatic fatty acid synthesis and enhanced β-oxidation and these effects are likely mediated through AMPK-mTOR-FoxO pathway. More research should be conducted with the Cobb 500 female broilers regarding BCAA needs as our work points to potential dietary BCAA needs considerably lower than currently used.

8.3 Branched-chain amino acids in stress, health, and disease

As we mentioned, poultry production is facing several challenges including emerging increase of metabolic disorders and environmental stress intensified by climate change. Heat stress is one of the most prevalent environmental challenges to poultry as modern broilers produce more heat due to high metabolic activity and they do not have sweat gland, thus they are not well adapted to high environmental temperatures. The adverse effect of heat stress on feed intake, growth performance, immunity and mortality is well documented (Cahaner and Leenstra, 1992; Dale and Fuller, 1980; Deeb *et al.*, 2002; Geraert *et al.*, 1996; Leenstra and Cahaner, 1992). Recently it has been reported that heat stress diminished protein and amino acid digestibility and thereby altered their circulating and intracellular profiles in chickens (Ito *et al.*, 2015). Moreover, amino acid deprived cells exhibited alteration in heat shock protein 70 and its transcription factor, HSF1 (Hensen *et al.*, 2012), which are also regulated by heat stress. BCAA supplementation prolongs exercise performance in human (Mittleman *et al.*, 1998). In chickens, several studies have been conducted to investigate effects of amino acid supplementation in low crude protein-diet, but most data were generated under thermo-neutral conditions, whereas data specific to hot and humid conditions are scarce. Therefore, further mechanistic studies focused on BCAAs such as L-valine supplementation during environmental stress are warranted.

Another emerging research area that needs more intense investigation is the use of BCAA in preventing metabolic disorders in broilers. For instance, the muscle disorder 'woody breast', characterized by muscle degeneration and immune cells infiltration (Sihvo *et al.*, 2014), is emerging on a global scale and constitutes a major animal health, welfare, and economic burden. Although the etiology of this disease is still unknown, our observation of an upregulated expression of hypoxia-inducing factor 1 (HIF-1) gene in woody breast compared with healthy chickens (unpublished data) indicates that hypoxia may play a key role in the development of this disorder. We hypothesize that breast muscle fibres of the fast growing modern broilers are characterized by large diameters which in turn may reduce available connective tissue spacing and blood supply leading to hypoxia. If this is the case, the use of arginine, the precursor of nitic oxide, which is a potent vasodilator, may help reducing or preventing development of woody breast. Again, to gain better insights in the beneficial effect of arginine, further mechanistic and molecular studies are needed. These studies will help to develop mechanism-based strategies (nutrition) to reduce the incidence of metabolic disorders in poultry.

8.4 Molecular pathways involved in phosphorus metabolism

Poultry production directly contributes to climate change by the release of greenhouse gasses. One major contributor in this phenomenon is the use of inorganic phosphate to meet the P requirement of the animal because two-thirds of the total P in plants (cereals) is in the form of phytate P (PP), which is poorly utilized by chickens due to very low phytase activity in their proventriculus and small intestine. To overcome this problem, poultry diets are often supplemented with inorganic phosphates which is mainly derived from phosphate rock that is a non-renewable resource and becoming increasingly scarce and expensive (Shastak *et al.*, 2012). This, in turn, poses the challenges of (1) substantial increase in feed cost, (2) securing future P supply for the global animal feed industry, and (3) contributing to environmental pollution and surface water eutrophication via P excretion in the manure. Thus accurate determination of P requirements and optimisation of its use are major thrusts of research. To do so, mechanistic understanding of P bioavailability, which encompasses digestion, absorption, and utilisation is needed.

Coordination of three organ systems, intestine, kidney, and bone is central to P homeostasis. In addition, there is a close association of P homeostasis with vitamin D (VD) and calcium (Ca) metabolism. Reduced circulating concentrations of Ca causes increased secretion of parathyroid hormone (PTH) by the parathyroid glands. Parathyroid hormone increased bone resorption and renal Ca reabsorption and increased renal excretion of P. In the kidneys, PTH also stimulates formation of the hormonal form of VD, which increases intestinal absorption of Ca and, to a lesser extent, the absorption of P. Elevated blood Ca concentrations as a result of these processes created a negative feedback loop to maintain homeostasis.

In poultry production, diet supplementation with microbial phytase is a practical method for improving P utilization by hydrolysing the phytate to inositol and inorganic phosphate (Viveros *et al.*, 2002). Results of previous studies have shown enhanced utilization of PP with additional VD or its metabolites by increasing the phytase activity, phytate hydrolysis and the absorption of Ca and P (Biehl and Baker, 1997; Qian *et al.*, 1997; Shafey *et al.*, 1990). A synergistic interaction between VD and phytase on broiler performance and bone status has also been reported (Green, 2012; Snow *et al.*, 2004). However the underlying molecular mechanisms are still not well defined. We recently demonstrated that 25-hydroxycholecalciferol enhanced breast muscle yield and increased the fractional rate of protein synthesis through activation of the mTOR-S6K pathway. As VD and P metabolism are associated, it is reasonable to hypothesise that the mTOR pathway may be involved in P homeostasis and utilisation. Results of recent research in humans indicated that fibroblast growth factor 23 (FGF23) play a key role in P metabolism (Takeda *et al.*, 2014), but similar effects have not been demonstrated in poultry.

8.5 Conclusions

1. To meet the substantial challenges the poultry industry is facing, intensive applied- and fundamental-based research is needed. New techniques involving nutrigenomics, genomics, proteomics, transcriptomics, mobilomics and metabolomics are now making their ways to solve the intervening puzzle between nutrients, genes and performances.
2. A personalised nutrition approach based on identification, selection and optimisation of nutrients, specifically BCAA and P, fine-tuned with animal genetic profiles will improve performance with minimal environmental impact.
3. These approaches have the potential to change the future of dietary guidelines and formulation.

References

Alexandratos, N.B.J., 2012. World agriculture towards 2030/2050. The 2012 revision. In: FAO (ed.). ESA Working paper No. 12-03. FAO, Rome, Italy.

Arakawa, M., Masaki, T., Nishimura, J., Seike, M. and Yoshimatsu, H., 2011. The effects of branched-chain amino acid granules on the accumulation of tissue triglycerides and uncoupling proteins in diet-induced obese mice. Endocrine Journal 58:161-170.

Avruch, J., Long, X., Ortiz-Vega, S., Rapley, J., Papageorgiou A. and Dai, N., 2009. Amino acid regulation of TOR complex 1. American Journal of Physiology. Endocrinology and Metabolism 296:E592-602.

Bai, J., Greene, E., Li, W., Kidd, M.T. and Dridi, S., 2015. Branched-chain amino acids modulate the expression of hepatic fatty acid metabolism-related genes in female broiler chickens. Molecular Nutrition and Food Research 59:1171-1181.

Biehl, R.R. and Baker, D.H., 1997. Utilization of phytate and nonphytate phosphorus in chicks as affected by source and amount of vitamin D3. Journal of Animal Science 75:2986-2993.

Blankenship, K., Gilley, A., Piekarski, A., Orlowski, S., Greene, E, Bottje, W. and Dridi, S., in press. Differential expression of feeding-related hypothalamic neuropeptides in the first generation of quails divergently selected for low or high feed efficiency. Neuropeptides. DOI: http://doi.org/10.1016/j.npep.2015.12.007.

Bouvarel, I., Vallee, C., Chagneau, A. M., Constantin, P., Lescoat, P., Ferreira, G. and Leterrier, C., 2008. Effects of various energy and protein levels during sequential feeding on feed preferences in meat-type chickens. Animal: An International Journal of Animal Bioscience 2:1674-1681.

Cahaner, A. and Leenstra, F., 1992. Effects of high temperature on growth and efficiency of male and female broilers from lines selected for high weight gain, favorable feed conversion, and high or low fat content. Poultry Science 71:1237-1250.

Calder, P.C., 2006. Branched-chain amino acids and immunity. The Journal of Nutrition 136:288S-293S.

Carling, D., Thornton, C., Woods, A. and Sanders, M.J., 2012. AMP-activated protein kinase: new regulation, new roles? The Biochemical Journal 445:11-27.

Chen, I.C., Hill, J.K., Ohlemuller, R., Roy, D.B. and Thomas, C.D., 2011. Rapid range shifts of species associated with high levels of climate warming. Science 333:1024-1026.

Combaret, L., Dardevet, D., Rieu, I., Pouch, M.N., Bechet, D., Taillandier, D., Grizard, J. and Attaix, D., 2005. A leucine-supplemented diet restores the defective postprandial inhibition of proteasome-dependent proteolysis in aged rat skeletal muscle. The Journal of Physiology 569:489-499.

Cota, D., Proulx, K., Smith, K.A., Kozma, S.C., Thomas, G., Woods, S.C. and Seeley, R.J., 2006. Hypothalamic mTOR signaling regulates food intake. Science 312:927-930.

Dale, N.M. and Fuller, H.L., 1980. Effect of diet composition on feed intake and growth of chicks under heat stress. II. Constant vs. cycling temperatures. Poultry Science 59:1434-1441.

De Simone, R., Vissicchio, F., Mingarelli, C., De Nuccio, C., Visentin, S., Ajmone-Cat, M.A. and Minghetti, L., 2013. Branched-chain amino acids influence the immune properties of microglial cells and their responsiveness to pro-inflammatory signals. Biochimica et Biophysica Acta 1832: 650-659.

Deeb, N., Shlosberg, A. and Cahaner, A., 2002. Genotype-by-environment interaction with broiler genotypes differing in growth rate. 4. Association between responses to heat stress and to cold-induced ascites. Poultry Science 81:1454-1462.

Du, Y., Meng, Q., Zhang, Q. and Guo, F., 2012. Isoleucine or valine deprivation stimulates fat loss via increasing energy expenditure and regulating lipid metabolism in WAT. Amino Acids 43:725-734.

Efeyan, A., Zoncu, R. and Sabatini, D.M., 2012. Amino acids and mTORC1: from lysosomes to disease. Trends in Molecular Medicine 18:524-533.

Elmquist, J.K., Elias, C.F. and Saper, C.B., 1999. From lesions to leptin: hypothalamic control of food intake and body weight. Neuron 22:221-232.

Everaert, N., Swennen, Q., Coustard, S.M., Willemsen, H., Careghi, C., Buyse, J., Bruggeman, V., Decuypere, E. and Tesseraud, S., 2010. The effect of the protein level in a pre-starter diet on the post-hatch performance and activation of ribosomal protein S6 kinase in muscle of neonatal broilers. The British Journal of Nutrition 103:206-211.

Farran, M.T. and Thomas, O.P., 1992. Valine deficiency. 1. The effect of feeding a valine-deficient diet during the starter period on performance and feather structure of male broiler chicks. Poultry Science 71:1879-1884.

Fingar, D.C. and Blenis, J., 2004. Target of rapamycin (TOR): an integrator of nutrient and growth factor signals and coordinator of cell growth and cell cycle progression. Oncogene 23:3151-3171.

Friedman, J.M. and Halaas, J.L., 1998. Leptin and the regulation of body weight in mammals. Nature 395:763-770.

Geraert, P.A., Padilha, J.C. and Guillaumin, S., 1996. Metabolic and endocrine changes induced by chronic heat exposure in broiler chickens: growth performance, body composition and energy retention. The British Journal of Nutrition 75:195-204.

Goichon, A., Chan, P., Lecleire, S., Coquard, A., Cailleux, A.F., Walrand, S., Lerebours, E., Vaudry, D., Dechelotte, P. and Coeffier, M., 2013. An enteral leucine supply modulates human duodenal mucosal proteome and decreases the expression of enzymes involved in fatty acid beta-oxidation. Journal of Proteomics 78:535-544.

Goodridge, A.G. and Ball, E.G., 1967. Lipogenesis in the pigeon: *in vivo* studies. The American Journal of Physiology 213:245-249.

Green, P.M., 2012. The effects of feeding high concentrations of cholecalciferol, phytase, or their combination on broiler chicks fed various concentrations of nonphytate phosphorus. Journal of Applied Poultry Research 21:579-587.

Hensen, S. M., Heldens, L., Van Enchevort, C.M., Van Genesen, S.T., Pruijin, G.J. and Lubsen, N.H., 2012. Heat shock factor 1 is inactivated by amino acid deprivation. Cell Stress and Chaperones 17:743-755.

Hood, R.L., 1982. The cellular basis for growth of the abdominal fat pad in broiler-type chickens. Poultry Science 61:117-121.

Ito, K., Bahry, M.A., Hui, Y., Furuse, M. and Chowdhury, V.S., 2015. Acute heat stress up-regulates neuropeptide Y precursor mRNA expression and alters brain and plasma concentrations of free amino acids in chicks. Comparative Biochemistry and Physiology. Part A, Molecular and Integrative Physiology 187:13-19.

Izumi, T., Kawamura, K., Ueda, H. and Bungo, T., 2004. Central administration of leucine, but not isoleucine and valine, stimulates feeding behavior in neonatal chicks. Neuroscience Letters 354:166-168.

Joshi, M.A., Jeoung, N.H., Obayashi, M., Hattab, E.M., Brocken, E.G., Liechty, E.A., Kubek, M.J., Vattem, K.M., Wek, R.C. and Harris, R.A., 2006. Impaired growth and neurological abnormalities in branched-chain alpha-keto acid dehydrogenase kinase-deficient mice. The Biochemical Journal 400:153-162.

Jousse, C., Averous, J., Bruhat, A., Carraro, V., Mordier, S. and Fafournoux, P., 2004. Amino acids as regulators of gene expression: molecular mechanisms. Biochemical and Biophysical Research Communications 313:447-452.

Kimball, S.R. and Jefferson, L.S., 2006. New functions for amino acids: effects on gene transcription and translation. The American Journal of Clinical Nutrition 83:500S-507S.

Krause, U., Bertrand, L., Maisin, L., Rosa, M. and Hue, L., 2002. Signalling pathways and combinatory effects of insulin and amino acids in isolated rat hepatocytes. European Journal of Biochemistry/FEBS 269:3742-3750.

Kroemer, G., Marino, G. and Levine, B., 2010. Autophagy and the integrated stress response. Molecular Cell 40:280-293.

Krzysik-Walker, S.M., Ocon-Grove, O.M., Maddineni, S.R., Hendricks, G.L. and Ramachandran, R., 2008. Is visfatin an adipokine or myokine? Evidence for greater visfatin expression in skeletal muscle than visceral fat in chickens. Endocrinology 149:1543-1550.

Lansard, M., Panserat, S., Plagnes-Juan, E., Dias, K., Seiliez, I. and Skiba-Cassy, S., 2011. L-leucine, L-methionine, and L-lysine are involved in the regulation of intermediary metabolism-related gene expression in rainbow trout hepatocytes. The Journal of Nutrition 141:75-80.

Lansard, M., Panserat, S., Plagnes-Juan, E., Seiliez, I. and Skiba-Cassy, S., 2010. Integration of insulin and amino acid signals that regulate hepatic metabolism-related gene expression in rainbow trout: role of TOR. Amino Acids 39:801-810.

Leenstra, F. and Cahaner, A., 1992. Effects of low, normal, and high temperatures on slaughter yield of broilers from lines selected for high weight gain, favorable feed conversion, and high or low fat content. Poultry Science 71:1994-2006.

Leeson, S., Caston, L. and Summers, J.D., 1996. Broiler response to diet energy. Poultry Science 75:529-535.

Leveille, G.A., O'Hea, E.K. and Chakbabarty, K., 1968. *In vivo* lipogenesis in the domestic chicken. Proceedings of the Society for Experimental Biology and Medicine. Society for Experimental Biology and Medicine 128:398-401.

Mao, X., Zeng, X., Huang, Z., Wang, J. and Qiao, S., 2013. Leptin and leucine synergistically regulate protein metabolism in C2C12 myotubes and mouse skeletal muscles. The British Journal of Nutrition 110:256-264.

Mittleman, K.D., Ricci, M.R. and Bailey, S.P., 1998. Branched-chain amino acids prolong exercise during heat stress in men and women. Medicine and Science in Sports and Exercise 30:83-91.

Morgane, P.J., 1961. Distinct 'feeding' and 'hunger motivating' systems in the lateral hypothalamus of the rat. Science 133:887-888.

Neset, T.S. and Cordell, D., 2012. Global phosphorus scarcity: identifying synergies for a sustainable future. Journal of the Science of Food and Agriculture 92:2-6.

Plavnik, I., Wax, E., Sklan, D. and Hurwitz, S., 1997. The response of broiler chickens and turkey poults to steam-pelleted diets supplemented with fat or carbohydrates. Poultry Science 76:1006-1013.

Qian, H., Kornegay, E.T. and Denbow, D.M., 1997. Utilization of phytate phosphorus and calcium as influenced by microbial phytase, cholecalciferol, and the calcium: total phosphorus ratio in broiler diets. Poultry Science 76:37-46.

Said, A.K. and Hegsted, D.M., 1970. Response of adult rats to low dietary levels of essential amino acids. The Journal of Nutrition 100:1363-1375.

Scheuermann, G.N., Bilgili, S.F., Hess, J.B. and Mulvaney, D.R., 2003. Breast muscle development in commercial broiler chickens. Poultry Science 82:1648-1658.

Seki, Y., Sato, K., Kono, T., Abe, H. and Akiba, Y., 2003. Broiler chickens (Ross strain) lack insulin-responsive glucose transporter GLUT4 and have GLUT8 cDNA. General and Comparative Endocrinology 133:80-87.

Shafey, T.M., McDonald, M.W. and Pym, R.A., 1990. Effects of dietary calcium, available phosphorus and vitamin D on growth rate, food utilisation, plasma and bone constituents and calcium and phosphorus retention of commercial broiler strains. British Poultry Science 31:587-602.

Shastak, Y., Witzig, M., Hartung, K. and Rodehutscord, M., 2012. Comparison of retention and prececal digestibility measurements in evaluating mineral phosphorus sources in broilers. Poultry Science 91:2201-2209.

Sihvo, H.K., Immonen, K. and Puolanne, E., 2014. Myodegeneration with fibrosis and regeneration in the pectoralis major muscle of broilers. Veterinary Pathology 51:619-623.

Simon, J., Freychet, P. and Rosselin, G., 1977. A study of insulin binding sites in the chicken tissues. Diabetologia 13:219-228.

Snow, J.L., Baker, D.H. and Parsons, C.M., 2004. Phytase, citric acid, and 1alpha-hydroxycholecalciferol improve phytate phosphorus utilization in chicks fed a corn-soybean meal diet. Poultry Science 83:1187-1192.
Takeda E, Y.H., Yamanaka-Okumura, H. and Taketani, Y., 2014. Increasing dietary phosphorus intake from food additive: potential for negative impact on bone health. Advances in Nutrition 5:92-97.
Tavernari, F.C., Lelis, G.R., Vieira, R.A., Rostagno, H.S., Albino, L.F. and Oliveira Neto, A.R., 2013. Valine needs in starting and growing Cobb (500) broilers. Poultry Science 92:151-157.
Thornton, S.A., Corzo, A., Pharr, G.T., Dozier, W.A., Miles, D.M. and Kidd, M.T., 2006. Valine requirements for immune and growth responses in broilers from 3 to 6 weeks of age. British Poultry Science 47:190-199.
Viveros, A., Brenes, A., Arija, I. and Centeno, C., 2002. Effects of microbial phytase supplementation on mineral utilization and serum enzyme activities in broiler chicks fed different levels of phosphorus. Poultry Science 81:1172-1183.
Wullschleger, S., Loewith, R. and Hall, M.N., 2006. TOR signaling in growth and metabolism. Cell 124:471-484.

Chapter 9
Phytase effects on protein and phosphorus bioavailability in fish diets

G.A. Morales[1,2], L. Marquez[3], A.J. Hernández[3] and F.J. Moyano[4]*
[1]Consejo Nacional de Investigaciones Científicas y Técnicas INPA-CONICET, Buenos Aires, Argentina; [2]Departamento de Producción Animal, Facultad de Agronomía, Universidad de Buenos Aires, Av. San Martín 4453, Cap. Fed. C1417DSE, Buenos Aires, Argentina; [3]Núcleo de Investigación en Producción Alimentaria, Universidad Católica de Temuco, Temuco, Chile; [4]Grupo de Nutrición y Alimentación Animal, Universidad de Almería, La Cañada de San Urbano, 04120 Almería, Spain; moralesg@agro.uba.ar

Abstract

In a global context of a fast growing aquaculture, the reduction of the aquatic pollution through the optimisation of aquafeeds and their nutritional value is a major concern for farmers, researchers and governments. More than 50% of phosphorus (P) reserves in seeds of most legumes are stored as phytate-bound P which is not bioavailable for monogastric animals including most fish species. Therefore, the presence of phytate in the aquafeed is traduced directly or indirectly in negative effects on growth performance and on minerals, protein and energy utilisation by fish. Phytases have been used in poultry and pig productions during the last decades, mainly to reduce the environmental impact and P loads from farm effluents, as well as to reduce feeding costs through nutrient sparing. Moreover, in recent years, researchers and companies are paying increasing attention to the use of phytases in high doses to obtain extra-phosphoric benefits on dietary nutrient utilisation and growth performance in monogastrics. The use of phytases in aquafeeds is still in an initial stage, but it is presented as one of the most effective tools used by the aquaculture to include alternative plant protein ingredients in diet formulas, improve fish growth and control the diet-related environment pollution from aquaculture operations. This chapter presents a discussion about the intervening factors that influence the potential benefits of phytases in fish plant based diets and how the inclusion of this additive might improve their protein and P bioavailability. The properties of phytic acid under the gastrointestinal environment, its interaction with other minerals, nutrients and digestive enzymes, as well as the use of phytases to avoid the negative effects of phytate on nutrients digestibility and water pollution are also discussed.

Keywords: aquaculture, enzymes, digestibility, nutrients, water pollution

9.1 Introduction

Annually, 120 million metric tons (MMT) of gaseous nitrogen (N_2) become reactive forms of nitrogen (N) around the world, while 20 MMT of phosphorus (P) are extracted from

DOI 10.3920/978-90-8686-836-0_9, © Wageningen Academic Publishers 2016

global reserves of phosphate rock for various industrial and agricultural uses. A large portion of these inputs end up in the environment (Smit *et al.*, 2015), slowly eroding the resilience of terrestrial and aquatic ecosystems (Rockström *et al.*, 2009). As agriculture and land livestock, aquaculture is an anthropic activity contributing to the release of nutrients into the aquatic environment. According to Verdegem (2013), in 2008 the global aquaculture production of fish and shellfish led to a discharge of 1.7 MMT of N to the aquatic environment, this accounts to 0.86% of human intervention in the global N cycle. In the same year, P released from aquaculture to the environment was 0.46 MMT, that represents around 2.3% of the world supply of P as fertilizer. Because P is a limiting nutrient for primary production in most aquatic ecosystems (Mainston and Parr, 2002), its discharge to the water contributes directly to eutrophication and excessive algal blooms degrading natural water bodies.

On the other hand, the world is now facing the decline in the overall availability and increased price of the mineral P, mainly due to the reduction in world reserves of phosphate rock (Abelson, 1999). In this context, the optimal use of these nutrients by aquaculture is presented as a key issue that will determine the sustainable growth of the activity in the coming decades.

Aquaculture has grown considerably in the last thirty years. Whereas it provided only 7% of fish for human consumption in 1974, this share had increased to 39% in 2004. Production of aquatic animals from aquaculture in 2014 amounted to 73.8 MMT in 2014 with an estimated first-sale value of US$ 160.2 billion, consisting of 49.8 million tonnes of finfish (US$ 99.2 billion) (FAO, 2016). Nevertheless, the demand of protein ingredients to produce aquafeeds, currently suffers from a great uncertainty about the projection of this industry in the future. One of the main inputs used since the beginning of the industry of feeds for aquaculture was the fishmeal (FM) from pelagic fisheries which for more than a decade are at their highest level of exploitation. This has led to a gradual decline in the FM world production over the last decade. By contrast, global demand continues growing, having increased FM prices by 150% from 2005 to 2015 (IndexMundi, 2015). Since in the same period other plant protein sources maintained their prices, or as in the case of soybean meal increased by 56% (Figure 9.1), aquafeeds industry was forced to include plant protein sources in aquafeed formulations at a higher rate. In this scenario, there has been a clear tendency to reduce the content of FM in feed formulations for farmed fish in recent years, particularly in those feeds used for salmonid species (Mallison, 2013) (Figure 9.2). However, the anti-nutritional factors (ANFs) present in most plant ingredients are considered the main cause limiting the inclusion rate of these ingredients in aquafeeds (Gatlin *et al.*, 2007).

ANFs are defined as innate components present in plant ingredient with a negative effect on feed consumption, digestion, nutrient uptake and animal welfare (Francis *et al.*, 2001). Different plant sources such as soybean meal, rapeseed, lupin, sunflower, among others, have a wide range of ANFs such as protease inhibitors, non-digestible carbohydrates, lectins, saponins, non-starch polysaccharides (NSP) and phytic acid. Despite some ANFs being thermolabile, such as trypsin inhibitors that can be thermally destroyed during the

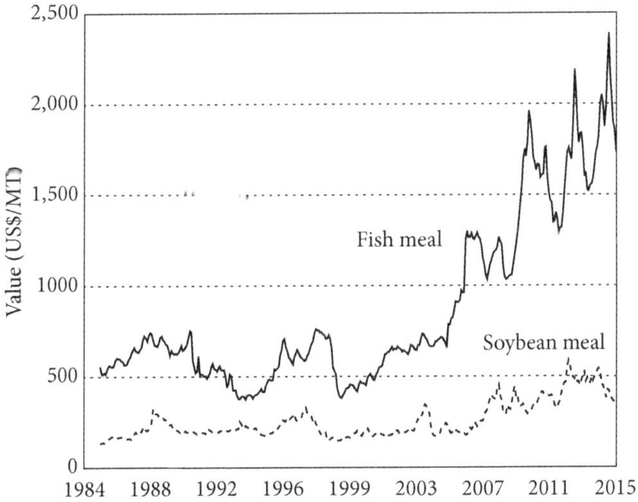

Figure 9.1. *Evolution of the price of fish meal and soybean meal in the last 30 years (IndexMundi, 2015).*

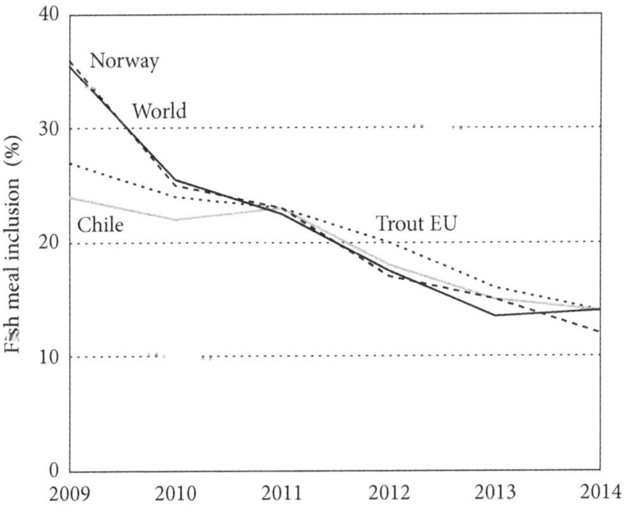

Figure 9.2. *Inclusion rate of fish meal in diets for salmonids (% of complete diet; Mallinson, 2013).*

extrusion of the diet, phytic acid is thermostable (Storebakken *et al.*, 2000). In this sense, the use of enzymes of microbial origin as a dietary additive, is one of the greatest advances in animal nutrition science to inactivate certain ANF present in diets with high content of plant ingredients.

9.2 Phytic acid

Phytic acid is the hexaphosphoric ester of the hexahydric cyclic alcohol meso-inositol. The chemical description of phytic acid is *myo*-inositol 1,2,3,4,5,6-hexakis dihydrogen phosphate (IUPAC–IUB, 1977). Phytic acid ($C_6H_{18}O_{24}P_6$; molecular mass: 660.04 g/mol), also known as phytate (IP_6), is the main form of P source in plant ingredients. Between 50% and 80% of P content in the seeds of most legumes is stored as IP_6 (Ravindran, 1995). The main function of IP_6 is as a P reservoir for growth of plants during different stages of their development. During the development of the grains, IP_6 is accumulated in the aleurone layer and germ of cereals and globoid crystals of legumes. The synthesis of IP_6 starts with the formation of *myo*-inositol, followed by a phosphorylation step to monophosphate (IP_1), with the last step involving phosphorylation of pentakisphosphate (IP_5) to IP_6 (Loewus, 2002).

According to Pontoppidan *et al.* (2007), in mature seeds of plants and most ingredients used in animal feeds, inositol phosphate is in the form of IP_6 (90-95%), with relatively small amount of IP_5 and tetrakisphosphate (IP_4) esters. Several physiological roles have been suggested for IP_6 in seeds and grains. These include energy store, source of cations and of *myo*-inositol cell wall precursor, and for initiation of dormancy. Phytic acid occurs primarily as salts of mono- and divalent cations (Zhao *et al.*, 2014) (e.g. potassium (K)-magnesium (Mg) salt in rice and calcium (Ca)- magnesium- potassium salt in soybeans) in discrete regions of cereal grains and legumes. Phytate is accumulated in seeds and grains during ripening, accompanied by other storage substances such as starch and lipids. Figure 9.3 shows the phytic P content in some of the plant ingredients commonly used in animal nutrition. The IP_6-P in plants may vary depending on the plant species, the part of

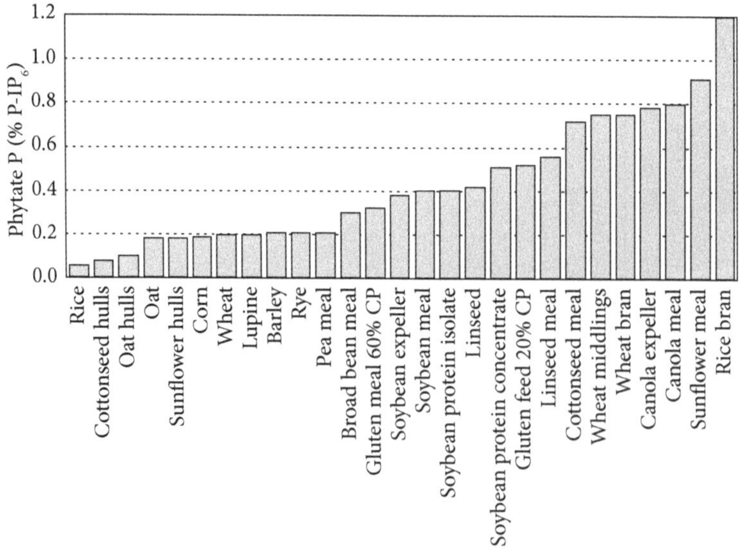

Figure 9.3. *Content of phytate phosphorus (P; in %; P-IP_6) in several plant ingredients used in animal feeds (De Blas* et al., *2010).*

the plant, the phenological stage and soil fertility, varying between 0.05% (rice) and 1.2% (rice bran).

9.3 Phytic acid within digestive tract

9.3.1 Phytic acid and minerals

Due to its chemical structure, IP_6 is a very stable compound. It differs from other organic molecules in its large number of phosphate moieties, which confers a high negative charge over a wide pH range. The six phosphate groups of the molecule can generate a total of 12 replaceable reactive sites with negative charges in complete dissociation, where between 3 and 9 are dissociated under the usual range of pH in the digestive tract. Thus, IP_6 is able to bind di- and trivalent minerals forming very stable insoluble complexes. Under normal physiological conditions, IP_6 can form insoluble complexes with multivalent cations such as zinc (Zn^{2+}), copper (Cu^{2+}), nickel (Ni^{2+}), cobalt (Co^{2+}), manganese (Mn^{2+}), Ca^{2+} and iron (Fe^{2+}) in that order of stability (Cheryan, 1980), reducing the digestibility of these minerals.

The solubility of IP_6 salts is largely pH dependent, decreasing as pH increases and forming insoluble salts of phytic acid at neutral pHs (Cheryan, 1980). Animal diets contain high amounts of Ca compared to other cationic minerals; thus the phytic acids complex mainly with Ca in the small intestine. According to Grynspan and Cheryan (1983) phytate salts are more soluble at low pH, tending to precipitate as IP_6-Ca complex at pH above 4.0. The magnitude of this effect depends on the Ca:IP_6 molar ratio. This suggests that only a fraction of the native IP_6 present in plant ingredients is soluble and available as phytase substrate when gastrointestinal pH is higher than 4.0. In this sense, a low solubility of IP_6 at pH near neutrality could be particularly disadvantageous for IP_6 dephosphorylation by phytases in fish species that are not capable to strongly acidify the digesta, as reported by several authors (Marquez *et al.*, 2012; Sugiura *et al.*, 2006; Yufera *et al.*, 2004). The pH in the gastrointestinal tract in carnivorous fish species is generally higher than in pigs and poultry. In salmonids, stomach pH in constantly fed fish can range from 3.5-5.6 (Austreng *et al.*, 2000; Kroghdahl *et al.*, 2015; Nordrum *et al.*, 2000) and intestinal pH can be as high as 9. The pH is generally higher in saltwater than in freshwater fish species, and similarly to terrestrial animals, HCl secretions lowers stomach pH in fish after feed intake (Marquez *et al.*, 2012). On the other hand, the concentration of minerals is generally higher in saltwater than in freshwater fish, in which waterborne minerals are ingested with feed and mainly absorbed in the gut and via the gills.

The negative effect of phytic acid on mineral utilisation by fish has been largely studied. Phytate can form ionic complexes with di- and trivalent mineral ions such as, $Mg2^{+}$ (Denstadli *et al.*, 2006), Zn^{2+}, Ca^{2+} (Fredlund *et al.*, 2006) reducing their bioavailability for fish growth. Feeding Atlantic salmon (*Salmo salar*) with high content of phytate from concentrated soybean meal (0.93%) led to a reduction in minerals (P, Ca, Mg and Zn) bioavailability and retention (Storebakken *et al.*, 1998). Papatryphon *et al.* (1999) also reported reduction in plasma levels of Ca and Zn in striped bass (*Morone saxatilis*) when feed was supplemented with a high dietary content of phytate (>1%).

9.3.2 Phytic acid and proteins

In addition to its known effects on the digestibility of P and others minerals, increasing evidence suggests that IP_6 may affect the digestibility of dietary protein and amino acids in pigs, poultry and even fish (Cowieson *et al.*, 2006; Kumar *et al.*, 2012). Phytic acid may interact with proteins to form insoluble IP_6-protein complexes over a wide pH range. Whether or not IP_6 affects the solubility of different nutrients depends greatly of the pH in the digestive tract. Several *in vitro* studies suggest that IP_6 interact in different ways with proteins depending mainly on the pH of the environment (Kies *et al.*, 2006; Konishi *et al.*, 1999; Okubo *et al.*, 1976; Wise, 1983).

When feed enters into stomach, IP_6 will either remain under this form, or dissociates and binds *de novo*, forming new complexes with proteins. Phytic acid can creates binary (protein-IP_6) or ternary (protein-mineral-IP_6) complexes (Kies *et al.*, 2006). IP_6-proteins interaction by binary complexes occurs through ionic bonds between IP_6 and basic amino acid residues at pH values below than the isoelectric point of proteins. Cosgrove (1966) suggested that as a result of complex formation, protein molecules become closely packed around the relatively small and highly charged IP_6 anion leading to the formation of macromolecular aggregations or insoluble coacervates. Selle *et al.* (2000) proposed that the *de novo* formation of binary protein-IP_6 complexes in animal gastrointestinal tract is a key mechanism by which IP_6 influences the protein digestibility. The main modes of forming IP_6-protein complexes that have been suggested are related with the presence of protein-IP_6 complexes in feed ingredients, the formation of binary and ternary complexes in the gastrointestinal tract and the inhibition of proteolytic enzymes (Kies *et al.*, 2006; Selle *et al.*, 2006).

The *de novo* formation of loose electrostatic association between IP_6 and proteins in the gastrointestinal tract occurs only under optimal pH conditions (Maenz, 2001). At pH below the isoelectric point of protein, the anionic phosphate groups of IP_6 binds strongly to the cationic groups of the protein to form binary complexes. Provided the high values of pKa of basic amino acids, these are highly susceptible to chelation, as they become polar and positively charged at a wide range of pH below their pKa. The binding potential of IP_6 to proteins increases as pH decreases, but maximizes at pH 2.5 (Adeola and Sands, 2003). Several factors influence the intensity of binary protein-IP_6 interactions in addition to pH. For example, dietary Ca^{2+} has a very high acid-binding capacity (Lawlor *et al.*, 2005). Thus, sufficient Ca^{2+} may dissociate protein-IP_6 complexes (Okubo *et al.*, 1974), which is attributed to Ca^{2+} competing with protein for access to negatively charged IP_6 molecules.

Once gastric content enters the intestine the pH of the chyme changes from proximal to distal intestine as it is impelled by peristaltic contractions. During the transition between stomach and proximal intestine, where most proteins reach their isoelectric point, IP_6 is gradually released from the binary complexes to form ternary complexes protein-cation-IP_6. These complexes are formed at the intermediate pH values (pH 5.0 and above), usual in the intestinal environment (Riche *et al.*, 2001), where IP_6 and proteins have a net negative charge. Ternary protein-mineral-IP_6 complexes can be formed *de novo* in the small intestine where the major components are linked via a cationic bridge, usually Ca^{2+}

which the most prevalent divalent cation present in the digesta. Nevertheless, the relative importance of such complex is difficult to assess, as IP_6 could form either mineral-IP_6 or ternary protein-cation-IP_6 complexes, with an equilibrium between these two forms depending on the concentration of the cation (Champagne, 1988).

Selle *et al.* (2012) discussed the possibility of an additional or alternative mechanism where IP_6 interacts with protein by acting as a 'Hofmeister anion'. Hofmeister or lyotropic series is a classification of cations and anions based on their capacity to stabilise or destabilise proteins (Baldwin, 1996), and it is postulated that ions influence protein solubility mainly by changing the hydrogen-bonding properties of water in the surrounding medium. Ions may break hydrogen bonds in water systems and are classified as chaotropes that destabilise proteins; alternatively, ions may form hydrogen bonds in water and stabilise proteins and these ions are kosmotropes (Zhang *et al.*, 2005). IP_6 contains six phosphate (HPO_4^{2-}) moieties and this suggests that the polyanionic IP_6 molecule has strong kosmotropic properties under the Hofmeister series that would tend to reduce protein solubility.

In an *in vitro* investigation (Morales *et al.*, 2011) simulating the gut pH values of the rainbow trout (2.5 in the stomach and 8.5 in the intestine) and using digestive extracts from the same species, neither the addition of IP_6 to a casein solution, nor the addition of phytase to a soybean protein concentrate, showed any relevant role of phytate ternary complexes on the bioavailability of dietary proteins. This study agrees with the general consensus that ternary complexes can be present during the digestion process but are not important in respect of IP_6 reducing protein availability (Selle *et al.*, 2012).

The theoretical diagram of the beneficial effect of phytase on P and protein bioavailability through binary protein-IP_6 and ternary protein-cation-IP_6 complexes disruption is shown in Figure 9.4.

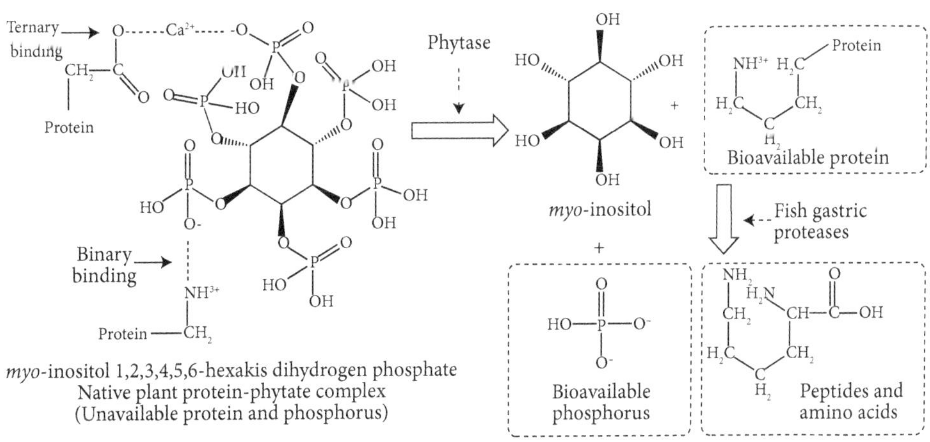

Figure 9.4. *Hypothesised beneficial effects of phytase on P and protein bioavailability through disruption of the binary protein-IP_6 and ternary protein-cation-IP_6 complexes. Based on the model proposed by Anderson (1985).*

Similarly to animal feeds used in pigs and poultry, aquafeeds contain a number of plant ingredients which have different types of proteins with variable affinity to be complexed by IP_6. The effect of enzymatic dephosphorylation of IP_6 on the solubility of different plant ingredients along a wide pH range was studied by Morales *et al.* (2013). The authors observed distinctive responses between leguminous (soybean meal, lupin meal, peas, broad bean meal, chickpea meal) and non-leguminous (canola meal, wheat middlings, wheat flour) ingredients in their solubility reaction to IP_6 dephosphorylation by phytase. While the increment in soluble protein after phytase treatment in broad bean meal, pea meal and chickpea meal was measured at both acid and basic pH, probably as a result of the partial degradation of either the binary and ternary complexes, such an effect was detected in soybean meal only at acid pH and slight or no effect was observed in the rest of ingredients (Figure 9.5).

It was suggested that differences in the solubility and further hydrolysis by gastric proteases could be related to the molecular structure of the species-specific storage proteins present in each plant ingredient which should present alterations in their charge at different pH's and hence in their affinity to bind to IP_6. Proteins with high contents in basic AA (i.e. arginine, lysine, histidine) would be particularly prone to form binary insoluble complexes with IP_6 under the acidic environment of fish stomach (Anderson, 1985). To get a better insight into this issue, the effect of phytase on the solubility of different protein fractions in several plant ingredients was evaluated by SDS-PAGE electrophoresis by Morales *et al.* (2013). Among protein fractions, legumin and vicilin showed the higher increase in their solubilisation in response to IP_6 dephosphorylation by phytase. These proteins, found mainly in peas and broad beans, contain a high proportion of basic AA (17.4 and 19.9%, respectively). In contrast, no significant effect was observed in wheat protein, of which the main storage proteins gliadins and glutenins contain a low amount (4.1%) of basic AA.

These findings support the hypothesis that the proportion of basic AA present in different plant proteins could play an important role in their interaction with phytic acid. The relation between the content of basic AA of each protein fraction and the observed increase in their solubility is shown in Figure 9.6. While in most proteins, their relative contents in basic AA appears to be a good predictor of the expected increase in solubility as a result of the enzymatic hydrolysis of IP_6, this was not the case for others such as conglutins present in *Lupinus* sp. (Morales *et al.*, 2013). More research on this topic would contribute to the interpretation of the binding effect of phytic acid on dietary proteins.

9.3.3 Phytic acid and digestive proteases

The formation of protein-IP_6 complexes may affect not only structural proteins in diets but also another kind of functional proteins like proteases (Inagawa *et al.*, 1987; Singh and Krikorian, 1982). Several ways of formation of these complexes affecting the protein hydrolysis in the digestive tract have been suggested: the existence of endogenous protein-IP_6 complexes in feed, the *de novo* formation of complexes in the gastrointestinal tract, and the inhibition of proteolytic enzymes, such as pepsin, trypsin and chymotrypsin (Kies *et al.*, 2006; Selle *et al.*, 2006), i.e. the interaction with protein substrates or with proteolytic enzymes.

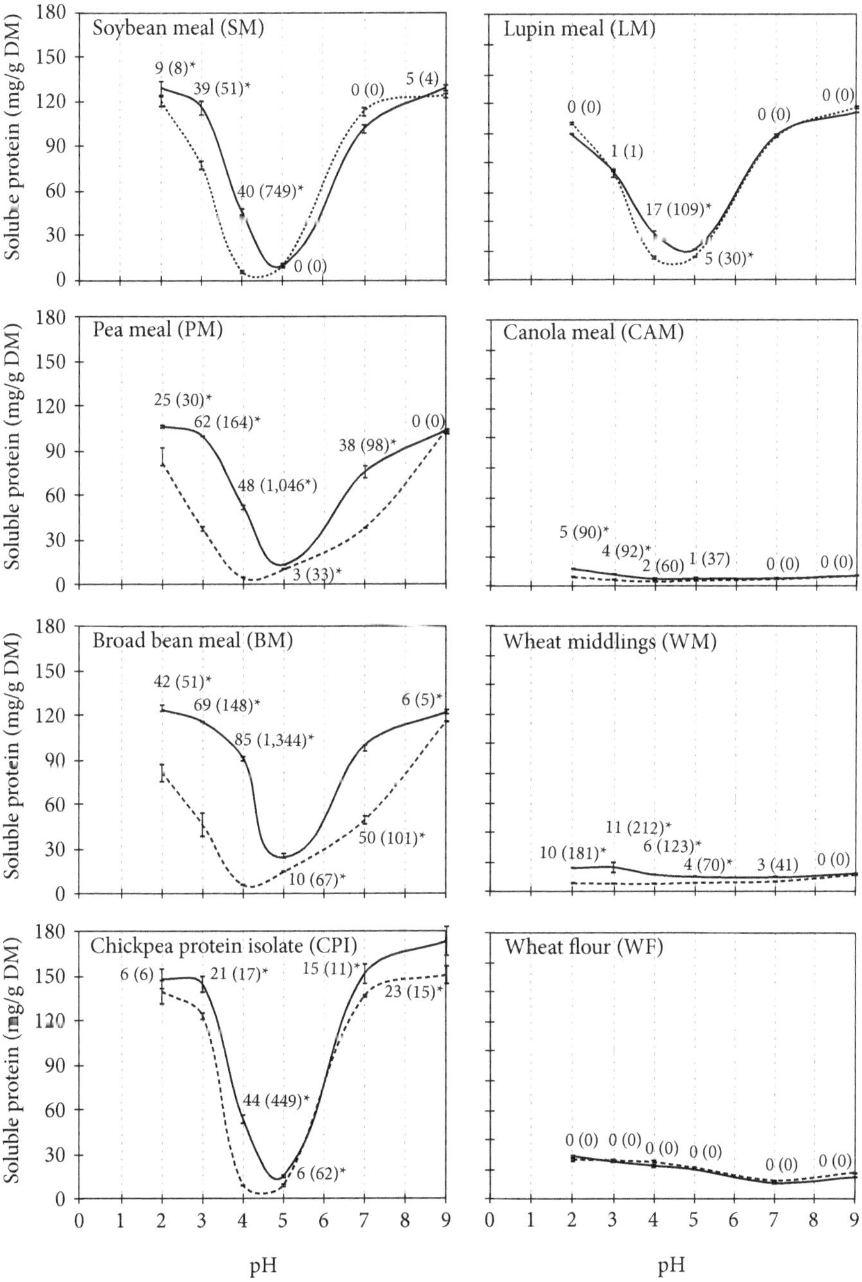

Figure 9.5. *Soluble protein (mg/g substrate) exposed at different simulated pH after a pre-incubation (pH 2.5; 25 °C, 60 min) of ingredients with (filled line) or without (dotted line) 2,500 FTU/kg DM. Values above curves indicate the increment in soluble protein as an effect of phytase treatment expressed as mg/g DM and (%) of the maximum value (Morales* et al.*, 2013).*

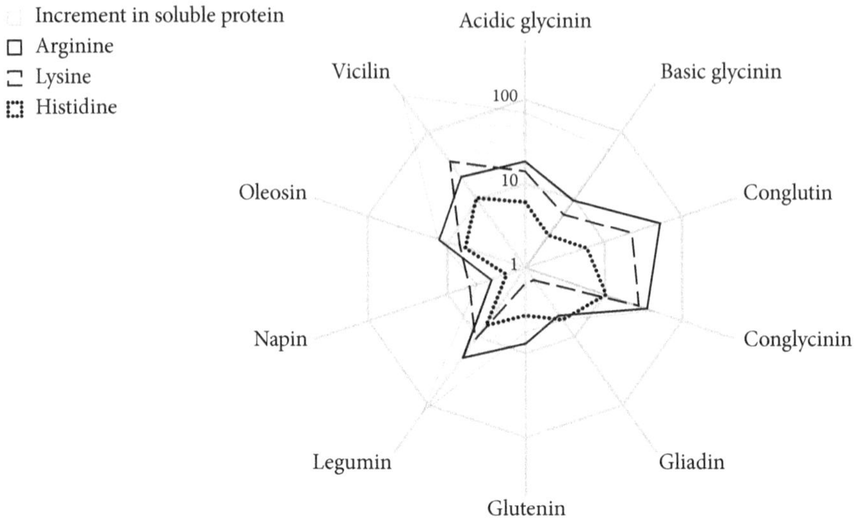

Figure 9.6. *Increase of the solubility of different protein fractions of plant ingredients after phytase treatment with 2,500 FTU/kg ingredient and the content of the three basic AA (logarithmic scale). Composition of AA data from Expassy (Magrane and UniProt Consortium, 2011); illustration adapted from data reported by Morales* et al. *(2013).*

Pepsin has a dual function in the gastrointestinal tract, firstly initiates the digestion of proteins, and secondly, the peptides generated by pepsin partially regulate the pancreatic stage of the digestive process by activating the release of gut hormones, as cholecystokinin and gastrin (Einarsson *et al.*, 1996). Consequently, the complexing of gastric pepsin by IP_6 can interfere with both digestive functions, reducing the global digestibility of dietary protein. It has been demonstrated that IP_6 can inhibit *in vitro* pepsin activity by the formation of the binary protein-IP_6 complexes (Vaintraub and Bulmaga, 1991). In agreement with this, Kies *et al.* (2006) observed that pepsin hydrolysed native protein-IP_6 complexes from soybean meal at a higher rate and to a larger extent when phytase was added to the mixture. More recently, an *in vitro* study has been suggested that native IP_6 can reduce the bioavailability of the dietary protein via the formation of insoluble IP_6-protein complexes with gastric pepsin from fish (Morales *et al.*, 2011). The simulation of the gastrointestinal condition of rainbow trout demonstrated that incubation of IP_6 in a concentration equivalent to that present in soybean meal with extracts prepared with either pure porcine pepsin or fish stomach content, produced a significant reduction in their soluble protein contents (68 and 99%, respectively) and as a consequence, in their protease activity (82 and 58%, respectively) (Morales *et al.*, 2011). Moreover, Morales *et al.* (2014) suggested that the activity of gastric protease measured in sea bream (*Sparus aurata*) fed on a diet supplemented with 2,500 phytase units (FTU)/kg diet was 60% higher than that of fish receiving the control diet. This study was the first evidence of the effect of IP_6 on the pepsin activity in fish stomach measured *in vivo*.

On the other hand, Singh and Krikorian (1982) have suggested that phytic acid may inhibit intestinal proteolysis by altering the protein configuration of alkaline proteases. They reported reduced activity of trypsin by IP_6 at pH 7.5 during the *in vitro* digestion of casein when trypsin was preincubated with IP_6. As both trypsinogen and trypsin require calcium ions to maintain their functionality, they hypothesized that the possible competitive binding of Ca^{2+} by IP_6 may partially reduce the activity of the enzyme. Moreover, it was suggested that IP_6 might bind with trypsin via Ca^{2+}, forming a ternary complex and thus reducing its activity. This hypothesis was confirmed by Caldwell (1992) who observed that addition of IP_6 to a reaction mixture containing Ca^{2+} reduced the activation of trypsinogen by about 90% at pH 8. It was concluded that Ca-IP_6 complexes increased the formation of inactive trypsin and substantially reduce the stabilising effect of Ca^{2+} on trypsin integrity. More recently, Morales *et al.* (2011) reported a 14% reduction in the activity of rainbow trout trypsin in the presence of IP_6, but no effect on chymotrypsin. Despite these studies demonstrating a significant negative effect of IP_6 on alkaline proteases activity, several investigations (Inagawa *et al.*, 1987; Kanaya *et al.*, 1976; Knuckles *et al.*, 1989; Reddy *et al.*, 1988; Vaintraub and Bulmaga, 1991) did not observe *in vitro* IP_6 inhibition of trypsin.

Similarly to the contradictory results obtained in different *in vitro* studies, *in vivo* studies are not completely consistent with the hypothesis that IP_6 affects the activity of alkaline proteases activity. Mroz *et al.* (1995) found that phytase increased both ileal digestibility of N (3.7%) and trypsin activity (10.9%) significantly in sows. In fish, Sajjadi and Carter (2004b) did not find any effect of the inclusion of dietary IP_6 on the trypsin activity of Atlantic salmon. In contrast, in the same species, Denstadli *et al.* (2006) observed that the cumulative trypsin activity in the pyloric intestine chyme was significantly lowered in fish fed a high dose of IP_6 (20.7 g/kg diet) compared to a control group fed a diet without IP_6 addition. However, this effect was not evident when fish were fed on diets with a content of IP_6 more similar to that present in plant-based diets (4.7 g/kg diet). Vandenberg *et al.* (2011) reported that the supplementation of 3,000 FTU/kg in a plant-based diet with 7.5 g/kg IP_6 showed no effect on proteolytic intestinal enzymes activity of rainbow trout. In addition, Morales *et al.* (2014) did not observe any positive effect of 2,500 FTU/kg added to a plant-based diet with 4.9 g/kg IP_6 on the activity of trypsin or chymotrypsin in sea bream.

Taking into account the aforementioned results, *in vivo* studies suggest that the formation of ternary IP_6-cation-protein complexes does not significantly affect the activity of intestinal proteases. *In vitro* evidence of the negative effects of IP_6 on the solubility and residual activity of intestinal proteases, from terrestrial animals and fish, would suggest possible compensatory mechanisms via intestinal protease secretion.

More studies evaluating spatio-temporal changes of lower groups of inositol phosphates (IP_{1-5}) within fish gastrointestinal tract should be conducted to get a better understanding of the dynamic of IP_6 dephosphorylation across the digestion, and then to know how IP esters interact with gastric and pancreatic proteases. In addition, *in vitro* assays using IP_6:proteases ratio and concentration of divalent cations close to those existing in live fish may help to understand how the intestinal proteases are inhibited, e.g. via direct ternary binding or indirectly by competition via sequestering cations (e.g. Ca^{2+} or Zn^{2+}).

9.4 Phytases

Phytases are chemically known as *myo*-inositol (1,2,3,4,5,6) hexakisphosphate phosphohydrolase, and catalyse the sequential release of phosphate from IP_6. Phytase sequesters orthophosphate groups from the inositol ring of IP_6 to produce free inorganic P, along with a chain of intermediate *myo*-inositol phosphates (inositol pentaphosphate to inositol monophosphate) (Debnath *et al.*, 2005a). The Enzyme Nomenclature Committee of the International Union of Biochemistry distinguishes two types of phytase: 3-phytase (EC 3.1.3.8) and 6-phytase (EC 3.1.3.26). This classification is based on the first phosphate group attacked by the enzyme. Phytases are widespread in nature and depending on their pH optima, phytases are often divided into acidic or alkaline types, indicating that their ideal 'work site' is in the stomach or the intestine. Phytase activity has been reported in plant and animal tissues and in a variety of microorganisms. Intestinal flora able to dephosphorylates IP_6 is present in pigs (colon/hindgut) and poultry (caeca), but not to the extent reported in ruminants (Schlemmer *et al.*, 2001; Selle and Ravindran, 2007).

There are antecedents of intestinal phytase activity in fish (Ellestad *et al.*, 2002a). These authors demonstrated that hybrid striped bass have a very low but existing intestinal phytase activity which could also be a non-specific phosphatase activity as it had very low affinity for IP_6. The same authors (Ellestad *et al.*, 2002b) compared the phytase activity in the intestinal brush border of three teleost fish species: the carnivorous hybrid striped bass and the omnivorous tilapia (*Oreochromis niloticus*) and carp (*Cyprinus carpio* L.). They concluded that tilapia could be capable of degrading substantial amounts of dietary IP_6, with observed apparent IP_6 disappearance of 50%, while hybrid bass and the carp could degrade only 1-2% of the IP_6 present in the diet. In a recent study, rainbow trout fed diets without phytase inclusion and a dietary IP_6 content of 13.2 g/kg showed apparent IP_6 digestibility values between 17 and 22% (Morales *et al.*, 2015). Although it is recognized that carnivorous species like rainbow trout have low ability to hydrolyse IP_6, some previous studies indicated that this species is capable to dephosphorylate the IP_6 to a variable extent. For example, Forster *et al.* (1999) and Denstadli *et al.* (2006) reported 5 and 15% IP_6 digestibility, respectively, in rainbow trout. According to Denstadli *et al.* (2006), the high apparent digestibility coefficient of IP_6 in the absence of an exogenous phytase could be explained by a triggering effect of IP_6 on the phosphatases present in the intestinal epithelium, but more are required in order to elucidate this mechanism.

Microbial produced phytases are other sources of IP_6 degrading enzymes and can mainly be divided into two categories: fungal phytases and bacterial phytases. Among fungal phytases, the first commercial phytase was produced from *Aspergillus niger* and was released in market in 1991. For several years, *A. niger* phytase was the only commercial enzyme used as feed additive. A second 3-phytase product was produced from *Aspergillus oryzae* containing the *A. niger* phytase gene. A third commercial 3-phytase was produced by a *Trichoderma reesei* expressing a gene originated from *Aspergillus awamori*. After that, other products based on microbial 6-phytase were produced by another *A. oryzae* strain transformed with the phytase gene of the basidiomycete strain *Peniophora lycii*.

Among bacteria, *Escherichia coli, Bacillus subtilis, Klebsiella terringa, Lactobacillus* sp., *Pseudomonas* sp. (Greiner *et al.*, 1993, 1997; Kerovuo *et al.*, 1998; Richardson and Hadobas, 1997) can degrade IP_6 during growth through production of extracellular phytases. However, the most important bacterial phytase developed commercially was an *E. coli* 6-phytase expressed in *Pichia pastoris* or *Schizosaccaromyces pombe*. In last years, new generations of improved 6-phytases were developed by companies, offering a range of enzyme products with variable response and efficacy depending on pH, temperature and substrate concentration: Phytases from *Buttiauxella* sp. expressed in *T. Reesei*; from *Citrobacter braakii* expressed in *A. oryzae* or modified *E. coli* phytases expressed in *T. reesei.*

The mode of action of the phytases starts by dephosphorylating fully phosphorylated IP_6 molecules, followed by penta- (IP_5), tetra- (IP_4), tri- (IP_3), di- and mono-esters of inositol in such an order. Thus, most of these phytases hydrolyse, in a first step, all of the available IP_6 to IP_5 before they hydrolyse IP_5 further to IP_4, IP_3 and lower esters (Yu *et al.*, 2012). Despite a theoretical complete hydrolysis should result in *myo*-inositol, phosphate and any other nutrients which are linked to phytic acid (minerals, proteins, amino acids, starch), hydrolysis under *in vivo* conditions is incomplete and result in a mixture of non-fully dephosphorylated inositol-phosphate esters. Depending on the dephosphorylation degree of the IP_6 molecule, the strength of the ionic interaction between different IP esters and other ionic molecules can vary in a large extent. Luttrell (1993) and Persson *et al.* (1998) demonstrated that IP_6 or IP_5 have a much greater chelation capacity for Ca^{2+} than IP_4 or IP_3. Indeed, IP_3 has only 10% of the chelation capacity of IP_6. Similarly, Yu *et al.* (2012) demonstrated that the capacity for binding to Fe^{3+} decreased proportionally from IP_6 to IP_3. In this last study, the authors observed that the aggregating capabilities of IP esters (IP_{1-6}) on soy protein and β-casein decreased dramatically from IP_6 to IP_5 and became negligible with IP_{1-4}.

In the same study, the effect of the *in vitro* sequential dephosphorylation of IP_6 on pepsin activity inhibition was evaluated. The reaction mixtures containing different IP esters (including IP_3, IP_4, IP_5, and IP_6) inhibited pepsin catalysed protein hydrolysis to varying degrees. These results show that IP_{3-6} (and not IP_{1-2}) still inhibited pepsin activity. Similar results were previously observed by Knuckles *et al.* (1989) who found that IP_5 and IP_6 have a more pronounced inhibition effect, whereas IP_1 and IP_2 does not seem to inhibit pepsin activity using casein and bovine serum albumen as substrates.

These studies suggest that IP_6 has the most relevant antinutritional effect on a number of dietary nutrients and represent the most important ester involved in the inhibition of the gastric digestion process of proteins. Therefore, as pointed out by Adeola and Cowieson (2011) the solution to these antinutritional effects of phytic acid is to dephosphorylate IP_6 by exogenous microbial phytases as quickly as possible in the proximal gastrointestinal tract to lower molecular weight esters. Because these lower IP esters have a reduced capacity to chelate divalent cations, they would be more soluble and hence more accessible for endogenous intestinal phytases and/or phosphatases.

9.5 Factors influencing phytase efficacy in fish

The net action of phytases on IP_6 degradation within animal digestive tract may be affected by a number of different factors including pH, temperature, substrate solubility, digestion time and presence of proteases (Dersjant-Li *et al.*, 2015). Most commercial microbial phytases act efficiently under the actual conditions present in the stomach. However, phytases do not always work equal to a certain pH, therefore IP_6 dephosphorylation efficacy will likely depend on the digestive conditions present in a given fish species.

9.5.1 Gastrointestinal pH

One of the main factors influencing phytases performance is the gastrointestinal pH, which can vary in a range from pH 2.0 to pH 8.0. However, most commercial phytases have a range of optimal activity under acidic condition, hence the main site of action within the animal digestive system is the stomach. The gastric pH and the gastric retention time are the main factors restricting the phytase efficiency. Once the chyme passes to the intestine, the enzyme, depending on the source, will become less active and the phytate less soluble.

The efficiency of different commercial phytases varies over a pH range of 2.5-6.5 (Figure 9.7) Among phytase sources, *E. coli* phytases are more active over a wide pH range with an optimum of pH 4.5, while *Citrobacter* and *Buttiauxella* sp. have a more acidic optimum (4.0 and 3.5, respectively). On the other hand, phytase activity profile can vary for different *E. coli* phytases due to the production technology and expression organism (e.g. *P. pastoris* and *T. reesei*). Taking into account the pH profile of the microbial phytases used in animal nutrition, the efficacy of phytases can differ among gastric and agastric fish. Gastrointestinal acidity in gastric fish such as rainbow trout, catfish and other carnivorous

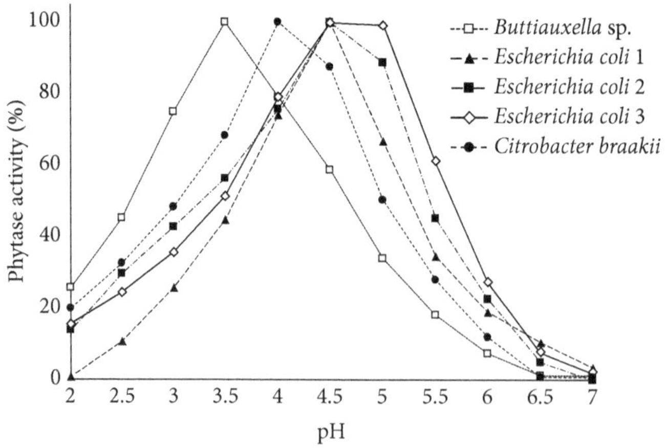

Figure 9.7. *Relative activity (%) of five commercial phytases in the pH range 2.0-7.0 using 5.1 mM sodium phytate as substrate, 37 °C, 30 min incubation time. Buffers used: 100 mM maleate buffer for pH 2.0-3.0; 100 mM acetate buffer pH 3.5-6.5, 100 mM Tris buffer for pH 7.0. pH was adjusted to eleven different values between pH 2.0 and 7.0. Maximum activity for each phytase was set to 100%.*

fish is more favourable for an efficient hydrolysis of IP_6 by microbial phytase. However, the pH in digestive tract of agastric fish such as common carp is around 6.5-8.4 (Ji, 1999). The large difference between the optimal pH of the phytase and the pH of the digestive tract of agastric fish, limits the efficacy of phytase in these fish species. The lack of an optimal efficacy of acid phytases in agastric fish supposes that the use of neutral phytases from genera as *Bacillus* and *Pedobacter* can increase the performance and nutrient digestibility in some species such as gibel carp (*Carassius gibelio*) (Liu *et al.*, 2012), grass carp (*Ctenopharyngodon idella*) (Liu *et al.*, 2014) and rohu (*Labeo rohita*) (Roy *et al.*, 2014). Alternatively to the commercially available phytases, which have an optimum function at acidic pH, neutral-alkaline phytases or phytases that are active over a wide pH range, would have a larger potential to be used in diets for fish, especially in agastric cultured fish. However, due the low solubility of IP_6 at neutral pH, the efficiency of phytases active at that pH could be limited. Additional research should be developed on this topic.

On the other hand, fish digestive pH can affect not only the relative efficacy of the phytase, but also the solubility of IP_6 and hence its accessibility to be enzymatically dephosphorylated by exogenous phytase. As pointed by Grynspan and Cheryan (1983), IP_6 solubility is largely pH-dependent, being more soluble at low pH but precipitating under the form of IP_6-cation at pH above 4. This suggests that only a fraction of the native IP_6 is soluble and available as phytase substrate when the pH of the chyme is not low enough. This hypothesis is supported by the results obtained by Morales *et al.* (2013), who showed that in most plant ingredients the relatively high values of solubility of IP_6-P at pH 2.0 dropped dramatically at pH 4.0. Recently, an *in vivo* study evaluating the spatio-temporal changes of the IP_6 levels within the gastrointestinal tract of sea bream suggests that only a low proportion of the ingested IP_6 (14%) remained in the soluble fraction of the stomach content, which reached a minimum pH after six hours post-feeding nearly 4.5 (Morales *et al.*, 2014). The low solubility of IP_6 at pH near to neutrality could severely limit the potential dephosphorylation of IP_6 by phytase in agastric fish species that are not able to acidify the digesta, as well as in the initial developmental stages of carnivorous fish with no fully functional stomach (Yufera and Darias, 2007).

9.5.2 Feed processing and rearing water temperature

In addition to pH, another key factor influencing the phytase activity is temperature. The vast majority of phytases are thermolabile, with an optimal functional temperature between 40 and 60 °C (Igbasan *et al.*, 2000). During the extrusion process, feed ingredients are exposed to high temperatures; hence the enzyme is partially destroyed. Thermostability of phytase is considered to be a desirable feature since most animal feeds processing involves a brief exposure to high temperatures and eventually the activity of the enzyme need to be retained (Kumar *et al.*, 2012). Phytase pre-treatment of plant ingredients before the extrusion process could be an option to avoid denaturation of the enzyme by excess heat (Nwanna, 2007). Moreover, spraying the feed after extrusion with a liquid phytase suspension is another option to solve the problem of high extrusion temperatures (Vielma *et al.*, 2004). However, the progress of companies in the development of enzymes with endogenous thermostable capacity and coatings with thermo-protective functions allows the addition of the enzyme before the extrusion process with relative low activity losses.

As mentioned above, phytase activity is influenced by gastrointestinal pH and also by temperature. Fish are ectotherms, that means that the body temperature is similar to the water. The optimum temperature of most microbial phytases is typically 30 to 50 °C higher than the body temperature of many fish species, depending on whether the enzyme is used in warm-water fish species (20 to 30 °C) or cold-water fish species (5 to 20 °C). In a previous study, Morales *et al.* (2011) tested the effect of phytase on P bioavailability under different temperatures (fish farmed 6 °C for winter and 16 °C for summer) and gastric pH conditions (pH 2.0, 3.0 and 4.0) simulating the rainbow trout digestion process. The amount of soluble P released and the residual IP_6-P from native IP_6 from soybean meal during an *in vitro* digestion with a gastric extract from rainbow trout indicated that the phytase tested is more efficient at low pH, high dosages and high temperature (Figure 9.8 and 9.9).

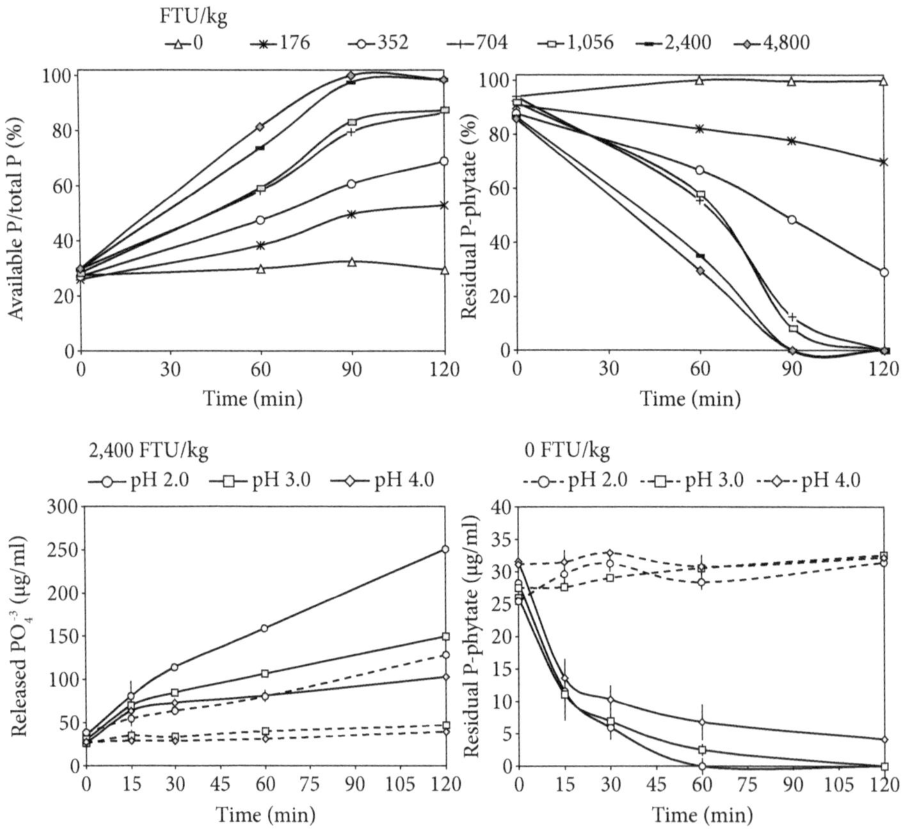

Figure 9.8. *Kinetics of soluble P released and residual IP_6-P content in soybean meal during* in vitro *gastric digestion with stomach enzymatic extract from rainbow trout simulation different gastric pHs (2.0, 3.0, 4.0) and different doses of* Escherichia coli *phytase (0-4,800 FTU/kg DM). Adapted from data reported by Morales* et al. *(2011).*

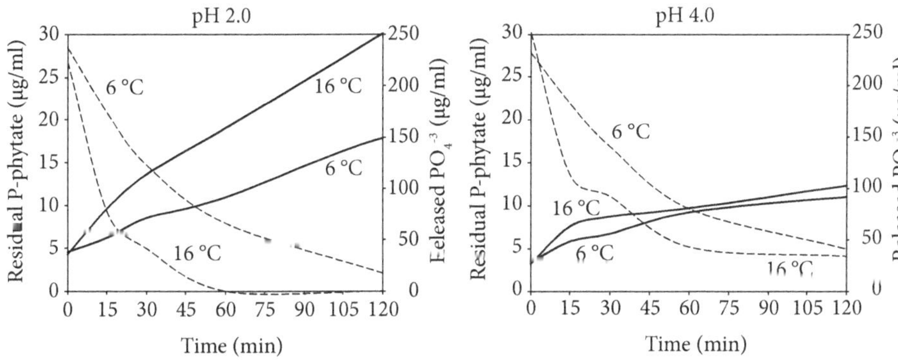

Figure 9.9. *Kinetics of soluble P released and residual IP_6-P content in soybean meal supplemented with 2,500 FTU/kg DM from* Escherichia coli *phytase during* in vitro *gastric digestion with stomach enzymatic extract from rainbow trout simulating different gastric pHs (2.0, 4.0) and different temperatures (winter: 6 °C and summer: 16 °C). Adapted from data reported by Morales* et al. *(2011).*

These results suggest that enzymatic degradation of IP_6 may be severely affected by low temperatures. Between 25 and 35% P release reduction was measured when simulating winter temperature in relation to those values obtained for summer temperatures. Although such a reduction may be compensated to a certain extent in the fish by a decrease in the digestive transit time (allowing a longer time for the action of phytase) (Fauconneau *et al.*, 1983), the net efficiency of the enzyme in cold-water species like salmonids during the winter may be quite low.

Denstadli *et al.* (2007) concluded that to increase mineral availability in Atlantic salmon reared in cold water (8 °C), pre-incubation of plant ingredients with phytase prior to extrusion might be a more rational method than the conventional addition of the enzyme by coating. The authors pre-incubated wheat and soy protein concentrate with phytase for one hour in the pre-conditioner at 45 °C prior to feed production and obtained a maximum IP_6 reduction of 86% in the non-acidified mixture, as well as significantly higher P digestibilities in comparison with those measured when the phytase was added through the conventional procedures. Nevertheless, such process is still experimental and needs further investigations in order to evaluate its economic feasibility.

Despite a positive effect of phytase pre-treatment of the ingredients (Cain and Garling, 1995; Denstadli *et al.*, 2007; Storebakken *et al.*, 1998) or phytase inclusion in the diet (Cheng and Hardy, 2003; Forster *et al.*, 1999; Lanari *et al.*, 1998; Vielma *et al.*, 1998, 2004), the selection of the more suitable method will depend on the water rearing temperature used for the target species. Limited information is available on phytases for cold-water fish species. A phytase which is active at low temperature may be more efficient for cold-water species such as salmon and trout. With the current available commercial microbial phytase products, it can be assumed that to produce a significant dephytination effect within the gastrotintestinal tract, a higher dose of enzyme will be needed in cold-water fish than in warm-water fish species. For these last species, as pointed by Forster *et al.* (1999), temperature does not seem to be a limiting factor for the IP_6 dephosphorylation by phytase, rather it is the pH. It could be speculated that, even when the rearing temperature is above 15 °C, the direct addition of the phytase to the ingredients mixture before extrusion would be a better option compared to the enzymatic pre-treatment of the ingredients as proposed Denstadli *et al.* (2007). The inclusion of phytase in diets for cold-water species, mainly when these species are fed during the winter season, represent a big challenge and comparative detailed studies to evaluate the technical and economic feasibility of the inclusion of phytases in aquafeeds and the way it should be used are still necessary.

9.5.3 Resistance against digestive proteases

Phytase is a functional protein molecule that can be hydrolysed by endogenous proteases within the digestive tract of animals. *In vitro* data reported by Elkhalil *et al.* (2007) suggest that *Aspergillus* phytase exposed to pepsin and pancreatin from pigs (60 min, 40 °C) can be inactivated to a high degree by both proteases. The reduction of *Aspergillus* phytase activity caused by incubation with pepsin amounted to 40.8% and by incubation with pancreatin to 34.2%. *Bacillus amyloliquefaciens* phytase was found to be susceptible to pepsin inactivation (32.4% residual activity), but was nearly completely resistant to pancreatin (98.2% residual activity). *E. coli* and *Klebsiella* phytase proved to be more resistant to pepsin (87-90% residual activity) and pancreatin (95-98% residual activity) inactivation compared to the other enzymes. Similarly, Kumar *et al.* (2003) evaluated the response of phytase derived from *P. lycii*, *A. niger* and *E. coli* to pepsin, trypsin and chymotrypsin for two hours pre-incubation. It was observed that the phytase derived from *E. coli* had greater pepsin resistance (76.7% residual activity) than *A. niger* and *P. lycii* phytases (31.4 and 5.4% residual activity, respectively). *E. coli* phytase also showed higher resistance to trypsin and chymotrypsin compared to the other phytases.

A recent interesting study reported by Menezes-Blackburn *et al.* (2015) investigated the performance of seven commercial phytases in an *in vitro* system simulating the poultry digestive tract. The authors evaluated the phytase inactivation by porcine pepsin determined by preincubating 20 mU of different phytase sources at pH 3.0 containing 3,000 U of pepsin for 45 min at 37 °C. *E coli* phytases resulted in the most resistant product (91-99% residual activity), followed by *C. braakii*, *Buttiauxella* sp., *A. niger* and *P. lycii* (Table 9.1).

Notwithstanding the general consensus that endogenous proteases from mammals can largely affect exogenous microbial phytase activity, the effect of endogenous fish proteases on phytase activity has not been deeply studied yet. In one study, Morales *et al.* (2011) evaluated the effect of gastric pepsin from stomach crude extracts from rainbow trout on the residual activity of two commercial phytases (*E. coli* and *P. lycii*). The authors observed that after one hour of gastric simulation (pH 2; 16 °C; 5,000 pepsin units per FTU) the activity of *P. lycii* phytase dropped rapidly, while *E. coli* phytase maintained its activity over 90% of the initial activity after four hours. In this study there was no difference between pepsin from porcine origin and the gastric crude extract from trout. The results indicated that under the acidic gastric environment of fish with a stomach, like that of rainbow trout, *E. coli* phytase would be more active and stable than phytase from *P. lycii*.

Table 9.1. *Optimal pH range, residual activity to gastric pepsin exposition of several commercial phytases used in animal nutrition.*

Donor organism	Expression organism	Type (initial cleavage)	Optimal pH range	Residual activity to pepsin (%)	Study[1]
Peniophora lycii	*Aspergillus oryzae*	6-phytase	4.5-5.5	6; 34	1,2
Aspergillus niger	*Aspergillus niger*	3-phytase	4.5-5.5	47	2
Escherichia coli	*Pichia pastoris*	6-phytase	4.0-5.0	90; 93	1,2
Escherichia coli	*Pichia pastoris*	6-phytase	4.0-5.0	95	3
Escherichia coli	*Pichia pastoris*	6-phytase	4.0-5.0	97	3
Escherichia coli	*Trichoderma reesei*	6-phytase	3.5-5.0	98; 91	2
Escherichia coli	*Schizosacch. pombe*	6-phytase	3.0-5.0	92	2
Buttiauxella sp.	*Trichoderma reesei*	6-phytase	3.0-3.5	85; 81	2,3
Citrobacter braakii	*Aspergillus oryzae*	6-phytase	3.0-4.5	92; 64	2,3

[1] 1 = Morales *et al.* (2011): pH 2.0, 16 °C, 180 min, trout pepsin:phytase ratio 5,000 PU:FTU. 2 = Menezes-Blackburn *et al.* (2015): pH 3.0, 37 °C, 45 min, porcine pepsin incubating 20 mFTU and 3,000 U of pepsin. 3 = Unpublished data (own elaboration): pH 2.0, 39 °C, 180 min, porcine pepsin:phytase ratio 9,000 PU:FTU.

9.6 Phytase responses in fish

Despite the use of phytase in diets for aquaculture is not as widespread as in diets for poultry and pigs, the effects of the use of these enzymes on nutrient utilisation and growth has been evaluated for numerous commercial aquatic species, mostly fish. The effects of supplemental phytase on nutrient utilisation and growth have been evaluated in: Rainbow trout (*Oncorhynchus mykiss*) (Cheng and Hardy, 2002, 2003, 2004; Cheng *et al.*, 2004; Dalsgaard *et al.*, 2009; Forster *et al.*, 1999; Lanari *et al.*, 1998; Morales *et al.*, 2015; Riche and Brown, 1996; Rodehutscord and Pfeffer, 1995; Sugiura *et al.*, 2000, 2001; Vandenberg *et al.*, 2011, 2012; Verlhac-Trichet *et al.*, 2014; Vielma *et al.*, 1998, 2002, 2004; Wang *et al.*, 2009; Yang *et al.*, 2011); Atlantic salmon (Carter and Sajjadi, 2011; Denstadli *et al.*, 2007; Sajjadi and Carter, 2004a,b; Storebakken *et al.*, 2000); Sea bass (*Dicentrarchus labrax*) (Oliva-Teles *et al.*, 1998); Sea bream (Morales *et al.*, 2014); Korean rockfish (*Sebastes schlegeli*) (Yoo *et al.*, 2005); Japanese flounder (*Paralichthys olivaceus*) (Masumoto *et al.*, 2001); Rohu (Roy *et al.*, 2014); Tiger puffer (*Takifugu rubripes*) (Laining *et al.*, 2011); Channel catfish (*Ictalurus punctatus*) (Eya and Lovell, 1997; Jackson *et al.*,1996; Li and Robinson, 1997); African catfish (*Clarias gariepinus*) (Schäfer *et al.*, 1995; VanWeerd *et al.*, 1999); Yellow catfish (*Pelteobagrus fulvidraco*) (Cheng *et al.*, 2015; Zhu *et al.*, 2014); Nile tilapia (Liebert and Portz, 2005; Portz and Liebert, 2004); Striped bass (Papatryphon and Soares, 2001; Papatryphon *et al.*, 1999; Powers-Hughes and Soares, 1998); Common carp (Hughes and Soares, 1998; Nwanna and Schwarz, 2007; Nwanna *et al.*, 2008; Schäfer *et al.*, 1995) and Grass carp (Liu *et al.*, 2013). For more data about comparative beneficial effects of phytase supplementation on nutrient digestibility and animal performance see the recent review reported by Lemos and Tacon (in press).

A comparative analysis of these studies suggests that the degree of effectiveness of phytases is considerably different from one investigation to another. Taking into account that effectiveness of microbial phytases may depend directly or indirectly on multiple factors, such as fish species, digestive physiology, water temperature, feeding protocol, dietary IP_6 levels, phytase inclusion rate and phytase source, among others, the range of optimal supplementation in fish is not yet entirely well defined. In general terms, previous studies evaluated the effect of phytases in a wide range of dietary levels (200-6,000 FTU/kg). Cao *et al.* (2007) who reviewed several studies, concluded that a phytase dose level of 250-2,000 FTU/kg feed is considered optimal for many fish species. However, this is a much wider range than that recommended for most diets designed for pigs or poultry (around 500 FTU/kg).

The large response variation among studies in aquaculture can be explained by the diversity of fish species and aquatic environments used in aquaculture industry. Gastric pH and the physiological temperature of the fish are the main factors to be considered for understanding how phytase works in a fish. The first depends on whether it is a carnivorous species with a developed stomach and defined gastric phase or an omnivorous fish species with a weak or absent gastric digestion phase. The capability or not to drop digesta pH determines the functional window of the phytase to hydrolase the dietary phytate.

Previous studies indicate that microbial phytase in salmonid diets based on plant ingredients significantly improved the digestibility of P, with a clear dose-dependent, positive response from 0 to 4,000 FTU/kg diet (Sugiura *et al.*, 2001; Vielma *et al*, 2004). However, most of the previous studies reported that the enzymatic dephosphorylation of IP_6 in fish digestive tract is still far to be complete. For example, the average P release of endogenous IP_6 present in diets with a high content of plant ingredients varies between 40 and 60% in rainbow trout (Cheng and Hardy, 2004; Cheng *et al.*, 2004; Forster *et al.*, 1999; Sugiura *et al.*, 2001; Vielma *et al.*, 1998, 2004). This implies that 40-60% of the phytate-P is still unavailable.

A recent study (Morales *et al.*, 2015) reported that by using a high dose of microbial phytase (4,000 FTU/kg feed) the endogenous IP_6 in a plant based diet (7% fish meal) for rainbow trout decreased 30% due to the added phytase already during the production process of the diet. Once fish ingested the diet, dephosphorylation of IP_6 progressed within the digestive tract reducing the IP_6 content in faeces from 32-34 in control treatments to 4-6 mg/g dry matter in phytase fed fish, indicating that phytase inclusion at such level can dephosphorylate up to 90% (IP_6 apparent digestibility coefficient) the native inositol phosphate from plant ingredients. In this study, phytase supplementation improved the digestibility of total dietary P from 38 to 76% and increased the apparent total tract digestibility of other minerals (Ca, Mg, Zn) and protein. In addition, 4,000 FTU/kg feed increased P retention (37 to 68%), Ca (9 to 76%), Mg (10 to 15%) and protein (49 to 52%), but no improvement was observed for Zn retention. These results are in agreement with those reported previously in an interesting study of Vielma *et al.* (2004) who evaluated phytase supplementation of 0, 500, 1000, 2,000 and 4,000 FTU/kg in a diet containing 50% soybean meal for rainbow trout. In this study, phytic acid content in the faecal dry matter of fish fed a diet without phytase was 34.9 mg/g, while faecal phytic acid concentration decreased in a non-linear pattern approaching the asymptote of 5.2 mg/g dry matter in fish fed phytase. As faecal phytic acid concentration decreased, apparent digestibility coefficients of P increased in a non-linear pattern with phytase supplementation. Apparent digestibility coefficient of P was 22.8% in fish fed no phytase, while it increased with phytase supplementation approaching the asymptote of 83.3%. From this study, it can be concluded that 2,000 FTU/kg feed, would be the optimal dose for the phytase tested to obtain a maximal phytate degradation and apparent digestibility coefficients for P, Zn and N.

In general terms, the positive effects of phytase on fish performance and nutrient digestibility in warm-water fish species can be considered comparable to the response observed in cold-water fish species. However, in most cases the phytase supplementation levels are lower in warm-water species compared to cold-water species, closely related to the effect of water temperature on phytase efficacy. Previous studies in warm-water species suggest a relative positive effect of dietary phytase supplementation on growth rate, feed conversion efficiency, P and other trace minerals digestibility. In addition, increased digestibilities of protein (2-18%), lipid (26-65%), energy (4-16%) and amino acids (5-12%) were also reported due to phytase supplementation from 500 to 1000 FTU/kg feed (Lemos and Tacon, in press).

Further *in vitro* and *in vivo* studies would be required to optimize dietary phytase doses, mainly in those fish species in which the gastric digestion phase is weak or inexistent so that, regardless of the physiological temperature of the fish, gut pH probably becomes the main determinant of the phytase efficacy. Taking into account the great variety of aquaculture fish species and rearing conditions, only an accurate study of the internal (e.g. dietary IP_6 content and source, phytase dose and source and dietary minerals) and external (e.g. pH, temperature, proteases and digestion rate) factors involved in fish digestion will contribute to the approximation of the optimal phytase dose needed to maximise nutrients retention and fish performance.

As pointed out above, pH can significantly affect the initial solubilisation of dietary IP_6 once feed is ingested by fish. Therefore, an efficient use of phytase as feed additive could only be achieved efficiently in fish able to decrease the pH of the digesta below 4.0. The pre-treatment of plant ingredients with phytase could be an alternative to increase IP_6-P bioavailability for agastric warm-water fish species, as suggested Denstadli *et al.* (2007) and for cold-water fish species where pH is not the limiting factor, but winter temperature reduces phytase efficiency within the fish gut.

However, another interesting approach could be the combination of phytase and organic acids that may reduce the pH of the digesta. The supplementation of diets with organic acids or their salts has been suggested for pig diets to improve growth through a better digestion, absorption and retention of a variety of nutrients (Partanen and Mroz, 1999). The suggested mode of action of organic acids in the digestive tract includes a reduction of pH in the stomach and in the small intestine, and the inhibition of Gram-negative bacteria by the undissociated form of the organic acids that passively diffuse through the bacteria cell wall, where it dissociates and inhibits the ability of the bacteria to multiply

(Partanen and Mroz, 1999). The reduction of pH in the stomach would increase pepsin activity improving protein digestion (Mroz *et al.*, 2000). Likewise, feed supplementation with organic acids has been shown to decrease duodenal pH, improve nitrogen retention and increase nutrient digestibility in pigs (Øverland *et al.*, 2000).

Regarding the enzymatic IP_6 dephosphorylation in the digestive tract, the use of organic acids such as citric acid, fumaric acid and formic acid could improve the effectiveness of microbial phytase through increased IP_6 solubility due to the reduction in stomach pH. In addition to their effect on intestinal pH, supplementary organic acids can be also a strong chelator of cations such as Ca^{2+}, removing these minerals from the phytate molecule and making it less stable and more susceptible to phytases (Khajepour and Hosseini, 2010). The synergic effect of organic acid such as citric acid and phytase in poultry and pigs is available in literature, but few studies have been conducted in fish (Shah *et al.*, 2015). Previous studies evaluated the combined effect of acidifiers and phytases in stomachless fish species (Baruah *et al.*, 2005, 2007a,b, 2009; Sugiura *et al.*, 2001). However, to the moment, there are no studies having evaluated the effect of such a combination on the solubility of IP_6 within the digestive tract of fish.

In addition to acidifiers, associated supplementation of phytases with non-starch polysaccharide degrading enzymes (glucanases, pentosanases, cellulases and xylanases) may facilitate IP_6 dephosphorylation by phytases. Fish have digestive enzymes capable of degrading starch present in the plant ingredients of the diet, being high in herbivorous fish, medium in omnivores and low in carnivores (Krogdahl *et al.*, 2005); however, they do not have non-starch polysaccharide degrading enzymes. This makes that non-starch polysaccharides present in the diet not digestible in the gastrointestinal tract of the fish, precluding its use as an energy source. The main disadvantages generated by non-starch polysaccharides in the gastrointestinal tract, are related to its high water holding capacity. This affects digestive transit, decreasing the interaction between digestive enzymes and macromolecules of dietary origin, resulting in a decreased bioavailability of nutrients (Storebakken and Austreng, 1987; Storebakken *et al.*, 1999).

On the other hand, some of the components of the plant cell wall (insoluble fractions) have a 'cage effect' encapsulating nutrients such as protein and minerals, affecting their digestion and absorption by fish, these being eliminated via the faeces. The addition of exogenous non-starch polysaccharide degrading enzymes in diets with a high content of plant sources is presented as a potential tool to increase the bioavailability of dietary nutrients in fish, as has been observed in terrestrial monogastric animals (Bedford, 2000; Khattak *et al.*, 2006). Phytase supplementation combined with non-starch polysaccharide degrading enzymes has not been studied in depth, but it is a potentially interesting application in aquaculture diets to improve the overall utilisation of nutrients by the animal as reported by Ai *et al.* (2007) and Signor *et al.* (2010).

9.7 Phytase super-dosing in fish

The concept of phytase super-dosing is based on the dietary addition of an enzyme dose high enough to hydrolyse dietary IP_6 quickly and thus eliminate its negative antinutritional effects in the early stages of the digestive process. In most terrestrial monogastrics the typically recommended phytase dose is in the order of 500 FTU/kg diet. The main reason for using this enzyme dosage, is to reduce the cost of inorganic P supplementation. With this typical dose it is assumed for new phytases that in pig and poultry diets, the dietary P availability can be increased by 1.5 g/kg available P compared to diets without phytase supplementation (Dersjant-Li *et al.*, 2015). It is assumed that higher doses of phytase result not only in a higher IP_6-P bioavailability but also in other, so called extra-phosphoric benefits for the animals.

One of the earliest studies of phytase super-dosing in monogastrics was conducted by Nelson *et al.* (1971) adding between 950 and 7,600 FTU/kg feed. In this work, the authors found that IP_6-P apparent disappearance was increased from 39% to 94% in broiler chickens. In another relevant study carried out by Shirley and Edwards (2003) with corn-based diets supplemented with phytase up to 12,000 FTU/kg a quadratic increase in phytate-P disappearance was observed with increasing phytase dose from approximately 42% at 93 FTU/kg to 95% at 12,000 FTU/kg feed.

Extra-phosphoric benefits of high doses of phytase supplementation are not only related to a higher digestibility of dietary amino acids and minerals, but also to energetic effects, both directly via improved solubility and digestibility of dietary nutrients (e.g. release of metal co-factors for endogenous enzymes or inositol liberation), and indirectly (via reduced endogenous loss, or net effects delivered via reduced synthesis of endogenous protein) (Cowieson *et al.*, 2011). In a recent study, Walk *et al.* (2013) concluded that the supplementation of phytase in excess of commercial recommendations (1,500 FTU/kg) may promote feed efficiency in broilers due to the alleviation of the antinutritional effects of phytate, rather than the release of P.

Several factors in relation to the use of phytase super-dosing in terrestrial monogastrics have been proposed, such as improving the utilisation of energy, amino acids, and trace minerals; improving nutrient intake because phytate might be acting as an appetite suppressant (Cowieson *et al.*, 2011); a higher release of phosphate or the restoration of the released P:Ca proportion (it is suggested that P and Ca are not necessarily released by phytase at a 1:1 ratio, so with high phytase doses P would continue to be released, whereas Ca would approach a maximum release) and generating *myo*-inositol with vitamin-like/lipotropic effects (Cowieson *et al.*, 2011). Despite these proposed factors for poultry and pigs, the mechanism of phytase super-dosing remains unknown for fish species.

In fish species, it has been observed that high levels of phytase not only result in an increased P utilisation, but also in a greater use of several dietary minerals and protein. For example, the effects of 1000 FTU/kg feed on growth, body composition, nutrient digestibility, retention and P release in effluent water were studied in rainbow trout by Lanari *et al.* (1998) with diets containing 33% soybean meal. Phytase supplementation increased apparent digestibility of P from 58.6 to 68.1%, daily weight gain and feed conversion ratio were improved in fish fed the phytase supplemented diet; and additionally body protein, ash and P contents also tended to be higher in the same group of fish. Vandenberg *et al.* (2011), feeding rainbow trout with a plant based diet (58% plant protein sources) supplemented with 3,000 FTU/kg feed reported an increment in apparent digestibility compared to the control diet without phytase for P (60%), Ca (70%), ash (68%), Mn (60%), Mg (45%), Fe (39%), Zn (12%), energy (3%) and protein (2%). Moreover, Vandenberg *et al.* (2012) also reported from a trial with rainbow trout fed a plant based diet (soybean meal 37%, corn gluten meal 22%, wheat middlings 13%) an increment of 100% in P digestibility and 3% in protein digestibility when 4,000 FTU/kg feed were fed compared to a control diet. A recent study conducted with the same species (rainbow trout) and the same phytase dose (4,000 FTU/kg feed) and a diet containing soy protein concentrate (37%) and sunflower expeller (20%) as the main IP_6 sources, showed an increment in P (92%), Ca (>100%), Mg (38%), Zn (60%) apparent digestibility and complete IP_6 degradability, as well as an increment in N retention (4%), weight gain (8%) and feed efficiency (6%) with respect to the fish fed the control diet without phytase supplementation (Morales *et al.*, 2015). Among other salmonid species, Carter and Sajjadi (2011) fed Atlantic salmon with a plant based diet, using phytase doses from 250 to 4,000 FTU/kg and found a positive dose-response with an increment in P digestibility (81%) up to 4,000 FTU/kg feed, in Mg (30%) and Zn (38%) up to 1000 FTU/kg feed and an increment in feed intake (18%) and weight gain (19%).

A complete dose-response study was carried out by Yan *et al.* (2002) who conducted a feeding trial to quantify the effects of phytase at 0, 500, 1000, 2,000, 4,000 and 8,000 FTU/ kg diet on the utilisation of dietary protein and minerals in channel catfish fingerlings fed an all plant protein diet composed of soybean meal, corn and wheat middlings. After 14 weeks, the authors found that fish fed phytase supplemented diets had higher concentration of ash, Ca, P and Mn in bones than fish fed the unsupplemented diet. Concentration of these minerals in bones did not differ between phytase supplemented groups. Bone Mg levels did not differ among fish fed ≥500 FTU/kg diet. The amount of Zn in bones of fish fed 8,000 FTU/kg feed was significantly higher than that in fish fed 0 or 500 FTU/kg feed. However, in the same study, phytase supplementation at levels of 8,000 FTU/kg diet did not increase weight gain or improve dietary protein utilisation of channel catfish fed an all-plant-protein diet.

As mentioned before, the differences in the response of phytase supplementation (even at high doses) among different fish species may be due to the presence or absence of an acidic gastric digestion. Despite these differences, in stomached fish species, the use of phytase super-dosing can improve protein digestion primarily due to a number of changes during fish gastric digestion: the rapid rupture of indigestible IP_6-protein complexes present in the diet, starting by IP_6-IP_5 esters and further IP_4-IP_3 esters; avoiding the *de novo* formation of IP-protein complexes, increasing protein solubility, increasing protein bioavailability through more soluble protein and more active digestive proteases, decreasing the negative effect of IP_6 on the protease activity of the gastric pepsin and reducing IP-Ca complexes with potential inhibiting capability of intestinal proteases (Kies *et al.*, 2006; Morales *et al.*, 2011, 2013, 2014; Selle *et al.*, 2000). Based on these data, an early and thorough hydrolysis of phytate by phytase in the upper digestive tract is essential for an improved digestion of P, minerals (e.g. Ca, Fe) and protein. These data imply also that to eliminate the antinutritional effects of phytate, IP_6 needs to be hydrolysed as completely as possible by phytase in the upper part of the digestive tract.

Further investigations on phytase super-dosing in fish is necessary to understand the mechanisms involved in fish digestion; however, the positive extra-phosphoric effects of high phytase doses on nutrient utilisation and growth performance is quite clear in the literature. The quick elimination of the antinutritional effects of most reactive IP esters (IP_6-IP_5), dephosphorylating these molecules as completely as possible by phytase in the upper part of the digestive tract, may be an accurate strategy to replace expensive animal protein meals by cheaper alternative plant protein ingredients with relative high levels of native phytic acid, to improve dietary nutrient utilisation and reduce the excretion of undigested nutrients to the water.

9.8 Phytases and water pollution

In animal production of terrestrial monogastrics, feeds are traditionally supplemented with inorganic phosphate to cover the nutritional requirements. However, feeds rich in plant ingredients have significant contents of IP_6-P that is excreted through the faeces and spread with the manure to the soil. This source of non-digestible nutrient may cause the potential eutrophication of superficial fresh water causing cyanobacterial blooms, hypoxia and death of aquatic animals and the production of nitrous oxide as potential green-house gas (Vats *et al.*, 2005). Similarly, aquaculture uses aquafeeds that contain relevant amounts of non-digestible IP_6-P. Therefore, inorganic P sources must be added to meet the fish requirements for this nutrient. However, excessive content of non-digestible dietary IP_6-P in aquafeed may cause eutrophication of rivers, lakes and reservoirs. P is a limiting nutrient for the primary production in most aquatic ecosystems (Mainston and Parr, 2002; Tyrrell, 1999). IP_6-P input to the ecosystems can directly contribute to eutrophication, causing algal bloom and hypoxia of natural waters.

Commercial diets for monogastrics supplemented with exogenous phytase gained recognition in the last decades as a nutritional strategy to increase dietary P digestibility and to reduce environmental pollution like the eutrophication of surface waters caused by excessive P output. Almost all of these products are market-oriented to chicken and pigs, species in which the enzyme has been highly successful due to the economic and environmental advantages generated.

However, in recent years, companies are promoting the use of phytases in aquaculture feeds. Many reports have demonstrated that phytase supplementation improved phytate-P availability in fish (Cao *et al.* 2007; Kumar *et al.*, 2012; Lemos and Tacon, in press). Thus, P digestibility and bone mineralisation can be considerably enhanced by phytase. Since supplementation of phytase can improve P bioavailability in plant based diets, it is possible to improve the P availability in diets and reduce the P excretion into water. Different studies suggested that addition of phytase to plant-based fish diets improves the utilisation of phytate-P by 20-60% and decreases the total-P load to the environment by 30-50% in rainbow trout (Forster *et al.*, 1999; Lanari *et al.*, 1998; Rodehutscord and Pfeffer, 1995; Sugiura *et al.* 2001), salmonids (Storebakken *et al.*, 1998), tilapias (Goda, 2007) and carps (Schaefer and Koppe, 1995). These studies suggest that phytase enzymes can be viewed as a solution to reduce the potential environmental impact caused by the discharge of nutrients from aquaculture to the aquatic environment.

As an example of the significance and potential application of phytase supplementation of commercial aquafeeds, in a recent study Morales *et al.* (2015) demonstrated that the use of 4,000 FTU/kg diet increased the retention of dietary P in rainbow trout from 36 to 67%, reducing almost 50% of the discharge of that nutrient into the environment (Figure 9.10).

The P loading in the solid output was the fraction most affected by phytase treatments, reducing the P waste loading from 3.8 to 1.8 g/kg BW gain (Morales *et al.*, 2015). Despite the fact that total P discharges were reduced by phytase, the soluble P fraction was relatively higher when fish were fed diets supplemented with phytase by increasing from 0.1 g to 0.3-0.5 g/kg BW gain.

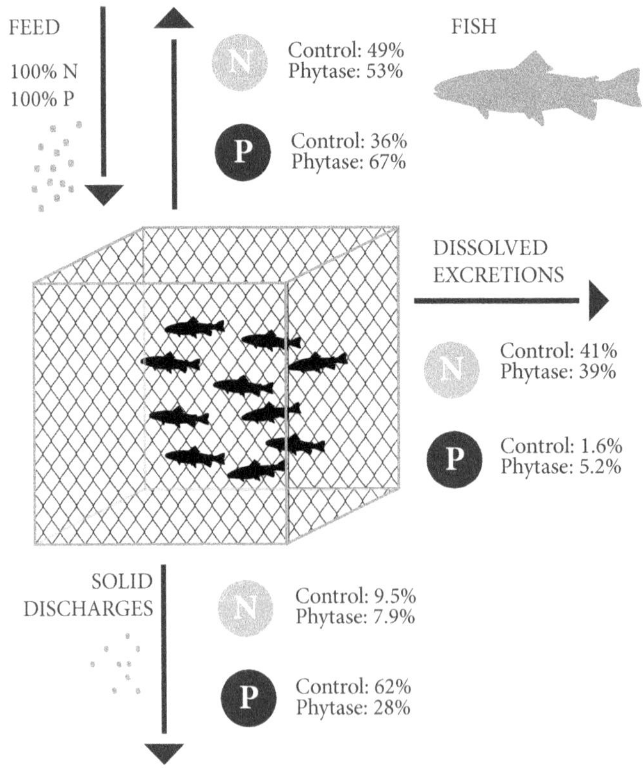

Figure 9.10. *Effect of phytase supplementation (4,000 FTU/kg feed) in a plant based diet for rainbow trout on the dissolved and solid discharged fractions of nitrogen (N) and phosphorus (P) into the water. Illustration adapted from data reported by Morales* et al. *(2015).*

Similar results were reported by Dalsgaard *et al.* (2009) who observed that the soluble fraction of the P discharged from rainbow trout fed a diet containing 1,400 FTU/kg was higher than that observed in fish fed a diet without phytase. This indicates that an as accurate as possible assumption of the P bioavailability of the diet formulation is essential to cover fish P requirements for maximum growth and mineral deposition (Rodehutscord, 1996) but, at the same time, reduce the excretion of P to the water.

The use of phytase as dietary supplement resulted not only in a lower release of P into the water, but also increased fish growth rate, improved feed efficiency and mineral retention (Morales *et al.*, 2015). Moreover, in the same study, the use of a high phytase application resulted in a higher N retention compared to fishes fed non-phytase diets, reducing the total N excretion by 13% (from 30 to 26 g/kg BW gain) to the environment compared to fish fed diets without phytase.

These results suggest that phytase has a high relevance from an environmental point of view. For example, if the results reported in literature are extrapolated to a medium-scale cage farming operation producing 1000 MT/year, the implementation of phytase supplementation in aquafeeds formulas could reduce the P discharges to the aquatic environment in the order of 2,000 kg/year. This is equivalent to approximately 8,800 kg/year of a monocalcium phosphate (MCP). If these values are extrapolated to the industry, only the Chilean salmonid culture (900 MMT; FAO, 2016) could reduce the P excretion to the environment in the order of 1,800 MMT of P, equivalent to almost 8,000 MMT of MCP per year released to the fjords.

Although the use of phytase as a feed additive in aquaculture feeds is not as common as in poultry and pigs nutrition, the available literature supports the benefits of phytases in terms of nutrient retention and reduction of discharges of indigestible nutrients in fish, demonstrating that this enzyme supplement is a key tool for a suitable management of feed formulations and environmentally sustainable aquaculture.

9.9 Outlook

In the last decades, aquaculture has been responsible for the impressive growth in the supply of fish for human consumption. The projections published in the OECD-FAO (FAO, 2016) suggest that the share of aquaculture in total fishery production will grow from 44% on average in 2013-15 to surpass capture fisheries in 2021. In 2025, this share will reach 52%. According to the global estimation reported by Tacon and Metian (2015), about 70% of fish and crustacean aquaculture production are direct-fed species (35.7 MMT in 2012), including carp, tilapia, shrimp, catfish, salmon, marine fish, other miscellaneous freshwater and diadromous fishes, freshwater crustaceans, milkfish and eel. The global aquafeed production was estimated to amount to 35.5 MMT for 2015 (Alltech, 2016).

Despite the aquaculture sector has been the largest consumer of fishmeal (68%) in 2012 (Naylor *et al.*, 2009), in a context of continuous global growth of aquaculture, the limiting availability and high cost of fishmeal opens an opportunity for phytases to be used as a feed additive giving the possibility to use cheaper alternative plant protein ingredients in aquafeed formulations, as well as to improve the bioavailability of several dietary nutrients.

On the other hand, farmers are facing increasing pressure concerning the discharges into the surrounding ecosystems. The environmental impact of intensive aquaculture operations is increasingly reviewed and environmental lobbyists as well as governments are restricting this industry (Debnath *et al.*, 2005b).

Aquaculture is presented as one of the main sources of food production to supply the growing demand of the global population. Like agriculture and terrestrial animal production, aquaculture is necessary for the increasing population. While in many cases the environmental impacts of aquaculture may be negligible, in many others it might not and technologies to optimize the use of resources should be developed, minimizing the ecological footprint of its activity. The discharges of nutrients from fish farms to the environment and their effects on the eutrophication of freshwater and marine resources, are considered one of the most concerning issues facing aquaculture today worldwide.

Currently, commercially available microbial phytases allow providing nutrient resources from plant ingredients to aquatic species that otherwise would not be capable for fish feeds (e.g. for salmonids, marine fish species). Moreover, among the benefits associated with the use of phytase in aquafeeds, a lower dependency of aquaculture on fishmeal from capture fisheries can be highlighted, as well as a reduced demand for inorganic P sources, the increase in the availability and retention of minerals, energy and protein in most studied fish species, and the smaller discharges of undigested P fractions, and to some extent reduce N excretion to the environment, which is especially relevant in areas of intensive aquaculture production.

9.10 Conclusions

1. The knowledge about phytase and phytate interactions, as well as the discovery of new phytase sources produced by biotechnology, greatly supported cost-effective production methods to improve poultry and pig nutrition. However, further investigations are necessary to optimize the functionality of the enzymes under the physiological conditions of fish.
2. Phytases currently used in terrestrial animal nutrition differ in their enzymatic features, in respect to their pH profiles, their resistance to digestive proteases, their intrinsic thermostabilities and their functionality at low temperatures and low substrate concentrations.
3. Further investigation of these enzyme properties, the characterisation of native IP_6 in all plant ingredients with a potential use in aquafeed formulations, as well as the efficiency of different phytases when interacting with dietary cations, is essential to understand the modes of action and to determine an optimal supplementation dose in feeds for the different aquaculture fish species.

References

Abelson, P.H., 1999. A potential phosphorus crisis. Science 283:2015.

Adeola, O. and Cowieson, A.J., 2011. Board-invited review: opportunities and challenges in using exogenous enzymes to improve nonruminant animal production. Journal of Animal Science 89:3189-3218.

Adeola, O. and Sands, J.S., 2003. Does supplementary microbial phytase improve amino acid utilization? A perspective that it does not. Journal of Animal Science 81:78-85.

Ai, Q., Mai, K., Zhang, W., Xu, W., Tan, B., Zhang, C. and Li, H., 2007. Effects of exogenous enzymes (phytase, non-starch polysaccharide enzyme) in diets on growth, feed utilization, nitrogen and phosphorus excretion of Japanese seabass, *Lateolabrax japonicus*. Comparative Biochemistry Physiology part A: Molecular and Integrative Physiology 147:502-508.

Alltech, 2016. Alltech Global Feed Survey summary. Alltech, Nicholasville, Kentucky, USA, 8 pp. Available at: http://stories.alltech.com/global-feed-survey-2016.html.

Anderson, P.A., 1985. Interactions between proteins and constituents that affect protein quality. In: Finley, J.W. and Hopkins, D.T. (eds.) Digestibility and amino acid availability in cereals and oilseeds. American Association of Cereal Chemists, St. Paul, MN, USA, pp. 31-45.

Austreng, E., Storebakken, T., Thomassen, M.S., Refstie, S. and Thomassen, Y., 2000. Evaluation of selected trivalent metal oxides as inert markers used to estimate apparent digestibility in salmonids. Aquaculture 188:65-78.

Baldwin, R.L., 1996. How Hofmeister ion interactions affect protein stability. Biophysical Journal 71:2056-2063.

Baruah, K., Pal, A.K., Sahu, N.P., Jain, K.K., Mukherjee, S.C. and Debnath, D., 2005. Dietary protein level, microbial phytase, citric acid and their interactions on bone mineralization of *Labeo rohita* (Hamilton) juveniles. Aquaculture Research 36:803-810.

Baruah, K., Sahu, N.P., Pal, A.K., Jain, K.K., Debnath, D., Sona, Y. and Mukherjee, S.C., 2007a. Interactions of dietary microbial phytase, citric acid and crude protein level on mineral utilization by rohu, *Labeo rohita* (Hamilton), Juveniles. Journal of World Aquaculture Society 38:238-249.

Baruah, K., Sahu, N.P., Pal, A.K., Jain, K.K., Debnath, D., Sona, Y. and Mukherjee, S.C., 2007b. Dietary microbial phytase and citric acid synergistically enhances nutrient digestibility and growth performance of *Labeo rohita* (Hamilton) juveniles at sub-optimal protein level. Aquaculture Research 36:803-812.

Baruah, K., Pal, A.K., Sahu, N.P., Debnath, D., Yengkokpam, S., Norouzitallab, P. and Sorgeloos, P., 2009. Dietary crude protein, citric acid and microbial phytase interacts to influence the hematoeimmunological parameters of rohu, *Labeo rohita*, juveniles. Journal of World Aquaculture Society 40:824-831.

Bedford, M.R., 2000. Exogeneous enzymes in monogastric nutrition – their current value and future benefits. Animal Feed Science and Technology 86:1-13.

Cain, K.D. and Garling, D.L., 1995. Pretreatment of soybean meal with phytase for salmonid diets to reduce phosphorus concentrations in hatchery effluents. The Progressive Fish Culturist 57:114-119.

Caldwell, R.A., 1992. Effect of calcium and phytic acid on the activation of trypsinogen and the stability of trypsin. Journal of Agriculture and Food Chemistry 40:43-46.

Cao, L., Wang, W., Yang, C., Yang, Y., Diana, J. and Yakupitiyage, A., 2007. Application of microbial phytase in fish feed. Enzyme Microbial Technology 40:497-507.

Carter, C.G. and Sajjadi, M., 2011. Low fishmeal diets for Atlantic salmon, *Salmo salar* L., using soy protein concentrate treated with graded levels of phytase. Aquaculture International 19:431-444.

Champagne, E.T., 1988. Effects of pH on mineral-phytate, protein-mineral-phytate and mineral-fiber interactions. Possible consequences of atrophic gastritis on mineral bioaccessibility from high fiber foods. Journal of the American College of Nutrition 7:499-508.

Cheng, Z.J. and Hardy, R.W., 2002. Effect of microbial phytase on apparent nutrient digestibility of barley, canola meal, wheat and wheat middlings, measured *in vivo* using rainbow trout (*Oncorhynchus mykiss*). Aquaculture Nutrition 8:271-277.

Cheng, Z.J. and Hardy, R.W., 2003. Effects of extrusion and expelling processing, and microbial phytase supplementation on apparent digestibility coefficients of nutrients in full-fat soybeans for rainbow trout (*Oncorhynchus mykiss*). Aquaculture 218:501-514.

Cheng, Z.J. and Hardy, R.W., 2004. Effects of microbial phytase supplementation in corn distiller's dried grain with soluble on nutrient digestibility and growth performance of rainbow trout, *Oncorhynchus mykiss*. Journal of Applied Aquaculture 15:83-100.

Cheng, Z.J., Hardy, R.W., Verlhac, V. and Gabaudan, J., 2004. Effects of microbial phytase supplementation and dosage on apparent digestibility coefficients of nutrients and dry matter in soybean product-based diets for rainbow trout *Oncorhynchus mykiss*. Journal of the World Aquaculture Society 35(1):1-15.

Cheng, N., Chen, P., Lei, W., Feng, M. and Wang, C., 2015. The sparing effect of phytase in plant-protein-based diets with decreasing supplementation of dietary NaH_2PO_4 for juvenile yellow catfish *Pelteobagrus fulvidraco*. Aquaculture Research 2015:1-12.

Cheryan, M., 1980. Phytic acid interactions in food systems. CRC Critical Reviews in Food Science and Nutrition 13:297-335.

Cosgrove, D.J., 1966. The chemistry and biochemistry of inositol polyphosphates. Reviews of Pure and Applied Chemistry 16:209-224.

Cowieson, A.J., Acamovic, T. and Bedford, M.R., 2006. Phytic acid and phytase: implications for protein utilization by poultry. Poultry Sciences 85:878-885.

Cowieson, A.J., Wilcock, P. and Bedford, M.R., 2011. Super-dosing effects of phytase in poultry and other monogastrics. World's Poultry Science Journal 67:225-235.

Dalsgaard, J., Ekmann, K.S., Pedersen, P.B. and Verlhac, V., 2009. Effect of supplemented fungal phytase on performance and phosphorus availability by phosphorus-depleted juvenile rainbow trout (*Oncorhynchus mykiss*), and on the magnitude and composition of phosphorus waste output. Aquaculture 286:105-112.

De Blas, C., Mateos, G.G. and García-Rebollar, P., 2010. Tablas FEDNA de composición y valor nutritivo de alimentos para la fabricación de piensos compuestos (3ª edición). Fundación Española para el Desarrollo de la Nutrición Animal. Madrid, Spain, 502 pp.

Debnath, D., Pal, A.K., Narottam, P.S., Jain, K.K., Yengkokpam, S. and Mukherjee, S.C., 2005a. Effect of dietary microbial phytase supplementation on growth and nutrient digestibility of *Pangasius pangasius* (Hamilton) fingerlings. Aquaculture Research 36:180-187.

Debnath, D., Sahu, N.P., Pal, A.K., Baruah, K., Yengkokpam, S. and Mukherjee, S.C., 2005b. Present scenario and future prospects of phytase in aqua feed – review. Asian Australasian Journal of Animal Sciences 18:1800-1812.

Denstadli, V., Skrede, A., Krogdahl, Å., Sahlstrøm, S. and Storebakken, T., 2006. Feed intake, growth, feed in Atlantic salmon (*Salmo salar* L.) fed graded levels of phytic acid. Aquaculture 256:365-376.

Denstadli, V., Storebakken, T., Svihus, B. and Skrede, A., 2007. A comparison of online phytase pre-treatment of vegetable feed ingredients and phytase coating in diets for Atlantic salmon *Salmo salar* L. reared in cold water. Aquaculture 269:414-426.

Dersjant-Li, Y., Awati, A., Schulze, H. and Partridge, G., 2015. Phytase in non-ruminant animal nutrition: a critical review on phytase activities in the gastrointestinal tract and influencing factors. Journal of the Science of Food and Agriculture 95(5):878-896.

Einarsson, S., Davies, P.S. and Talbot, C., 1996. The effect of feeding on the secretion of pepsin, trypsin and chymotrypsin in the Atlantic salmon, *Salmo salar* L. Fish Physiology and Biochemistry 15:439-446.

Elkhalil, E.A.I., Männer, K., Borriss, R. and Simon, O., 2007. *In vitro* and *in vivo* characteristics of bacterial phytases and their efficacy in broiler chickens. British Poultry Science 48:64-70.

Ellestad, L.E., Angel, R. and Soares Jr., J.H., 2002a. Intestinal phytase I: detection of preliminary characterization of activity in the intestinal brush border membrane of hybrid striped bass *Morone saxatilis* × *M. chrysops*. Fish Physiology and Biochemistry 26:249-258.

Ellestad, L.E., Angel, R. and Soares Jr., J.H., 2002b. Intestinal phytase II: a comparison of activity and *in vivo* phytate hydrolysis in three teleost species with differing digestive strategies. Fish Physiology and Biochemistry 26:259-273.

Eya, J.C. and Lovell, R.T., 1997. Net absorption of dietary phosphorus from various inorganic sources and effect of fungal phytase on net absorption of plant phosphorus by channel catfish *Ictalurus punctatus*. Journal of the World Aquaculture Society 28(4):386-391.

Food and Agriculture Organisation (FAO), 2016. The state of world fisheries and aquaculture 2016. Contributing to food security and nutrition for all. FAO, Rome, Italy, 200 pp.

Fauconneau, B.G., Choubert, G, Blanc, D., Breque, J. and Luquet, P., 1983. Influence of environmental temperature on flow rate of foodstuffs through the gastrointestinal tract of rainbow trout. Aquaculture 34:27-39.

Forster, I., Higgs, D.A., Dosanjh, B.S., Rowshandeli, M. and Parr, J., 1999. Potential for dietary phytase to improve the nutritive value of canola protein concentrate and decrease phosphorus output in rainbow trout (*Oncorhynchus mykiss*) held in 11 °C fresh water. Aquaculture 179:109-125.

Francis, G., Makkar, H.P.S. and Becker, K., 2001. Antinutritional factors present in plant derived alternate fish feed ingredients and their effects in fish. Aquaculture 199:197-227.

Fredlund, K., Isaksson, M., Rossander-Hulthen, L., Almgren, A. and Sandberg, A.S., 2006. Absorption of zinc and retention of calcium: dose-dependent inhibition by phytate. Journal of Trace Elements in Medicine and Biology 20:49-57.

Gatlin III, D.M., Barrows, F.T., Bellis, D., Brown, P., Campen, J., Dabrowski, K., Gaylord, T.G., Hardy, R.W., Herman, E., Hu, G., Krogdahl, Å., Nelson, R., Overturf, K., Rust, M., Sealey, W., Skonberg, D., Souza, E., Stone, D., Wilson, R. and Wurtele, E., 2007. Expanding the utilization of sustainable plant products in aquafeeds – a review. Aquaculture Research 38:551-579.

Goda, A.M.A.S., 2007. Effect of dietary soybean meal and phytase levels on growth, feed utilization and phosphorus discharge for Nile tilapia (*Oreochromis niloticus* L.). Journal of Fisheries and Aquatic Science 2:248-263.

Greiner, R., Konietzny, U. and Jany, K.D., 1993. Purification and characterization of two phytases from *Escherichia coli*. Archives of Biochemistry and Biophysics 303:107-113.

Greiner, R., Haller, E., Konietzny, U. and Jany, K.D., 1997. Purification and characterization of a phytase from Klebsiella terrigena. Archives of Biochemistry and Biophysics 341:201-206.

Grynspan, F. and Cheryan, M., 1983. Calcium phyate: effect of pH and molar ratio on *in vitro* solubility. Journal of the American Oil Chemists' Society 60:1761-1764.

Hughes, K.P. and Soares, J.H., 1998. Efficacy of phytase on phosphorus utilization in practical diets fed to striped bass *Morone saxatilis*. Aquaculture Nutrition 4:133-140.

Igbasan, F.A., Männer, K., Miksch, G., Borriss, R., Farouk, A. and Simon, O., 2000. Comparative studies on the *in vitro* properties of phytases from various microbial origins. Archives of Animal Nutrition 53:353-373.

Inagawa, J., Kiyosawa, I. and Nagasawa, T., 1987. Effects of phytic acid on the digestion of casein and soybean protein with trypsin, pancreatin and pepsin. Nippon Eiyo Shokuryo Gakkaishi 40:367-373.

IndexMundi, 2015. Commodity price index. Available at: http://www.indexmundi.com.

IUPAC-IUB, 1977. Nomenclature of phosphorus containing compounds of biochemical importance. European Journal of Biochemistry 79:1-9.

Jackson, L.S., Li, M.H. and Robinson, E.H., 1996. Use of microbial phytase in channel catfish *Ictalurus punctatus* diets to improve utilization of phytate phosphorus. Journal of World Aquaculture Society 27:309-313.

Ji, H., 1999. Anti-nutritional factors in plant based fish feed. Fish Reserve 19:22-24.

Kanaya, K., Yasumoto, K. and Mitsuda, H., 1976. Pepsin inhibition by phytate contained in rice bran. Japan Society of Nutrition and Food Science 29:341-346.

Kerovuo, J., Lauraeus, M., Nurminem, P., Kalkkinen, N. and Apajalahti, J., 1998. Isolation, characterization, molecular gene cloning, and sequencing of a novel phytase from *Bacillus* subtilis. Applied and Environmental Microbiology 64:2079-2085.

Khajepour, F. and Hosseini, S.A., 2010. Mineral status of juvenile Beluga (*Huso huso*) fed citric acid supplemented diets. World Applied Sciences Journal 11(6):682-686.

Khattak, F.M., Pasha, T.N., Hayat, Z. and Mahmud, A., 2006. Enzymes in poultry nutrition. Journal of Animal and Plant Sciences 16:1-7.

Kies, A.K., De Jonge, L.H., Kemme, P.A. and Jongbloed, A.W., 2006. Interaction between protein, phytate, and microbial phytase *in vitro* studies. Journal of Agriculture and Food Chemistry 54:1753-1758.

Knuckles, B.E., Kuzmicky, D.D., Gumbmann, M.R. and Betschart, A.A., 1989. Effect of *myo*-inositol phosphate esters on *in vitro* and *in vivo* digestion of protein. Journal of Food Science 54:1348-1350.

Konishi, C., Matsui, T., Park, H., Yano, H. and Yano, F., 1999. Heat treatment of soyabean and rapeseed meals suppress rumen degradation of phytate phosphorus in sheep. Animal Feed Science and Technology 80:115-122.

Krogdahl, A., Hemre, G.I. and Mommsen, T.P., 2005. Carbohydrates in fish nutrition: digestion and absorption in postlarval stages. Aquaculture Nutrition 11:103-122.

Krogdahl, A., Sundby, A. and Holm, H., 2015. Characteristics of digestive processes in Atlantic salmon (*Salmo salar*). Enzyme pH optima, chyme pH, and enzyme activities. Aquaculture 449:27-36.

Kumar, V., Miasnikov, A., Sands, J.S. and Simmins, P.H., 2003. *In vitro* activities of three phytases under different pH and protease challenges. In: Paterson, J.E. (ed.) Proceedings of the Australian Pig Science Association, November 23-26, 2003. Werribee, Australia, pp. 164.

Kumar, V., Sinha, A.K., Makkar, H.P.S., De Boeck, D. and Becker, K., 2012. Phytate and phytase in fish nutrition. Journal of Animal Physiology and Animal Nutrition 96:335-364.

Laining, A., Ishikawa, M., Kyaw, K., Gao, J., Binh, N.T. and Koshio, S., 2011. Dietary calcium/phosphorus ratio influences the efficacy of microbial phytase on growth, mineral digestibility and vertebral mineralization in juvenile tiger puffer, *Takifugu rubripes*. Aquaculture Nutrition 17:267-277.

Lanari, D., D'Agaro, E. and Turri, C., 1998. Use of nonlinear regression to evaluate the effects of phytase enzyme treatment of plant protein diets for rainbow trout, *Oncorhynchus mykiss*. Aquaculture 161:345-356.

Lawlor, P.G., Lynch, P.B., Caffrey, P.J., O'Reilly, J.J. and O'Connel, M.K., 2005. Measurements of the acid binding capacity of ingredients used in pig diets. Irish Veterinary Journal 58:447-452.

Lemos, D. and Tacon, A.G.J., in press. Use of phytases in fish and shrimp feeds: a review. Reviews in Aquaculture. DOI: http://doi.org/10.1111/raq.12138.

Li, M.H. and Robinson, E.H., 1997. Microbial phytase can replace inorganic phosphorus supplements in channel catfish *Ictalurus punctatus* diets. Journal of the World Aquaculture Society 28(4):402-406.

Liebert, F., Portz, L., 2005. Nutrient utilization of Nile tilapia *Oreochromis niloticus* fed plant based low phosphorus diets supplemented with graded levels of different sources of microbial phytase. Aquaculture 248:111-119.

Liu, L.W., Su, J. and Luo, Y., 2012. Effect of partial replacement of dietary monocalcium phosphate with neutral phytase on growth performance and phosphorus digestibility in gibel carp, *Carassius auratus gibelio* (Bloch). Aquaculture Research 43:1404-1413.

Liu, L.W., Luo, Y.W., Hou, H.L., Pan, J. and Zhang, W., 2013. Partial replacement of monocalcium phosphate with neutral phytase in diets for grass carp, *Ctenopharyngodon idellus*. Journal Applied Ichthyology 29:520-525.

Liu, L., Zhou, Y., Wu, J., Zhang, W., Abbas, K., Fang, L.X. and Luo Y., 2014. Supplemental graded levels of neutral phytase using pretreatment and spraying methods in the diet of grass carp, *Ctenopharyngodon idellus*. Aquaculture Research 45:1932-1941.

Loewus, F., 2002. Biosynthesis of phytate in food grains and seeds. In: Reddy, N.R. and Sathe, S.K. (eds.) Food phytates. CRC Press, Boca Raton, FL, USA, pp. 53-61.

Luttrell, B.M., 1993. The biological relevance of the binding of calcium ions by inositol phosphates. Journal of Biological Chemistry 268:1521-1524.

Maenz, D.D., 2001. Enzymatic characteristics of phytases as they relate to their use in animal feeds. In: Bedford, M.R. and Partridge, G.G. (eds.) Enzymes in farm animal nutrition. CABI Publishing, New York, NY, USA, pp. 61-84.

Magrane, M. and UniProt Consortium, 2011. UniProt knowledgebase: a hub of integrated protein data. Database 11.

Mainston, C.P. and Parr, W., 2002. Phosphorus in rivers – ecology and management. Science of the Total Environment 282-283:25-47.

Mallison, A., 2013. Marine ingredients overview. Intrafish investment forum. London, UK. Available at: http://www.intrafishevents.com.

Marquez, L., Robles, R., Morales, G.A. and Moyano, F.J., 2012. Gut pH as a limiting factor for digestive proteolysis in cultured juveniles of the gilthead sea bream (*Sparus aurata*). Fish Physiology and Biochemistry 38:859-869.

Masumoto, T., Tamura, B. and Shimeno, S., 2001. Effects of phytase on bioavailability of phosphorus in soybean meal-based diets for Japanese flounder *Paralichthys olivaceus*. Fisheries Science 67:1075-1080.

Menezes-Blackburn, D., Gabler, S. and Greiner, R., 2015. Performance of seven commercial phytases in an *in vitro* simulation of poultry digestive tract. Journal of Agricultural and Food Chemistry 63(27):6142-6149.

Morales, G.A., Moyano, F.J. and Marquez, L., 2011. *In vitro* assessment of the effects of phytate and phytase on nitrogen and phosphorus bioaccessibility within fish digestive tract. Animal Feed Science and Technology 170:209-221.

Morales, G.A., Saenz de Rodrigañez, M.A., Marquez, L., Diaz, M. and Moyano, F.J., 2013. Solubilisation of protein fractions induced by *Escherichia coli* phytase and its effects on *in vitro* fish digestion of plant proteins. Animal Feed Science and Technology 181:54-64.

Morales, G.A., Marquez, L., Saenz de Rodrigañez, M.A., Bermúdez, L., Robles, R. and Moyano, F.J., 2014. Effect of phytase supplementation of a plant-based diet on phosphorus and nitrogen bioavailability in sea bream *Sparus aurata*. Aquaculture Nutrition 20:172-182.

Morales, G.A., Denstadli, V., Collins, S.A., Mydland, L.T., Moyano, F.J. and Øverland, M., 2015. Phytase and sodium diformate supplementation in a plant-based diet improves protein and minerals utilization in rainbow trout (*Oncorhynchus mykiss*). Aquaculture Nutrition. DOI: http://dx.doi.org/10.1111/anu.12340.

Mroz, Z., Krasuki, W. and Grela, E., 1995. Physiological responses of lactating sows to feeding rapeseed '00' and microbial phytase. In: Hennessy, D.P. and Cranwell, P.D. (eds.) Manipulating pig production. Vol. 5. Australasian Pig Science Association, Werribee, Australia, pp. 185.

Mroz, Z., Jonbloed, A.W., Partanen, K.H., Vreman, K., Kemme, P.A. and Kogut, J., 2000. The effects of calcium benzoate in diets with or without organic acids on dietary buffering capacity, apparent digestibility, retention of nutrients, and manure characteristics in swine. Journal of Animal Sciences 78:2622-2632.

Naylor, R.L., Hardy, R.W., Bureau, D.P., Chiu, A., Elliott, M., Farrell, A.P., Forster, I., Gatlin, D.M., Goldburg, R.J., Hua, K. and Nichols, P.D., 2009. Feeding aquaculture in an era of finite resources. Proceedings of the National Academy of Sciences 106:15103-15110.

Nelson, T.S., Shieh, T.R., Wodzinski, R.J. and Ware, J.H., 1971. Effect of supplemental phytase on the utilization of phytate phosphorus by chicks. Journal of Nutrition 101:1289-1293.

Nordrum, S., Krogdahl, Å., Røsjø, C., Olli, J.J. and Holm, H., 2000. Effects of methionine, cysteine and medium chain triglycerides on nutrient digestibility, absorption of amino acids along the intestinal tract and nutrient retention in Atlantic salmon (*Salmo salar* L.) under a pair-feeding regime. Aquaculture 186:341-360.

Nwanna, L.C., 2007. Effect of dietary phytase on growth, enzyme activities and phosphorus load of Nile tilapia (*Oreochromis niloticus*). Journal of Engineering and Applied Sciences 2:972-976.

Nwanna, L.C. and Schwarz, F.J., 2007. Effect of supplemental phytase on growth, phosphorus digestibility and bone mineralization of common carp (*Cyprinus carpio* L). Aquaculture Research 38:1037-1044.

Nwanna, L.C., Kolahsa, M., Eisenreich, R. and Schwarz, F.J., 2008. Pre-treatment of dietary plant feedstuffs with phytase and its effect on growth and mineral concentration in common carp (*Cyprinus carpio* L.). Journal of Animal Physiology and Animal Nutrition 92:677-682.

Okubo, K., Iacobucci, G.A. and Myers, D.V., 1974. Effect of Ca^{2+} on phytate binding to glycinin. Cereal Science Today 19:401.

Okubo, K., Myers, D.V. and Iacobucci, G.A., 1976. Binding of phytic acid to glycinin. Cereal Chemistry 53:513-524.

Oliva-Teles, A., Pereira, J.P., Gouveia, A. and Gomes, E., 1998. Utilisation of diets supplemented with microbial phytase by seabass (*Dicentrarchus labrax*) juveniles. Aquatic Living Resources 11(4):255-259.

Øverland, M., Granli, T., Kjos, N.P., Fjetland, O., Steien, H. and Stokstad, M., 2000. Effect of dietary formates on growth performance, carcass traits, sensory quality, intestinal microflora, and stomach alterations in growing-finishing pigs. Journal of Animal Science 78:1875-1884.

Papatryphon, E., Howell, R.A. and Soares, J.H.J., 1999. Growth and mineral absorption by striped bass *Morone saxatilis* fed a plant feedstuff based diet supplemented with phytase. Journal of the World Aquaculture Society 30:161-173.

Papatryphon, E. and Soares, J.H., 2001. The effect of phytase on apparent digestibility of four practical plant feedstuffs fed to striped bass, *Morone saxatilis*. Aquaculture Nutrition 7:161-167.

Partanen, K.H. and Mroz, Z., 1999. Organic acids for performance enhancement in pig diets. Nutrition Research Reviews 12:117-145.

Persson, H., Turk, M., Nyman, M. and Sandberg, A.S., 1998. Binding of Cu^{2+}, Zn^{2+} and Cd^{2+} to inositol tri-, tetra-, penta- and hexa-phosphates. Journal of Agriculture and Food Chemistry 46:3194-3200.

Pontoppidan, K., Pettersson, D. and Sandberg, A.S., 2007. *Peniophora lycii* phytase is stable and degrades phytate and solubilises minerals *in vitro* during simulation of gastrointestinal digestion in the pig. Journal of the Science of Food and Agriculture 87:2700-2708.

Portz, L. and Liebert, F., 2004. Growth, nutrient utilization and parameters of mineral metabolism in Nile tilapia *Oreochromis niloticus* (Linnaeus, 1758) fed plant-based diets with graded levels of microbial phytase. Journal Animal Physiology and Animal Nutrition 88:311-320.

Powers-Hughes, K. and Soares, J.H., 1998. Efficacy of phytase on phosphorus utilization in practical diets fed to striped bass *Morone saxatilis*. Aquaculture Nutrition 4:133-140.

Ravindran, V., 1995. Phytases in poultry nutrition: an overview. Proceedings of the Australian Poultry Science Symposium 7:135-139.

Reddy, N.R., Sathe, S.K. and Pierson, M.D., 1988. Removal of phytate from great northern beans (*Phaseolus vulgaris* L.). Journal of Food Science 53:107-147.

Richardson, A.E. and Hadobas, P.A., 1997. Soil isolates of *Pseudomonas* spp. that utilize inositol phosphates. Canadian Journal of Microbiology 43:509-516.

Riche, M. and Brown, P.B., 1996. Availability of phosphorus from feedstuffs fed to rainbow trout, *Oncorhynchus mykiss*. Aquaculture 142:269-282.

Riche, M., Trottier, N.L., Ku, P.K. and Garling, D.L., 2001. Apparent digestibility of crude protein and apparent availability of individual amino acids in tilapia (*Oreochromis niloticus*) fed phytase pretreated soybean meal diets. Fish Physiology and Biochemistry 25:181-194.

Rockström, J., Steffen, W., Noone, K., Persson, A., Chapin, F.S., Lambin, E.F., Lenton, T.M., Scheffer, M., Folke, C., Schellnhuber, H.J., Nykvist, B., De Wit, C.A., Hughes, T., Van der Leeuw, S., Rodhe, H., Sörlin, S., Snyder, P.K., Costanza, R., Svedin, U., Falkenmark, M., Karlberg, L., Corell, R.W., Fabry, V.J. Hansen, J., Walker, B., Liverman, D., Richardson, K., Crutzen, P. and Foley, J.A., 2009. A safe operating space for humanity. Nature 461:472-475.

Rodehutscord, M. and Pfeffer, E., 1995. Effects of supplemental microbial phytase on phosphorus digestibility and utilization in rainbow trout, *Oncorchynchus mykiss*. Water Science and Technology 31:143-147.

Rodehutscord, M., 1996. Response of rainbow trout (*Oncorhynchus mykiss*) growing from 50 to 200 g to supplements of dibasic sodium phosphate in a semipurified diet. Journal of Nutrition 126:324-331.

Roy, T., Banerjee, G., Dan, S.K., Ghosh, P. and Ray, A.K., 2014. Improvement of nutritive value of sesame oilseed meal in formulated diets for rohu, *Labeo rohita* (Hamilton), fingerlings after fermentation with two phytase producing bacterial strains isolated from fish gut. Aquaculture International 22:633-652.

Sajjadi, M. and Carter, C.G., 2004a. Dietary phytase supplementation and the utilization of phosphorus by Atlantic salmon (*Salmo salar* L.) fed a canola-meal-based diet. Aquaculture 240:417-431.

Sajjadi, M. and Carter, C.G., 2004b. Effect of phytic acid and phytase on feed intake, growth, digestibility and trypsin activity in Atlantic salmon (*Salmo salar*, L.). Aquaculture Nutrition 10:135-142.

Schaefer, A. and Koppe, W.M., 1995. Effect of a microbial phytase on utilization of native phosphorus by carp in a diet based on soybean meal. Water Science and Technology 31:149-155.

Schäfer, A., Koppe, W.M., Meyer-Burgdorff, K.H. and Günter, K.D., 1995. Effects of a microbial phytase on the utilization of native phosphorus by carp in a diet based on soybean meal. Water Science and Technology 31:149-155.

Schlemmer, U., Jany, K.D., Berk, A., Schulz, E. and Rechkemmer, G., 2001. Degradation of phytate in the gut of pigs – pathway of gastrointestinal inositol phosphate hydrolysis and enzymes involved. Archives of Animal Nutrition 55:255-280.

Selle, P.H. and Ravindran, V., 2007. Microbial phytase in poultry nutrition. Animal Feed Science and Technology 135:1-41.

Selle, P.H., Ravindran, V., Caldwell, R.A. and Bryden, W.L., 2000. Phytate and phytase: consequences for protein utilization. Nutrition Research Reviews 13:255-278.

Selle, P.H., Ravindran, V., Bryden, W.L. and Scott, T., 2006. Influence of dietary phytate and exogenous phytase on amino acid digestibility in poultry: a review. Journal of Poultry Science 43:89-103.

Selle, P.H., Cowieson, A.J., Cowieson, N.P. and Ravindran, V., 2012. Protein-phytate interactions in pig and poultry nutrition: a reappraisal. Nutrition Research Reviews 6:1-17.

Shah, S.Z.H., Afzal, M., Akmal, A., Fatima, M. and Hussain, S.M., 2015. Effect of citric acid and phytase on growth performance and mineralization of *Labeo rohita* juveniles fed soybean meal based diet. International Journal of Agriculture and Biology 18:111-116.

Shirley, R.B. and Edwards Jr., H.M., 2003. Graded levels of phytase past industry standards improves broiler performance. Poultry Science 82:671-680.

Signor, A.A., Boscolo, W.R., Bittencourt, F., Feiden, A., Goncalves, G.S. and Azambuja de Freitas, J.M., 2010. Performance of juvenile Nile tilapia fed diets with enzymatic complex. Revista Brasileira de Zootecnia 39:977-983.

Singh, M. and Krikorian, A.D., 1982. Inhibition of trypsin activity by phytate. Journal of Agricultural and Food Chemistry 30:799-800.

Smit, A.L., Van Middelkoop, J.C., Van Dijk, W. and Van Reuler, H., 2015. A substance flow analysis of phosphorus in the food production, processing and consumption system of the Netherlands. Nutrient Cycling in Agroecosystems 103:1-13.

Storebakken, T. and Austreng, E., 1987. Binders in fish feeds: II. Effect of different alginates on digestibility of macronutrients in rainbow trout. Aquaculture 60:121-131.

Storebakken, T., Shearer, K.D. and Roem, A.J., 1998. Availability of protein, phosphorus and other elements in fish meal, soy-protein concentrate and phytase-treated soy-protein-concentrate-based diets to Atlantic salmon, *Salmo salar*. Aquaculture 161:365-379.

Storebakken, T., Kvien, I.S., Shearer, K.D., Grisdale-Helland, B. and Helland, S.J., 1999. Estimation of gastrointestinal evacuation rate in Atlantic salmon (*Salmo salar*) using inert markers and collection of faeces by sieving: evacuation of diets with fish meal, soybean meal or bacterial meal. Aquaculture 172:291-299.

Storebakken, T., Shearer, K.D. and Roem, A.J., 2000. Growth, uptake and retention of nitrogen and phosphorus, and absorption of other minerals in Atlantic salmon *Salmo salar* fed diets with fish meal and soy-protein concentrate as the main sources of protein. Aquaculture Nutrition 6:103-108.

Sugiura, S., Babbitt, J., Dong, F. and Hardy, R., 2000. Utilization of fish and animal by-product meals in low-pollution feeds for rainbow trout (*Oncorhynchus mykiss*) (Walbaum). Aquaculture Research 31:585-593.

Sugiura, S.H., Gabaudan, J., Dong, F.M. and Hardy, R.W., 2001. Dietary microbial phytase supplementation and the utilization of phosphorus, trace minerals and protein by rainbow trout *Oncorhynchus mykiss* (Walbaum) fed soybean meal-based diets. Aquaculture Research 32:583-592.

Sugiura, S.H., Roy, P.K. and Ferraris, R.P., 2006. Dietary acidification enhances phosphorus digestibility but decreases H^+/K^+-ATPase expression in rainbow trout. Journal of Experimental Biology 209:3719-3728.

Tacon, A.G.J. and Metian, M., 2015. Feed matters: satisfying the feed demand of aquaculture. Reviews in Fisheries Science and Aquaculture 23:1-10.

Tyrrell, T. 1999. The relative influences of nitrogen and phosphorus on oceanic primary product. Nature 400:525-531.

Vaintraub, I.A. and Bulmaga, V.P., 1991. Effect of phytate on the *in vitro* activity of digestive proteinases. Journal of Agricultural and Food Chemistry 39:859-861.

Vandenberg, G.W., Scott, S.L., Sarker, P.K., Dallaire, V. and Noüe, J., 2011. Encapsulation of microbial phytase: effects on phosphorus bioavailability in rainbow trout (*Oncorhynchus mykiss*). Animal Feed Science and Technology 169:230-243.

Vandenberg, G.W., Scott, S.L. and Noüe, J., 2012. Factors affecting nutrient digestibility in rainbow trout (*Oncorhynchus mykiss*) fed a plant protein-based diet supplemented with microbial phytase. Aquaculture Nutrition 18:369-379.

VanWeerd, J.H., Khalaf, K.H., Aartsen, E.J. and Tijssen, P.A., 1999. Balance trials with African catfish *Clarias gariepinus* fed phytase-treated soybean meal-based diets. Aquaculture Nutrition 5:135-142.

Vats, P., Bhattacharyya, M.S. and Banerjee, U.C., 2005. Use of phytases (*myo*-inositolhexakisphosphate phosphohy-drolases) for combatting environmental pollution: a biological approach. Critical Reviews in Environmental Science and Technology 35(5):469-486.

Verdegem, M.C.J., 2013. Nutrient discharge from aquaculture operations in function of system design and production environment. Reviews in Aquaculture 4:1-14.

Verlhac-Trichet, V., Vielma, J., Dias, J., Rema, P., Santigosa, E., Wahli, T. and Vogel, K., 2014. The efficacy of a novel microbial 6-phytase expressed in *Aspergillus oryzae* on the performance and phosphorus utilization of cold- and warm-water fish: rainbow trout, *Oncorhynchus mykiss*, and Nile tilapia, *Oreochromis niloticus*. Journal of the World Aquaculture Society 45(4):367-379.

Vielma, J., Lall, S.P. and Koskela, J., 1998. Effects of dietary phytase and cholecalciferol on phosphorus bioavailability in rainbow trout (*Oncorhynchus mykiss*). Aquaculture 163:309-323.

Vielma, J., Ruohonen, K. and Peisker, M., 2002. Dephytinization of two soy proteins increases phosphorus and protein utilization by rainbow trout, *Oncorhynchus mykiss*. Aquaculture 204:145-156

Vielma, J., Ruohonen, K., Gabaudan, J. and Vogel, K., 2004. Top-spraying soybean meal-based diets with phytase improves protein and mineral digestibilities but not lysine utilization in rainbow trout, *Oncorhynchus mykiss* (Walbaum). Aquaculture Research 35:955-964.

Walk, C.L., Bedford, M.R., Santos, T.S., Paiva, D., Bradley, J.R., Wladecki, H., Honaker, C. and McElroy, A.P., 2013. Extra-phosphoric effects of superdoses of a novel microbial phytase. Poultry Science 92:719-725.

Wang, F., Yang, Y.H., Han, Z.Z., Dong, H.W., Yang, C.H. and Zou, Z.Y., 2009. Effects of phytase pretreatment of soybean meal and phytase-sprayed in diets on growth, apparent digestibility coefficient and nutrient excretion of rainbow trout (*Oncorhynchus mykiss* Walbaum). Aquaculture International 17:143-157.

Wise, A., 1983. Dietary factors determining the biological activity of phytates. Nutrition Abstracts and Reviews in Clinical Nutrition 53:791-806.

Yan, W., Reigh, R.C. and Xu, Z., 2002. Effects of fungal phytase on utilization of dietary protein and minerals, and dephosphorylation of phytic acid in the alimentary tract of channel catfish *Ictalurus punctarus* fed an all-plant-protein diet. Journal of the World Aquaculture Society 33:10-22.

Yang, Y.H., Wang, Y.Y., Lu, Y. and Li, Q.Z., 2011. Effect of replacing fish meal with soybean meal on growth, feed utilization and nitrogen and phosphorus excretion on rainbow trout (*Oncorhynchus mykiss*). Aquaculture International 19:405-419.

Yoo, G.Y., Wang, X., Choi, S., Hana, K., Kang, J.C. and Bai, S.C., 2005. Dietary microbial phytase increased the phosphorus digestibility in juvenile Korean rockfish *Sebastes schlegeli* fed diets containing soybean meal. Aquaculture 243:315-322.

Yu, S., Cowieson, A., Gilbert, C., Plumstead, P. and Dalsgaard, S., 2012. Interactions of phytate and *myo*-inositol phosphate esters (IP_{1-5}) including IP_5 isomers with dietary protein and iron and inhibition of pepsin. Journal of Animal Science 90:1824-1832.

Yufera, M., Fernandez-Diaz, C., Vidaurreta, A., Cara, J.B. and Moyano, F.J., 2004. Gastrointestinal pH and development of the acid digestion in larvae and early juveniles of *Sparus aurata* (Pisces: Teleostei). Marine Biology 144:863-869.

Yufera, M. and Darias, M.J., 2007. Changes in the gastrointestinal pH from larvae to adult in Senegal sole (*Solea senegalensis*). Aquaculture 267:94-99.

Zhang, Y., Furyk, S., Bergbreiter, D.E. and Cremer, P.S., 2005. Specific ion effects on the water solubility of macromolecules: PNIPAM and the Hofmeister series. Journal of the American Chemical Society 127:1405-1410.

Zhao, F., Moore, K.L. Lombi, E. and Zhu, Y., 2014. Imaging element distribution and speciation in plant cells. Trends in Plant Science 19:183-192.

Zhu, Y., Qiu, X., Ding, Q., Duan, M. and Wang, C., 2014. Combined effects of dietary phytase and organic acid on growth and phosphorus utilization of juvenile yellow catfish *Pelteobagrus fulvidraco*. Aquaculture 430:1-8.

Chapter 10
Interactions between minerals and phytate degradation in poultry – challenges for phosphorus digestibility assays

M. Rodehutscord
Institute of Animal Science, University of Hohenheim, Emil-Wolff-Str. 10, 70599 Stuttgart, Germany; inst450@uni-hohenheim.de

Abstract

Determination of precaecal digestibility of phosphorus (pcdP) has been put forward as the approach of choice to characterise the availability of phosphorus (P) from feed raw materials and mixed diets in poultry. Mixed diets usually contain P sources of plant and mineral or animal origin. In plant feed raw materials, pcdP is closely related to gastro-intestinal degradation of phytate (any salt of phytic acid (*myo*-inositol 1,2,3,4,5,6-hexakis dihydrogen phosphate; $InsP_6$)). The broiler has the potential to hydrolyse two third of dietary $InsP_6$ in low-calcium (Ca) P diets devoid of any phytase. However, supplements of Ca or P or both strongly reduce gastro-intestinal $InsP_6$ hydrolysis and thus digestibility of $InsP_6$-P. These effects can be compensated by concurrent supplementation of phytase, with compensatory effects being higher at dosages exceeding the industry standard level of supplementation. The standard protocol for the determination of pcdP of the World's Poultry Science Association contains limitations in regard to dietary P and Ca levels, but it does not consider interactions with $InsP_6$ hydrolysis. Thus, pcdP values of non-plant P sources are underestimated if determined in diets that contain $InsP_6$ and not corrected for $InsP_6$ disappearance. Likewise, pcdP values of plant P sources are overestimated in diets that contain non-plant P sources. It is suggested that chemical analysis in pcdP studies is extended by $InsP_6$ determination.

Keywords: inositol phosphates, calcium, phytase, evaluation

10.1 Introduction

All living organisms depend upon a continuous supply of phosphates to be absorbed from the intestine. Phosphorus (P) is bound in different forms in feed raw materials for poultry, this being a main reason for differences to exist in P availability. It is widely accepted that sustainable usage of P in the poultry industry means aiming at a precise-as-possible P supply to the bird, considering the bird's requirement, attempting to minimise P excretion and cost of feeding, and to save the scarce and finite rock phosphates stores we have on earth. Thus, characterisation of P availability and influencing factors is a key towards achieving this sustainability.

C.L. Walk *et al.* (ed.) **Phytate destruction – consequences for precision animal nutrition**
DOI 10.3920/978-90-8686-836-0_10, © Wageningen Academic Publishers 2016

It has been reviewed elsewhere that during the last decades different approaches have been developed and applied to characterise P availability in poultry, and that the plethora of approaches is disadvantageous for establishment of an evaluation system for industry application (Shastak and Rodehutscord, 2013). Aiming for global harmonisation, a working group of the World's Poultry Science Association has suggested a standard trial protocol for the determination of P availability based on the precaecal digestibility of P (pcdP) (WPSA, 2013). This protocol implies that graded dietary levels of a P source are used, and pcdP of the P source or P equivalency of an enzyme is calculated by regression analysis. Hence, unless semi-purified or purified diets are used, experimentation means using combinations of different sources (e.g. plant and non-plant) and levels of P in the diet. The approach assumes that, as long as threshold levels of calcium (Ca) and P in the diet are obeyed, pcdP values for different P sources are additive, meaning interactions between P and Ca sources not be relevant in a mixed diet. This assumption is challenged by recent studies reporting interactions to exist especially in regard to plant P.

This contribution will report the current knowledge about interactions and come up with ideas how they could be considered in an updated standard trial protocol.

10.2 Disappearance of inositol phosphates in the digestive tract of poultry

In plant seeds and feedstuffs obtained from the seeds, such as oilseed meals, most of the P is present as phytate-P (any salt of phytic acid (*myo*-inositol 1,2,3,4,5,6-hexakis dihydrogen phosphate; $InsP_6$)). In contrast to some textbook knowledge and recommendations made by past committees (GfE, 1999; NRC, 1994), recent broiler studies have found that more than two third of $InsP_6$ from maize-soybean meal-based diets devoid of phytase disappeared until the end of the ileum. This has been extensively reviewed by Rodehutscord and Rosenfelder in Chapter 1 (Rodehutscord and Rosenfelder, 2016). When discussing this high potential of the broiler chicken to hydrolyse $InsP_6$ it must be kept in mind that the underlying studies were conducted with diets devoid of any mineral P supplement and at a low to moderate level of Ca. While this is a condition that does not currently reflect a practical diet, economic and ecological constraints trigger development towards P and Ca reduction, especially in growing and finishing diets. However, the most common situation is that diets contain a variable mix of $InsP_6$, mineral P and/or Ca and often an exogenous phytase. Thus, it is of interest to know about potential effects that supplements can have on $InsP_6$ degradation in the digestive tract of poultry.

10.3 Relationship between precaecal $InsP_6$ disappearance and P digestibility

Gastro-intestinal $InsP_6$ hydrolysis can affect the metabolic P supply of the bird only when the released phosphate is absorbed. Recent studies conducted in our institute measured both, $InsP_6$ disappearance and P digestibility in the same birds. A compilation of data from the different experiments showed variability in values of $InsP_6$ disappearance and P digestibility up to the lower half of ileum, thus allowing for further calculation (Figure 10.1). Linear regression analysis applied to the data yielded an estimate for the slope of 0.78; thus, with each increment in $InsP_6$-P disappearance by 1 g, the amount of P digested increased by 0.78 g. This is a relatively rough estimate of what could be referred to as '$InsP_6$-P digestibility,' associated with a substantial variation as indicated by the standard error (SE) of the estimate (0.21), and it indicates several experimental variables contributing to variation. However, the slope of 0.78 is a good indication that a very high proportion of $InsP_6$-P can be absorbed by the animal. The 'non-absorbed' proportion (on average 22%) likewise is related to incomplete release of P from $InsP_6$ rather than incomplete absorption of released P. Phosphorus contained in lower inositol phosphates in the terminal ileum are assumed not absorbed by the bird, but is included in the measured $InsP_6$ disappearance. Isomers of $InsP_5$, $InsP_4$ and $InsP_3$ were found to exist in the lower ileum regardless of whether enzymes were supplemented to the diet or not (Zeller *et al.*, 2015a,b,c). The sum of $InsP_5$, $InsP_4$ and $InsP_3$ accounted for 9 to 37% of total InsP detected in the ileum, with an average proportion of 17% across all treatments of these experiments (Table 10.1).

10.4 Dietary variables affecting inositol phosphate disappearance

10.4.1 Supplements of P and Ca

$InsP_6$ hydrolysis in the digestive tract is subject to phytate solubility. Dietary factors influencing phytate solubility may therefore have an effect on $InsP_6$ disappearance. It has been shown that the level of intestinal $InsP_6$ disappearance in broiler chickens strongly depends upon the Ca concentration of the diet. The level of pc $InsP_6$ disappearance

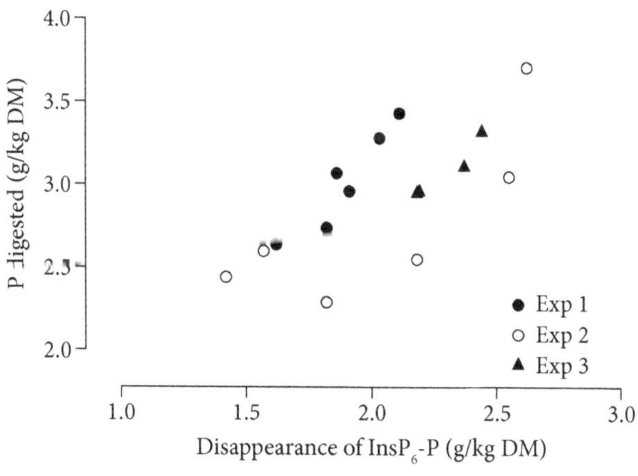

Figure 10.1. *Comparison of $InsP_6$-P disappearance and P digestibility until the end of ileum of broiler chickens in three experiments (Zeller* et al., *2015a,b,c). Each data point represents a treatment mean. Diets were based on maize or wheat; some were supplemented with phytase and some were not. A linear regression through all data had a slope of 0.78 (SE 0.21).*

Table 10.1. *Sum of* $InsP_3$*,* $InsP_4$ *and* $InsP_5$ *detected in digesta from the terminal ileum of broiler chickens.*

Diet type	$InsP_3 + InsP_4 + InsP_5$		Reference
	nmol/kg DM	Proportion of total InsP (%)[1]	
Maize-SBM[2] without Phy	2,289	15	Zeller *et al.* (2015c)
Maize-SBM + 500 FTU/kg	2,629	20	
Maize-SBM + 12,500 FTU/kg	1,299	26	
Maize-SBM + MCP without Phy	3,851	14	
Maize-SBM + MCP + 500 FTU/kg	7,183	27	
Maize-SBM + MCP + 12,500 FTU/kg	2,166	37	
Maize-SBM without Phy	1,193	9	Zeller *et al.* (2015b)
Maize-SBM + Phy A	1,605	12	
Maize-SBM + Phy E1	1,547	13	
Maize-SBM + Phy E2	1,444	14	
UT wheat without enzyme	2,228	17	Zeller *et al.* (2015a)
UT wheat + Phy	1,204	15	
UT wheat + Phy + Xyl	605	11	
MT wheat without enzyme	1,787	19	
MT wheat + Phy	569	11	
MT wheat + Phy + Xyl	646	10	

[1] Total InsP: $InsP_3 + InsP_4 + InsP_5 + InsP_6$.
[2] SBM: soybean meal, Phy: phytase, with A –*Aspergillus;* E1 and E2 modified, *Escherichia coli* originated, MCP: monocalcium phosphate, MT: microwave-treated, UT: untreated, Xyl: xylanase.

decreased from about 70 to about 20% with increasing Ca from 0.2 to 0.7% (Tamim and Angel, 2003; Tamim *et al.*, 2004). Only very recently it has been reported that pcdP of maize was increased from 25 to 57% when dietary Ca was reduced from 0.95 to 0.13% (Perryman *et al.*, 2016), which was perhaps caused by Ca effects on $InsP_6$ degradation. In another study, pc $InsP_6$ disappearance was about 25% and not significantly different when 0.7 and 1.0% Ca were used in the diet (Li *et al.*, 2016). This is an indication that Ca does not exhibit negative effects once a certain threshold Ca level is exceeded. When comparing results of the aforementioned studies it should be kept in mind that the diets used by Li *et al.* (2016) included P from meat and bone meal or monocalcium phosphate (MCP), meaning that Ca was included together with mineral P and raising the question whether mineral P can have an effect as well.

Phosphorus addition to the medium in an *in vitro* assay decreased synthesis of phosphatases, including phytase by *Aspergillus ficuum* (Shieh *et al.*, 1969). Such inhibition of enzymes by mineral P may also happen in the digestive tract, with the consequence of reduced $InsP_6$ hydrolysis. When mineral supplements containing both Ca and P were included in the diet, $InsP_6$ disappearance was repeatedly found to be significantly reduced (Figure 10.2). However, a diminishing effect on $InsP_6$ disappearance was also found when

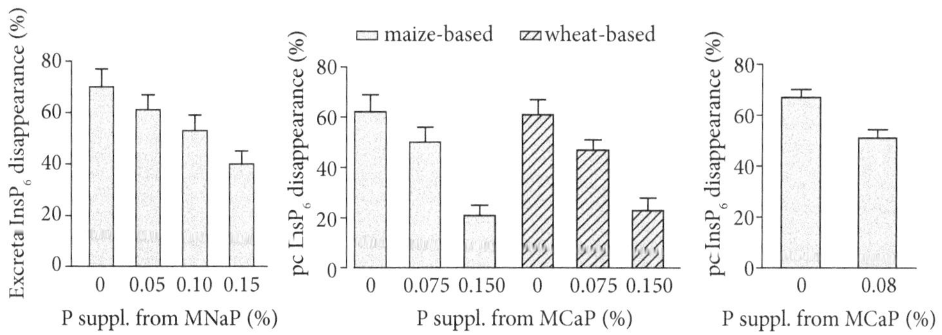

Figure 10.2. *$InsP_6$ disappearance in broiler chickens as affected by supplements of anhydrous monosodium phosphate (MNaP) and monocalcium phosphate monohydrate (MCaP) in maize- and wheat-based diets not containing an exogenous phytase (Shastak* et al.*, 2014; Zeller* et al.*, 2015c).*

P was supplemented without Ca, by monosodium phosphate inclusion. If the end product inhibition shown *in vitro* by Shieh *et al.* (1969) also happens in the digestive tract, then this would partially explain why pc $InsP_6$ breakdown is reduced upon supplementation of a (Ca-free) mineral phosphate. When measured on the excreta level, $InsP_6$ disappearance was about 67% with dietary concentrations of 0.6% Ca and 0.5% P (Delezie *et al.*, 2012). $InsP_6$ disappearance was significantly reduced to 53% upon moderate increase of Ca (0.8%) and P (0.6%) together, but reduction was more pronounced for P than for Ca when they were supplemented separately. This supports the findings reported for $InsP_6$ degradation up to the final two thirds of the ileum as shown in Figure 10.2.

In a very recent experiment it was found that when the diet did not contain phytase, a mineral P supplement reduced pc $InsP_6$ disappearance and that this reduction was even more when P and Ca were supplemented together (Figure 10.3). When diets contained a phytase supplement, a negative P effect was not found, but Ca tended to reduce $InsP_6$ disappearance. Perhaps, the modulating effects of P and Ca are different because phytases of different origin (endogenous mucosal, endogenous microbial, exogenous) or their production are targeted by the minerals in a different way. Taken the high variability of P and Ca levels in practical diets it is recommended that further research projects aim at quantifying the multiple interactions in order to make implementation in practical feed formulation possible.

10.4.2 Supplements of vitamin D_3

Supplementation of 1,25-dihydroxycholecalciferol (1,25-DHCC), alone or in combination with a phytase supplement, increased different criteria of P utilisation in broiler chickens than those previously addressed (Mitchell and Edwards Jr., 1996). The authors suggested that this effect is caused by less phytate precipitation due to an increase in Ca absorption, or by endogenous mucosal phytase being directly increased by the 1,25-DHCC supplement. The authors also mentioned that this effect could be unique to broiler chickens. In accordance with these findings, $InsP_6$ disappearance in another study and measured on

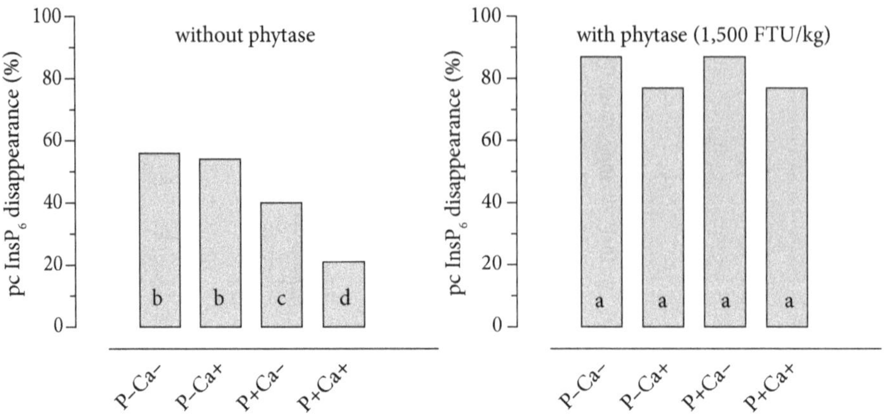

Figure 10.3. *InsP$_6$ disappearance in the terminal ileum of broiler chickens as affected by supplements of monosodium phosphate or limestone or both, to a maize-soybean meal diet. Concentrations of P and Ca were 4.1 (P-), 6.9 (P+), 6.2 (Ca-) and 10.4 (Ca+); the phytase was a modified,* Escherichia coli *derived phytase (Quantum Blue, AB Vista, Marlborough, UK) (Sommerfeld* et al., *2016).*

the excreta level, was slightly but significantly increased when the cholecalciferol level of the diet was increased from 1,250 to 2,500 IU/kg (Delezie *et al.*, 2012).

10.4.3 Feed processing

Feed processing can also have an effect on InsP$_6$ degradation through influencing the accessibility of InsP$_6$ for phytase. An *in vitro* study using wheat showed that grinding increased the amount of dialyzable phosphate (Żyła *et al.*, 1999), perhaps because phytase had better access to InsP$_6$, both located differently in the seed. Phytase from ground wheat can access InsP$_6$ from soybean meal when incubated together (Blaabjerg *et al.*, 2010). Microwave treatment of wheat reduced intrinsic phytase activity and InsP$_6$ disappearance in the crop of broilers, but increased InsP$_6$ disappearance in the terminal ileum (Zeller *et al.*, 2015a, 2016). The authors suggested this to be a consequence of disrupted wheat aleurone structures by microwaving that increased accessibility of InsP$_6$ by endogenous epithelial or microbial phytase in the small intestine. Maize ground through 1-mm screen compared to 3-mm showed higher InsP$_6$ hydrolysis rate upon phytase supplementation *in vitro* (Ton Nu *et al.*, 2014). In line with this observation, phytase supplementation increased pcdP and toe ash in broilers fed a medium but not when fed a coarse particle size maize-based diet (Amerah and Ravindran, 2009).

10.5 Interactions between supplements of P, Ca and phytase

The interactions between InsP$_6$ breakdown and dietary variables, especially P and Ca supplements as described before, carries on to phytase supplements also being viewed in light of interactions. Because using Ca and P supplements in poultry diets is common practice it can be assumed that birds cannot exploit their potential of P utilisation without added phytase.

In line with the high level of pc $InsP_6$ disappearance found in broiler chickens fed low-Ca P diets, the effects of added phytase at standard industry level are not high when such diets are used. However, the effects of phytase supplementation at dosages far exceeding the current standards (12,500 vs 500 FTU/kg of diet) caused a further improvement in pc $InsP_6$ disappearance (78-92%) and pcdP (58-71%), respectively (Zeller *et al.*, 2015c). Also of interest, standard phytase supplementation partly compensated for the inhibitory effects that the mineral P source had on $InsP_6$ disappearance and completely compensated for the inhibitory effects at the very high level of supplementation (Figure 10.3 and 10.4). In the same birds that had a pc $InsP_6$ disappearance of 92% with 12,500 FTU/kg in the diet, $InsP_6$ disappearance in the crop already was 80% (Zeller *et al.*, 2016), demonstrating the high relevance of the processes in the anterior section of the gastro-intestinal tract. When using different combinations of dietary P, $InsP_6$, Ca and phytase concentrations in a four-factorial arrangement of treatments, Li *et al.* (2016) also detected multiple interactions on pc $InsP_6$ disappearance.

Results indicate a dilemma: Broiler chickens are able to digest about two thirds of $InsP_6$-P when they are challenged by feeding low-CaP diets. But when mineral Ca P sources are added to meet the requirements, less $InsP_6$-P is available with the consequence that either mineral P must be further increased or exogenous phytase added. With the increasing pressure to reduce P excretion in the poultry industry it probably is one of the great challenges to solve the dilemma in finding the right balance between the mineral and enzyme supplements used in the different growth phases of the birds. Yet we do not have sufficient data to model the plethora of effects on a quantitative basis.

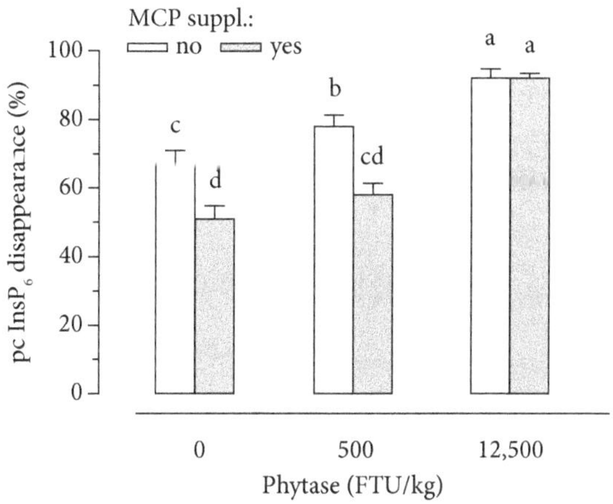

Figure 10.4. *Interactive effects of monocalcium phosphate (MCP) and phytase supplements on $InsP_6$ disappearance until the terminal ileum of broiler chickens fed maize-soybean meal-based diets (Zeller* et al.*, 2015c).*

10.6 Implications for P digestibility studies

The concept of P digestibility determination implies that graded levels of a test P source are used and digestibility of the test P source is calculated by regression analysis. This assumes interactions between the test P source and other ingredients in the diet not to exist. As explained in the previous sections this assumption was found to be wrong.

The interactions probably have the highest relevance when mineral and animal P sources are studied because P supplementation will reduce the degradation of $InsP_6$ contained in the basal diet and hence digestibility of the supplemented P source is underestimated. In three experiments from our institute, pc disappearance of $InsP_6$-P was reduced by approximately 0.4-0.5 g/kg DM with each gram of mineral P supplemented to the diet (Figure 10.5).

The effect shown in Figure 10.5 is ignored when experiments are only based on total P analysis. However, it can have a great influence on the calculated pcdP of the supplemented P source. The influence is demonstrated in Figure 10.6 as an example. The digestibility value of MCP, calculated without correction, was about 54% with only marginal effects of the basal diet used (Shastak *et al.*, 2014). However, when the data were corrected for the reduction in $InsP_6$ breakdown simultaneously measured, the calculated pcdP of MCP is much higher and reaches the range of digestibility usually observed in pigs (Rodehutscord *et al.*, 1994). The conclusion from this model calculation is that chemical analysis in P digestibility trials should be extended by $InsP_6$ analysis of diet and digesta samples so that $InsP_6$-P disappearance is traceable.

Interactions may also occur when P digestibility of plant raw materials is investigated using the regression technique. However, to date these interactions are not well investigated and not quantifiable.

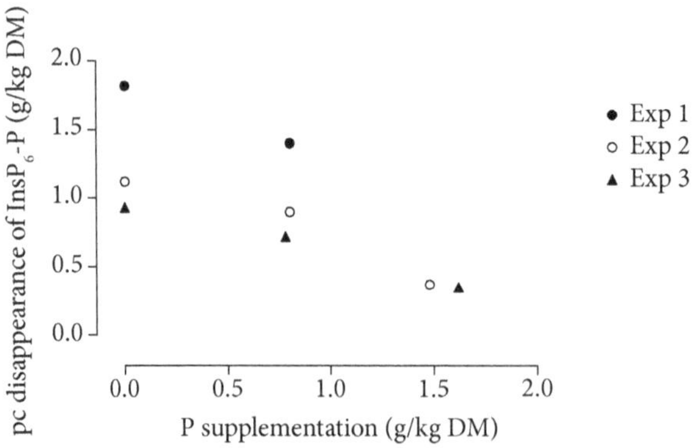

Figure 10.5. *Responses in precaecal $InsP_6$ disappearance to addition of mineral P to the diet in three experiments with broiler chickens (Shastak* et al.*, 2014; Zeller* et al.*, 2015c).*

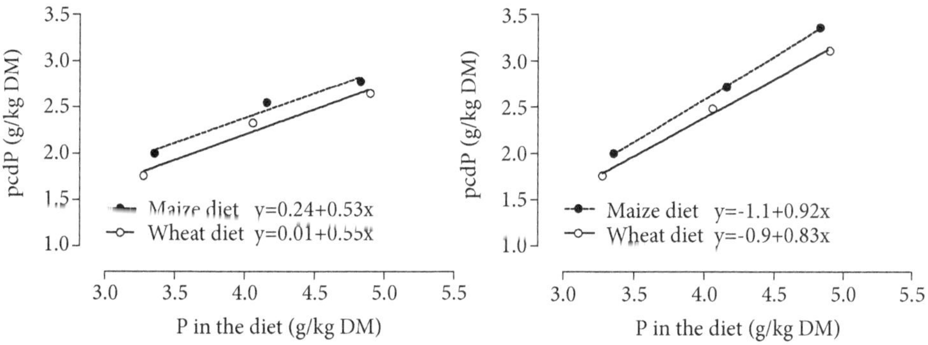

Figure 10.6. *Example for the calculation of precaecal digestibility of P (pcdP) of monocalcium phosphate in broilers by linear regression using either a maize-based or wheat-based diet. Values were without (left panel) or with (right panel) correction for* $InsP_6$*-P disappearance and assuming a pc digestibility of* $InsP_6$*-P of 78% as estimated in Figure 10.1. Original data are taken from Shastak* et al. *(2014).*

Interactive effects can be avoided if the basal diet is free of $InsP_6$. Thus, an alternative approach is to use a phytate-free semi-purified basal diet, such as recently done for testing pcdP of meat and bone meal (Mutucumarana *et al.*, 2015). Nevertheless it should be kept in mind that semi-purified diets might not allow for feed intake and growth as typical in the industry and might change digesta passage rate and microbiota composition; hence effects on the determined pcdP are possible.

10.7 Conclusions

1. The high potential for gastro-intestinal hydrolysis of $InsP_6$ in broilers is suppressed by feed phosphate and Ca supplementation. This suppression can be compensated by use of exogenous phytase.
2. At marginal level of P supply, about 78% of $InsP_6$-P disappearing until the end of ileum is absorbed. The remaining 22% is present as lower inositol phosphates.
3. Studies on the pcdP of non-plant P sources need additional determination of $InsP_6$ disappearance.

References

Amerah, A.M. and Ravindran, V., 2009. Influence of particle size and microbial phytase supplementation on the performance, nutrient utilisation and digestive tract parameters of broiler starters. Animal Production Science 49:704-710.

Blaabjerg, K., Carlsson, N.G., Hansen-Møller, J. and Poulsen, H.D., 2010. Effect of heat-treatment, phytase, xylanase and soaking time on inositol phosphate degradation *in vitro* in wheat, soybean meal and rapeseed cake. Animal Feed Science and Technology 162:123-134.

Delezie, E., Maertens, L. and Huyghebaert, G., 2012. Consequences of phosphorus interactions with calcium, phytase, and cholecalciferol on zootechnical performance and mineral retention in broiler chickens. Poultry Science 91:2523-2531.

GfE [Gesellschaft für Ernährungsphysiologie], 1999. Energie- und Nährstoffbedarf landwirtschaftlicher Nutztiere. 7. Empfehlungen zur Energie- und Nährstoffversorgung der Legehennen und Masthühner (Broiler). DLG-Verlag, Frankfurt am Main, Germany.

Li, W., Angel, R., Kim, S.-W., Brady, K., Yu, S. and Plumstead, P.W., 2016. Impacts of dietary calcium, phytate, and nonphytate phosphorus concentrations in the presence or absence of phytase on inositol hexakisphosphate (IP_6) degradation in different segments of broilers digestive tract. Poultry Science 95:581-589.

Mitchell, R.D. and Edwards Jr., H.M., 1996. Additive effects of 1,25-dihydroxycholecalciferol and phytase on phytate phosphorus utilization and related parameters in broiler chickens. Poultry Science 75:111-119.

Mutucumarana, R.K., Ravindran, V., Ravindran, G. and Cowieson, A.J., 2015. Measurement of true ileal phosphorus digestibility in meat and bone meal for broiler chickens. Poultry Science 94:1611-1618.

NRC [National Research Council], 1994. Nutrient requirements of poultry. National Academy Press, Washington, D.C., USA.

Perryman, K.R., Masey O'Neill, H.V., Bedford, M.R. and Dozier, W.A., 2016. Effects of calcium feeding strategy on true ileal phosphorus digestibility and true phosphorus retention determined with growing broilers. Poultry Science 95:1077-1087.

Rodehutscord, M. and Rosenfelder P., 2016. Update on phytate degradation pattern in the gastrointestinal tract of pigs and broiler chickens. In: Walk, C.L., Kühn, I., Stein, H.H., Kidd, M.T. and Rodehutscord, M. (eds.) Phytate destruction – consequences for precision animal nutrition. Wageningen Academic Publishers, Wageningen, the Netherlands, pp. 15-32.

Rodehutscord, M., Faust, M., Düngelhoef, M., Spiekers, H. and Pfeffer, E., 1994. Zur Messung der Verdaulichkeit des Phosphors aus mineralischen Phosphor-Trägern sowie aus Mineralfuttern, Eiweiβkonzentraten und Alleinfuttern für Schweine. Journal of Animal Physiology and Animal Nutrition 71:169-178.

Shastak, Y. and Rodehutscord, M., 2013. Determination and estimation of phosphorus availability in growing poultry and their historical development. World`s Poultry Science Journal 69:569-585.

Shastak, Y., Zeller, E., Witzig, M., Schollenberger, M. and Rodehutscord, M., 2014. Effects of the composition of the basal diet on the evaluation of mineral phosphorus sources and interactions with phytate hydrolysis in broilers. Poultry Science 93:2548-2559.

Shieh, T.R., Wodzinski, R.J. and Ware, J.H., 1969. Regulation of the formation of acid phosphatases by inorganic phosphate in *Aspergillus ficuum*. Journal of Bacteriology 100:161-1165.

Sommerfeld, V., Schollenberger, M., Kühn, I. and Rodehutscord, M., 2016. Interacitve effects of P, Ca and phytase supplements on $InsP_6$ degradation and net absorption of P and Ca up to the lower ileum of broiler chickens. Proceedings of the Society of Nutrition Physiology 25:45 (abstr.).

Tamim, N.M. and Angel, R., 2003. Phytate phosphorus hydrolysis as influenced by dietary calcium and micro-mineral source in broiler diets. Journal of Agricultural and Food Chemistry 51:4687-4693.

Tamim, N.M., Angel, R. and Christman, M., 2004. Influence of dietary calcium and phytase on phytate phosphorus hydrolysis in broiler chickens. Poultry Science 83:1358-1367.

Ton Nu, M.A., Blaabjerg, K. and Poulsen, H.D., 2014. Effect of particle size and microbial phytase on phytate degradation in incubated maize and soybean meal. Animal 8:534-541.

WPSA [Working Group No 2: Nutrition of the European Federation of Branches of WPSA], 2013. Determination of phosphorus availability in poultry. World`s Poultry Science Journal 69:687-698.

Zeller, E., Schollenberger, M., Kühn, I. and Rodehutscord, M., 2015a. Effect of diets containing enzyme supplements and microwave-treated or untreated wheat on inositol phosphates in the small intestine of broilers. Animal Feed Science and Technology 204:42-51.

Zeller, E., Schollenberger, M., Kühn, I. and Rodehutscord, M., 2015b. Hydrolysis of phytate and formation of inositol phosphate isomers without or with supplemented phytases in different segments of the digestive tract of broilers. Journal of Nutritional Science 4:e1.

Zeller, E., Schollenberger, M., Kühn, I. and Rodehutscord, M., 2016. Dietary effects on inositol phosphate breakdown in the crop of broilers. Archives of Animal Nutrition 70:57-71.

Zeller, E., Schollenberger, M., Witzig, M., Shastak, Y., Kühn, I., Hoelzle, L.E. and Rodehutscord, M., 2015c. Interactions between supplemented mineral phosphorus and phytase on phytate hydrolysis and inositol phosphates in the small intestine of broilers. Poultry Science 94:1018-1029.

Żyła, K., Gogol, D., Koreleski, J., Świątkiewicz, S. and Ledoux, D.R., 1999. Simultaneous application of phytase and xylanase to broiler feeds based on wheat: *in vitro* measurements of phosphorus and pentose release from wheats and wheat-based feeds. Journal of the Science of Food and Agriculture 79:1832-1840.

Chapter 11
Phosphorus metabolism and transportation in the intestine

O.A. Olukosi
Monogastric Science Research Centre, Scotland's Rural College, Edinburgh, EH9 3JG, United Kingdom; oluyinka.olukosi@sruc.ac.uk

Abstract

Phosphorus (P), absorbed in the form of inorganic phosphate (Pi), is required by animals for normal growth and development, but its oversupply is detrimental to the animal. There are controls both at the intestinal and renal levels to ensure maximal Pi absorption when lumen Pi content is low and prevent vascular calcification when lumen Pi content is high. This tight control is the form of a sodium-coupled transporter belonging to the solute carrier (SLC34) family. In the intestine the sodium-phosphate transporter (NaPi-IIb) is present very early in life in birds and is responsible for transcellular Pi uptake mainly from the apical surface of mature enterocytes. Studies in broilers show that the transporter is expressed in all the sections of the small intestine but is most highly expressed in the duodenum and jejunum. There are comparatively few studies on the dietary and non-dietary factors that influence NaPi-IIb mRNA or protein expression in the intestine of poultry species. This review discusses studies showing the regulation of NaPi-IIb in the intestine of, mainly, broilers by dietary levels of phosphorus, vitamin D_3 and phytase supplementation among others.

Keywords: broilers, NaPi-IIb

11.1 Introduction

Phosphorus (P), absorbed in the form of inorganic phosphate (Pi), plays many important roles in cellular metabolism, growth, development, bone formation/mineralisation, energy metabolism, nucleic acid synthesis, cell signalling, and acid-base regulation. In view of the many functions of Pi, its content in the body must be tightly controlled to avoid deficiency leading to poor growth and development or an excess leading to vascular calcification. The homeostasis of Pi is controlled by its flux in the extracellular and intercellular compartments. Most of body Pi is located in the bone (~85%) and the remaining are in the intracellular (~15%) and extracellular (<0.05%) fluids. In view of its many roles, the requirement for Pi cannot possibly be met by simple diffusional movement of the nutrient, and consequently there is need for a transporter system that is capable of accumulating Pi against its electrochemical gradient.

C.L. Walk *et al.* (ed.) **Phytate destruction – consequences for precision animal nutrition**
DOI 10.3920/978-90-8686-836-0_11, © Wageningen Academic Publishers 2016

11.2 Phosphate absorption

Phosphate is absorbed from the small intestine via both paracellular and transcellular pathways (Fuchs and Peterlik, 1979). Paracellular absorption occurs by means of electrochemical gradients across intestinal epithelium (this is quantitatively important immediately postprandial) and is therefore saturable. Phosphate absorption from the intestinal lumen however is largely via a transcellular pathway (Eto *et al.*, 2006) and is initiated by transportation of Pi at the brush border membrane. Phosphate entry into the enterocyte is the rate-limiting step in Pi absorption across the intestinal epithelium. Early studies showed that the epithelial transportation of Pi across the intestinal epithelium is coupled with Na (Berner *et al.*, 1976; Quamme, 1985). Phosphate transport is a secondary active transport characterised by the flux of Pi being coupled to the flux of another substance (in this case Na) whose transport is primary active. Berner *et al.* (1976) observed in rats that Pi was taken up by a saturable transport system at a pH of 7.4 and in the presence of a Na gradient across the membrane. Consequently the driving force for Pi shuttle across the epithelium is not Pi itself but, rather, an asymmetrical distribution of Na (or a Na gradient) across the plasma membrane. This transport system is important as it allows transport of Pi against its concentration gradient. The sodium-phosphate (Na-Pi) transporter is a plasma membrane-bound symporter that mediates the transport of extracellular Pi into cells coupled with Na transport. This intestinal absorption of Pi is enhanced by low P intake and vitamin D_3 (Larbier and Leclerq, 1992; Miyamoto *et al.*, 2005; Proszkoqiec-Weglar and Angel, 2013) and they are the most important regulators of Pi absorption in the intestine.

In birds, Pi is absorbed in all the sections of the small intestine with the most quantitative absorption taking place in the jejunum (Larbier and Leclerq, 1992). The greatest transporter activity for Pi is in the duodenum, followed by the jejunum but because of the comparatively shorter length of the duodenum, there is greater quantity of Pi absorption in the jejunum. Most studies present data on Pi utilisation at the ileal and total tract levels and virtually no studies have reported on P digestibility at pre-ileal level. The dearth of data may be due to the generally poor recovery of digestibility marker in pre-ileal sections of the digestive tract. Recently however, Zeller *et al.* (2015) reported the hydrolysis of *myo*-inositol hexakis-phosphate (IP_6) in different sections of the digestive tract as well net P absorption in the different sections of the small intestine of broilers.

In an experiment to study P digestibility and phytic acid disappearance along the digestive tract, it was observed that there was a drastic reduction in the concentration of Pi measured in the digesta at the crop and the distal end of the ileum both in diets with high and low levels of dietary P. The observations from the two studies show that there is an effective hydrolysis of *myo*-inositol hexakis dihydrogen phosphate and absorption of P (Figure 11.1) as well as clearance of total P (Figure 11.2) in digestive tract by the end of the ileum.

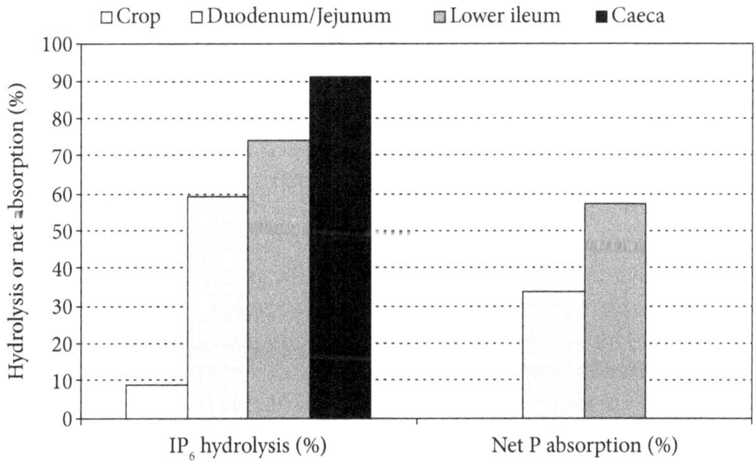

Figure 11.1. Myo-*inositol 1,2,3,4,5,6 hexakis (dihydrogen phosphate) hydrolysis along the digestive tract and net phosphorus absorption in the small intestinal sections of 25-day old broilers. Data based on broilers receiving diets with 7.5 and 5.2 g/kg DM of Ca and total phosphorus, respectively without phytase supplementation (Zeller* et al., *2015).*

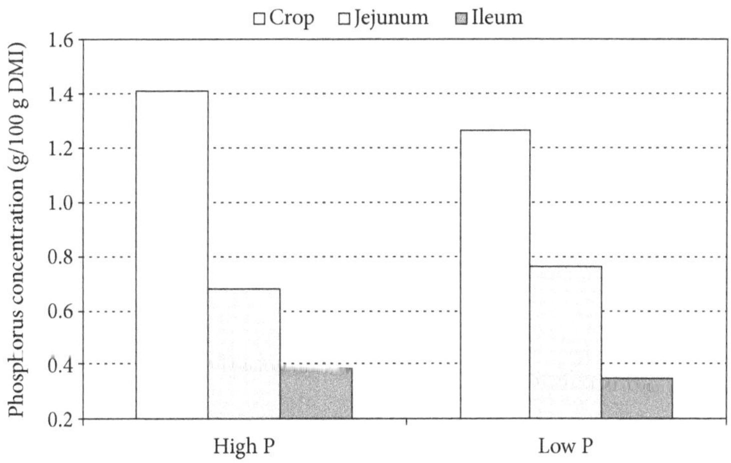

Figure 11.2. *Concentration of total phosphorus (g/100 g dry matter intake) in the digesta along the digestive tract of 21-day old broilers. Concentration was calculated as* $P_0 \times (C_i / C_0)$ *where* P_0 *is the concentration of total P in the digesta and* C_i *and* C_0 *are the concentrations of the marker in the diet and digesta, respectively (Olukosi* et al., *unpublished).*

11.3 Phosphate transporter – NaPi-IIb

The solute carrier family (SLC34) comprises three transporters namely NaPi-IIa, NaPi-IIb and NaPi-IIc, also known as SLC34A1, SLC34A2 and SLC34A3, respectively (Werner *et al.*, 1998). NaPi-IIa and NaPi-IIc are primarily localised to the kidney (Murer *et al.*, 2004; Wang and Yin, 2009). The transporters use the free energy of the downhill electrochemical Na^+ gradient across the luminal membrane. Intestinal Pi transportation is dependent on NaPi-IIb protein abundance and mRNA levels. The fact that the transporter is apically located (in the intestine) and that there is practically no Pi transportation in NaPi-IIb knock-out mouse indicate that the transporter is responsible for transportation of Pi from the intestinal lumen. Huber *et al.* (2006) showed that for hens, NaPi-IIa in the kidney plays the most dominant role in adaptation of hens to low dietary P intake. In their study, high dietary P reduced expression of NaPi-IIa and Vmax of renal Na-dependent Pi transport. These effects resulted in decreased renal Pi uptake and an increase in Pi excretion. Another Na-dependent transporter detected in the brush border of rodent small intestine belongs to SLC20 family and is known as NaPi-III or PiT1 (Giral *et al.*, 2009). Huber *et al.* (2015) reported identifying the band for NaPi-III in the small intestine of broilers and that it was influenced by phytase supplementation.

Human intestinal NaPi-IIb is the second mammalian phosphate transporter identified and was reported to have 76% homology to mouse intestinal NaPi-IIb (Xu *et al.*, 1999). Shibasaki *et al.* (2009) demonstrated that NaPi-IIb is expressed early during the embryonic development where it is possibly responsible for Pi absorption from maternal circulation. Mutation to produce NaPi-IIb gene deficient mice resulted in *in utero* mortality soon after implantation thus showing the obligatory requirement for the transporter for early embryonic development. We have shown previously (Olukosi *et al.*, 2010) that chicks immediately after hatch have high level of expression of NaPi-IIb mRNA in all sections of the small intestine. Thus it is reasonable to assume that, just as for mammals, the transporter is expressed early on during embryonic development where it arguably serves for the uptake of Pi from the yolk.

Protein for NaPi-IIb is expressed in a number of organs (Yan *et al.*, 2007) but glycosylated NaPi-IIb is localised in the brush border of the enterocytes. Huber *et al.* (2015) reported two bands of NaPi-IIb in the small intestine of broiler chickens. Because these bands were inversely regulated by dietary interventions, the authors suggested that they were isoforms. Protein Band1 increased whereas Band2 decreased in response to phytase supplementation but no effect of variation in total P on the protein expression was observed. Similar observation had been made in zebrafish (Graham *et al.*, 2003). The two possible isoforms of NaPi-IIb were described as a low affinity high capacity transporter which is activated by high P such as in diets supplemented with phytase or inorganic P. The second isoform was described as a high affinity, low capacity transporter that was reduced in diets supplemented with a source of inorganic P.

11.4 Ontogeny and expression of NaPi-IIb in the small intestinal sections

In mammals, the highest expression of NaPi-IIb in the intestine occurs during suckling and decreased rapidly at weaning. The protein is not fully glycosylated until after weaning. Indications are that maturation of NaPi-IIb occurs early in post-natal life in mammals (Wang and Yin, 2009). Xu *et al.* (2002) noted that NaPi-IIb gene expression in jejunum of rats decreased with age from 2 weeks to adult age. Others have reported that Pi absorption decreases with age in mammals (Borowitz and Granrud, 1992) and the decrease in NaPi-IIb agrees with that observation. At hatch and prior to feeding, NaPi-IIb expression was greatest in the duodenum and jejunum and lowest in the ileum. This profile was maintained in intestinal mucosa samples collected on days 7 and 14 (Olukosi *et al.*, 2010) (Figure 11.3). The overall trend is towards increasing NaPi-IIb mRNA in all the sections of the small intestine as the birds aged from 0 to 21 days. The dietary nPP level did not have any significant effect on NaPi-IIb expression. Total tract total P retention also generally increases with age from day 7 to 21 in the broiler chickens (Olukosi *et al.*, 2007).

Stefanello *et al.* (2016) reported that jejunal expression of NaPi-IIb increased with age in broiler chickens between 0 to 21 days. The largest increase was from weeks 1 to 2 and expression only marginally increased from weeks 2 to 3. Because the digestive and absorptive capacity largely reaches maturity (or peak) by day 15 of age in broiler chickens (Mahagna and Nir, 1996) the observed dramatic increase in NaPi-IIb expression from day 7 to 14 but much less difference from day 14 to 21 agrees with the report of digestive capacity reaching maximum at week 2.

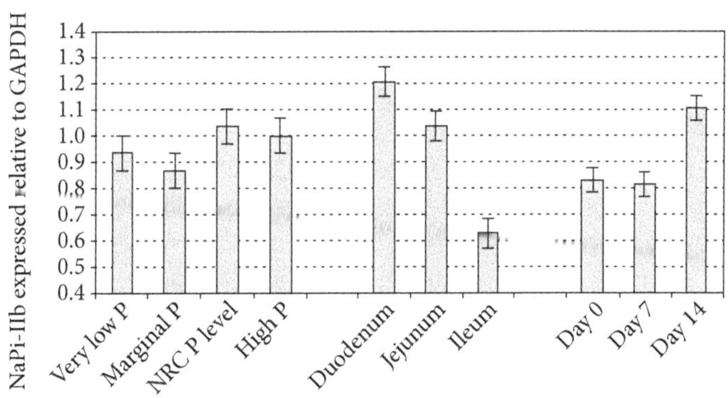

Figure 11.3. *Expression of NaPi-IIb in diets with various levels of dietary phosphorus (left bars), in different sections of the small intestine (middle bars) and at different ages (right bars). Dietary non-phytate phosphorus (P) levels were 0.25, 0.35, 0.45 and 0.55% for very low P, marginal P, NRC P level and high P diets, respectively. NaPi-IIb level was expressed relative to glyceraldehyde 3-phosphate dehydrogenase (GAPDH). Samples were collected from mucosa scrapings of the small intestine of broiler chickens (Olukosi* et al., *2010).*

It appears that there are species differences in the spatial distribution of NaPi-IIb in different sections of the small intestine. Giral *et al.* (2009) noted that NaPi-IIb is mostly located in the duodenum and jejunum but their expression is low in the ileum of the rat. On the other hand, Radanovic *et al.*, (2005) reported that in mouse, NaPi-IIb was mainly expressed in the ileum and less so in the duodenum and jejunum. Although Pi absorption took place throughout the small intestine, the authors noted that in mouse the transcellular Pi absorption is mainly in the ileum but low-P diet upregulated the NaPi-IIb expression throughout the sections of the small intestine. In the broiler chickens, expression of NaPi-IIb decreases from the duodenum to the ileum (Han *et al.*, 2009; Olukosi *et al.*, 2010, 2013).

11.5 Dietary factors affecting NaPi-IIb expression

Regulations of the membrane-bound NaPi-IIb occur at both transcriptional and translational levels. The changes at transcriptional level result in variation in gene expression which is exhibited as changes in mRNA abundance. Regulations at translational level result in changes in the number of functional proteins on the membrane. Both of these changes affect the transporter protein function. Table 11.1 provides a summary of some recent data on dietary effects on NaPi-IIb gene expression in poultry, mainly broilers. The data generally show that increasing dietary P level, or phytase supplementation, reduces the expression of NaPi-IIb and that the response of the transporter to dietary P levels is dependent on dietary calcium (Ca) levels.

11.5.1 Dietary phosphorus level effect

There have been observations by several authors that dietary low Pi increases the expression of NaPi-IIb in the intestine of different species (Hattenhauer *et al.*, 1999; Olukosi *et al.*, 2011) compared to similar animals that receive high dietary P. Hattenhauer *et al.* (1999) indicated that the response was confined to the mature enterocytes and were not observed in the crypts. Although acute low P increases expression of NaPi-IIa in renal tubules, there is no such response to acute P deprivation in the intestine of the mice and so it appears that it takes some time to elicit the response in the brush border membrane.

Giral *et al.* (2009) observed that rats fed chronically low P diet (0.1% P) had 67% increase in jejunal Pi uptake compared with those fed 0.6 or 1.2% dietary P. Along with increased Pi uptake, the authors also reported an increase in expression and protein of NaPi-IIb in the jejunum but not in the duodenum. The authors also observed that there is a transient, adaptive and acute overexpression of NaPi-IIb in brush border membrane of rats switched from high- to low-P diet in the duodenum but not jejunum and that this adaptive mechanism induced postprandial hyperphosphatemia. Their study suggested that a chronic Pi effect was observed in the jejunum but an acute response in the duodenum.

In chickens, whereas many studies have reported an increased expression of NaPi-IIb in response to low-P diets (Li *et al.*, 2012; Nie *et al.*, 2003; Olukosi *et al.*, 2010, 2011; Yan *et*

Table 11.1. *Effects of dietary interventions on gene expression of NaPi-IIb.*[1]

Age	Nutrition interventions	Effect on NaPi-IIb expression	Reference
Broiler chickens			
42 d	low nPP (0.35%) and phytase at 0, 500 and 1000 FTU/kg	decreased expression with phytase supplementation	Adedokun *et al.*, 2009
26 d	0.50% nPP (with or without coccidial vaccine challenge)	coccicidia challenge increased expression	Adedokun *et al.*, 2012
21 d	0.50% nPP (additional phytate salt, MYO or phytase)	phytate increased expression in diets without phytase; no phytase or MYO effect	Beeson, 2016 (unpublished)
21 d	low (0.35%) or adequate (0.45%) nPP with or without phytase	no significant effect	Beeson *et al.*, 2016
21 d	0.21% nPP (plus 5 μg/kg vit D_3)	increased expression	Han *et al.*, 2009
25 d	0.39 or 0.47% total P (in 2×3 factorial arrangement with 0, 500 or 12,500 FTU/kg phytase)	protein Band1 increased, Band2 decreased by phytase; no response to variation in tP	Huber *et al.* 2015
21 d	0.35, 0.40, 0.45 and 0.50% nPP with three levels of CP (factorial)	increased expression in duodenum with increased nutrient density; non-linear effect of nPP	Li *et al.*, 2011
21 d	0.6 or 1.1% Ca and 0.27 or 0.45% nPP (2×2 factorial arrangement)	reduced duodenal expression at 0.50% nPP in high-Ca diet	Li *et al.*, 2012
7 d	0.25, 0.35, 0.45, and 0.55% nPP	reduced expression with increasing P in jejunum and ileum	Olukosi *et al.*, 2011
21 d	0.18 nPP (control), supplemented with phytase, TCP or MCP	no effect	Olukosi *et al.*, 2013
7-21 d	total P (0.69, 0.99 and 1.29%)	no significant effect at any age	Stefanello *et al.*, 2016
21 d	0.32% tP (control), increasing tP and CP	increased expression due to CP, no effect of P	Xue *et al.*, 2016
4 d	1.2% Ca (0.55% nPP) or 0.6% Ca (0.25% nPP)	increased expression	Yan *et al.*, 2007
Laying hens			
24 w	0.073, 0.240 and 0.343% total P	increased expression in diet with highest tP level	Huber *et al.*, 2006
28 w	0.20, 0.25, 0.30, 0.35 and 0.40% nPP	linear decrease in duodenal expression	Nie *et al.*, 2003

[1] CP = crude protein; FTU = phytase units; MCP = mono-calcium phosphate; MYO = *myo*-inositol; nPP = non-phytate phosphorus; TCP = tri-calcium phosphate; tP = total phosphorus.

al., 2007) others have also reported no change in expression of the gene in response to variation in dietary P level (Beeson *et al.*, 2016; Olukosi *et al.*, 2013; Xue *et al.*, 2016) or inconsistent effect (Li *et al.*, 2011).

The differences in observations in the chickens studies seem to be due to various factors such as: (1) the age when the samples were taken (greater likelihood of response in younger birds); (2) the section of the small intestine that is sampled (expression greater in the proximal small intestine); (3) the extent of dietary P reduction and gaps in dietary P levels (more severe P deficiency and wider dietary P levels more likely to elicit response); and (4) whether the entire tissue or the mucosa scrapings are used for the gene expression study. Other factors that will play different roles may include dietary factors that will generally affect P utilisation such as Ca:P ratio and the dietary level of phytate P. Consequently these factors need into be taken to account when comparing data across studies.

11.5.2 Phytase supplementation effects

Because phytase is well known for its effect in increasing P digestibility especially in low P diets and because NaPi-IIb is mainly responsible for transcellular P absorption the response of the transporter to phytase supplementation is of interest. Adedokun *et al.* (2009) reported that NaPi-IIb expression decreased in the small intestine of 42-d old broilers receiving phytase-supplemented diet. This observation is in line with observation of reduction in NaPi-IIb expression with higher dietary P. Phytase supplementation is expected to increase luminal P content, hence a reduction in the intestinal expression of the transporter is the expected outcome because NaPi-IIb expression is generally reduced in the presence of abundant luminal P.

Huber *et al.* (2015) reported different responses of NaPi-IIb protein Band1 and 2 to phytase supplementation. The bands are possibly isoforms of NaPi-IIb that have been shown previously to respond differently to dietary P level. On the other hand, Olukosi *et al.* (2013) reported that phytase supplementation had no effect on NaPi-IIb expression in all the three sections of the small intestine in 21-d old broilers. Beeson *et al.* (2016) similarly observed no phytase effect on NaPi-IIb expression in diets deficient or adequate in P. As with the effect of dietary P on NaPi-IIb expression, the different factors that may influence phytase effect need to be considered in across-studies comparison of phytase supplementation effect on NaPi-IIb expression.

11.5.3 Dietary vitamin D level

Chronic feeding of a diet low in P leads to hypophosphatemia, which in turn induces an increase in vitamin D_3 by 1α-hydroxylase in the kidney and helps to increase P reabsorption in the kidney and stimulates intestinal P uptake (Segawa *et al.*, 2004). There is generally very poor P absorption when there is a deficiency of vitamin D (Rizzoli *et al.*, 1977; Segawa *et al.*, 2004). Giral *et al.* (2009) showed that the active form of vitamin D_3 activates vitamin D_3 receptor, which in turn induces transcriptional upregulation of NaPi-IIb in the small intestine although there is suggestion that low-P can also activate NaPi-IIb upregulation in pathways independent of vitamin D_3.

Han *et al.* (2009) reported that supplementation of 1,25-$(OH)_2D_3$ to diet low in P resulted in increased NaPi-IIb expression in duodenum, jejunum and ileum of 21-d old chickens. The diets also resulted in increased P absorption by the chickens. Similarly Hattenhauer *et al.*

(1999) demonstrated the up-regulation of NaPi-IIb by dietary low P or vitamin D_3 without transcriptional changes for the gene. The authors suggested that post-transcriptional and non-genomic effects are possible explanations for this observation.

11.5.4 Oral vaccine challenge

Although not a dietary intervention, disease challenges influence nutrient utilisation generally and can have specific effects on P absorption. Adedokun *et al.* (2012) observed an increase in jejunal NaPi-IIb expression in broilers following an oral challenge with coccidia vaccine. The diets used in the study were P-adequate, purified or semi-purified diets and P digestibility was not assessed in the experiment. The authors suggested that the increased expression likely indicated increased need for P absorption during coccidial challenge. It is difficult, and possibly ill-advised, to compare effects seen in purified diet with those seen when feeding practical diets. It is likely that the increased NaPi-IIb expression observed in the purified diets that already met P requirement is due to low feed (and consequently P) intake that is often observed in birds fed purified diets. The low nutrient intake may be further confounded by a disease challenge consequently leading to changes in NaPi-IIb expression. Clearly the possible effects of disease challenge on NaPi-IIb expression requires further elucidation.

11.5.5 Other dietary factors

Li *et al.* (2011) reported an increase in duodenal NaPi-IIb expression in broilers that received increasing nutrient density (i.e. increasing dietary metabolisable energy and crude protein levels) and non-phytate P. The authors suggested that increased NaPi-IIb expression in response to increased nutrient density is in line with the increase in growth performance of the birds as nutrient density increased. In addition, increased NaPi-IIb occurred simultaneously with an increase in serum level of parathyroid hormone and 1-α-hydoxylase levels.

Li *et al.* (2012) reported the influence of Ca:P on NaPi-IIb expression. The authors demonstrated that when adequate P diets were fed, low dietary Ca increased NaPi-IIb expression but the influence of nPP level on NaPi IIb expression was only significant in diets with adequate Ca. Because of the impact of Ca:P on P availability for absorption, it is conceivable that the most dramatic increase in NaPi-IIb in response to low P was observed in diets with high Ca. In such diets, the deficiency of P will be even more severe when diets already low in P also have high Ca. Beeson (2016) recently observed that dietary manipulation of phytate level using a phytate salt increased jejunal expression of NaPi-IIb. Xue *et al.* (2016) observed an increase in NaPi-IIb expression due to increasing dietary protein, but not total P, level.

11.6 Conclusions

1. It is apparent that animals have a robust system to ensure consistent supply of Pi to meet the animal need by regulation of Pi uptake or reabsorption at the intestinal and renal levels, respectively.
2. Preponderance of studies show that low P stimulates increased capacity for intestinal P absorption and renal P reabsorption (by regulating NaPi-IIb and NaPi-IIa, respectively).
3. On the other hand, in order to prevent vascular calcification, P absorption from the intestine is reduced when there is high supply of Pi. This capacity for regulating P is present early on during embryonic development and certainly at hatching for birds.
4. It will appear then that the most important regulations of NaPi-IIb will be those that contribute to reducing dietary P loss. It is unlikely that expression of NaPi-IIb per se will be the bottleneck for P clearance from the intestinal lumen.
5. In the absence of other physiological complications, birds will be able to regulate their P uptake from the intestinal lumen, but when there is P abundance in the lumen, a negative feedback will prevail to prevent entry of too much P into the cell because of its potentially damaging effects.

There is need to more fully understand the regulations of NaPi-IIb and the influence that dietary and non-dietary factors may have on the regulation of the transporter in non-ruminant animals. It stands to reason that where the regulation of the transporter may be more beneficial from animal nutrition point of view are to understand: (1) how the transporter can be regulated to improve P utilisation early in life when capacity for nutrient utilisation is inherently low and (2) later in life when animals quantitatively consume and excrete more nutrients.

References

Adedokun, S.A., Ajuwon, K.M., Romero, L.F. and Adeola, O., 2012. Ileal endogenous amino acid losses: Response of broiler chickens to fiber and mild coccidial vaccine challenge. Poultry Science 91:899-907.

Adedokun, S.A., Olukosi, O.A., Adeola, O. and Ajuwon, K.M., 2009. Exogenous phytase alters the expression of nutrient transporters and markers of inflammation in the gastrointestinal tracts of broiler chickens. Poultry Science 88 (Supp. 1):59.

Beeson, L.A., 2016. An investigation into the possible health-promotion modes of action of regular or high dosages of phytase. Ph.D. Thesis, University of Glasgow, UK.

Beeson, L.A., Walk, C.L., Hastie, P., Bedford, M.R. and Olukosi, O.A., 2016. Gene expression of nutrient transporters and markers of liver anti-oxidants, gut level immune response and cellular integrity in broilers fed diets containing regular or high doses of phytase. Book of Abstracts, World's Poultry Science Association UK Branch Spring Meeting. April 6-7, Chester, UK, pp. 12-13.

Berner, W., Kinne, R. and Murer, H., 1976. Phosphate transport into brush-border membrane vesicles. Biochemical Journal 160:467-474.

Borowitz, S.M. and Granrud G.S., 1992. Ontogeny of intestinal phosphate absorption in rabbits. American Journal of Physiology. Gastrointestinal and Liver Physiology 262:G847-G853.

Eto, N., Tomita, M. and Hayashi, M., 2006. NaPi-mediated transcellular permeation is the dominant route in intestinal inorganic phosphate route in intestinal inorganic phosphate absorption in rats. Drug Metabolism and Phamacokinetics 21:217-221.

Fuchs, R. and Peterlik, M., 1979. Pathways of phosphate transport in chick jejunum: influence of vitamin D and extracellular sodium. Pflugers Archiv 381:217-222.

Giral, H., Caldas, Y., Sutherland, E., Wilson, P., Breusegem, S., Barry, N., Blaine, J, Jiang, T., Wang, X.X. and Levi, M., 2009. Regulation of rat intestinal Na-dependent phosphate transporters by dietary phosphate. American Journal of Physiology. Renal Physiology 297:F1466-F1475.

Graham, C., Nalbant, P., Scholermann, B., Hentschel, H., Kinne, K.H. and Werner, A., 2003. Characterization of a type IIb sodium dependent cotransporter from zebrafish (*Danio rerio*) kidney. American Journal of Physiology. Renal Physiology. 284:F727-F736.

Han, J.C., Yang, X.D., Zhang, T., Li, H., Li, W.L., Zhang, Z.Y. and Yao J.H., 2009. Effects of 1α-hydroxycholecalciferol on growth performance, parameters of tibia and plasma, meat quality, and type IIb sodium phosphate cotransporter gene expression of one- to twenty-one-day-old broilers. Poultry Science 88:323-329.

Hattenhauer, O, Traebert, M., Murer, H. and Biber, J., 1999. Regulation of small intestinal Na-Pi type IIb cotransporter by dietary phosphate intake. American Journal of Physiology. Gastrointestinal and Liver Physiology 277:G756-G762.

Huber, K. Zeller, E. and Rodehutscord, M., 2015. Modulation of small intestinal phosphate transporter by dietary supplements of mineral phosphorus and phytase in broilers. Poultry Science 94:1009-1017.

Huber, K., Hempel, R. and Rodehutscord, M., 2006. Adaptation of epithelial sodium-dependent phosphate transport in jejunum and kidney of hens to variations in dietary phosphorus intake. Poultry Science 85:1980-1986.

Larbier, M. and Leclercq, B., 1992. Nutrition and feeding of poultry. Nottingham University Press, Loughborough, Leicester, UK, 305 pp.

Li, J., Yuan, J., Guo, Y., Sun, Q. and Hu, X., 2012. The influence of dietary calcium and phosphorus imbalance on intestinal NaPi-IIb and calbindin mRNA expression and tibia parameters of broilers. Asian-Australasian Journal Animal Science 25:552-558.

Li, J., Yuan, J., Guo, Y., Yang, Y., Bun, S., Hu, X. and Yi, D., 2011. The effect of dietary nutrient density on growth performance, physiological parameters, and small intestinal type IIb sodium phosphate co-transporter expression in broilers. Journal of Animal Science and Biotechnology 2:102-110.

Mahagna, M. and Nir, I., 1996. Comparative development of digestive organs, intestinal dissacharidases and some blood metabolites in broilers and layer-type chicks after hatching. British Poultry Science 37:359-371.

Miyamoto, K., Ito, M., Kuwahata, M., Kato, S. and Segawa, H., 2005. Inhibition of intestinal sodium-dependent inorganic phosphate transport by fibroblast growth factor 23. Therapeutic Apheresis and Dialysis 9:331-335.

Murer, H., Forster, I. and Biber, J., 2004. The sodium phosphate cotransporter family SLC34. Pflugers Archiv – European Journal of Physiology 447:763-767.

Nie, W., Yang, Y., Yuan, J., Wang, Z. and Guo, Y., 2003. Effect of dietary non-phytate phosphorus on laying performance and small intestinal epithelial phosphate transporter expression in Dwarf pink-shell laying hens. Journal of Animal Science and Biotechnology 4:34. Available at: www.jasbsci.com/content/4/1/34.

Olukosi, O.A., Adedokun, S.A., Ajuwon, K.M. and Adeola, O., 2010. The ontogeny of sodium-dependent phosphate transporter (NaPi-IIb) in broilers and its relationship with phosphorus excretion in broiler chicks. Poultry Science 89 (Supp. 1):59.

Olukosi, O.A., Adedokun, S.A., Ajuwon, K.M. and Adeola, O., 2011. Early responses of sodium-dependent phosphate transporter type IIb in broiler chicks to dietary phosphorus intervention. British Poultry Abstracts 7:39.

Olukosi, O.A., Cowieson, A.J. and Adeola, O., 2007. Age-related influence of a cocktail of xylanase, amylase and protease or phytase individually or in combination in broilers. Poultry Science 86:77-86.

Olukosi, O.A., Kong, C., Fru-Nji, F., Ajuwon, K.M. and Adeola, O., 2013. Assessment of a bacterial 6-phytase in the diets of broiler chickens. Poultry Science 92:2101-2108.

Proszkowiec-Weglarz, M. and Angel, R. 2013. Calcium and phosphorus metabolism in broilers: effect of homeostatic mechanism on calcium and phosphorus digestibility. Journal of Applied Poultry Research 22: 609-627.

Quamme, G.A., 1985. Phosphate transport in intestinal brush border membrane vesicles: effect of pH and dietary phosphate. American Journal of Physiology. 249:G168-G176.

Radanovic, T., Wagner, C.A., Murer, H. and Biber, J., 2005. Regulation of intestinal phosphate transport I. Segmental expression and adaptation to low-Pi diet of the type IIb Na-Pi cotransporter in mouse small intestine. American Journal of Physiology. Gastrointestinal and Liver Physiology 288:G496-G500.

Rizzoli, R., Fleisch, H. and Bonjour, J.P., 1977. Role of 1,25-dihydroxyvitamin D3 on intestinal phosphate absorption in rats with a normal vitamin D supply. The Journal of Clinical Investigation 60:639-647.

Segawa, H., Kaneko, I., Yamanaka, S., Ito, M., Kuwahata, M., Inoue, Y., Kato, S. and Miyamoto, K., 2004. Intestinal Na-Pi cotransporter adaptation to dietary Pi content in vitamin D receptor null mice. American Journal of Physiology. Renal Physiology 287:F39-F47.

Shibasaki, Y., Etoh, N., Hayasaka, M., Takahashi, M., Kakitani, M., Yamashita, T., Tomizuka, K. and Hanaoka, K., 2009. Targeted deletion of the tybe IIb Na^+-dependent Pi-co-transporter, *NaPi-IIb*, results in early embryonic lethality. Biochemical and Biophysical Research Communications 381:482-486.

Stefanello, C., Vieira, S.L., Xue, P., Ajuwon, K.M. and Adeola, O., 2016. Age-related energy values of bakery meal for broiler chickens determined using the regression method. Poultry Science 95:1582-1590.

Wang, B. and Yin, Y., 2009. Regulation of the type IIb sodium-dependent phosphate co-transporter expression in the intestine. Frontiers of Agriculture in China 3:226-230.

Werner, A., Dehmelt, L. and Nalbant, P., 1998. Na-dependent phosphate cotransporters: the NaPi protein families. Journal of Experimental Biology 201:3135-3142.

Xu, H., Bai, L., Collins, J.F. and Ghishan, F.K., 1999. Molecular cloning, functional characterization, tissue distribution, and chromosomal localization of a human, small intestinal sodium-phosphate (Na_-Pi) transporter (SLC34A2). Genomics 62:281-284.

Xu, H., Bai, L., Collins, J.F. and Ghishan, F.K., 2002. Age-dependent regulation of rat intestinal type IIb sodium-phosphate cotransporter by 1,25-(OH)2 vitamin D3. American Journal of Physiology. Cell Physiology 282:C487-C493.

Xue, P.C., Ajuwon, K.M. and Adeola, O., 2016. Phosphorus and nitrogen utilization responses of broiler chickens to dietary crude protein and phosphorus levels. Poultry Science. DOI: http://dx.doi.org/10.3382/ps/pew156.

Yan, F., Angel, R. and Ashwell, C.M., 2007. Characterization of the chicken small intestine type IIb sodium phosphate cotransporter. Poultry Science 86:67-76.

Zeller, E., Schollenberger, M., Kühn, I. and Rodehutscord, M., 2015. Hydrolysis of phytate and formation of inositol phosphate isomers without or with supplemented phytases in different segments of the digestive tract of broilers. Journal of Nutritional Science e4:1-12. DOI: http://dx.doi.org/10.1017/jns.2014.62.

Chapter 12
Measurement of calcium digestibility in feed ingredients for poultry – methodology and challenges

*M.N. Anwar and V. Ravindran**
Monogastric Research Centre, Institute of Veterinary, Animal and Biomedical Science, Massey University, Palmerston North 4442, New Zealand; v.ravindran@massey.ac.nz

Abstract

Determination of calcium (Ca) digestibility in feed ingredients for poultry has not received any attention in the past due to the abundance and low cost of limestone, the major inorganic Ca source, and the low Ca concentration in plant feed ingredients. The recent interest in the determination of phosphorus (P) digestibility in feed ingredients necessitates the measurement of Ca digestibility because of the close relationship between P and Ca metabolism. Currently there is no established method available for the determination of Ca digestibility in poultry. Three methods, namely direct, difference, and regression, have been used for the determination of nutrient digestibility in feed ingredients. In the research reported herein, assay methodologies were evaluated with the primary aim of contributing to future discussions on the development of a suitable method to measure Ca digestibility for poultry. In poultry, Ca digestibility needs to be measured at the ileal level because of the possible urine Ca contamination at the excreta level. Ileal digestible data generated for meat and bone meal and limestone indicate that Ca digestibility values are quite different from bioavailability values reported in the past and what has come to be generally accepted in the industry. Issues that need to be considered in the development of a consensus methodology for use across laboratories are identified.

Keywords: endogenous losses, ileal digestibility, limestone, meat and bone meal, particle size

12.1 Introduction

Calcium (Ca), phosphorus (P) and vitamin D_3 are essential minerals required for skeletal development. Almost 99% of the body Ca is present in the skeleton and the balance 1% is involved in a wide array of metabolic and physiological functions in the body. Calcium plays important roles *inter alia* in blood clotting, muscle contraction, transmission of nerve impulse, hormone secretion, and regulation of heart function (Suttle, 2010). Deficiency of Ca can cause poor growth performance, rickets and tibial dyschondroplasia in broiler chickens. On the other hand, high dietary Ca levels can cause reduction in availability of

DOI 10.3920/978-90-8686-836-0_12, © Wageningen Academic Publishers 2016

several important and costly nutrients, including P, protein and lipids (Ballam *et al.*, 1984; Mutucumarana *et al.*, 2014a; Tamim and Angel, 2003), and bird performance.

Measurement of Ca digestibility in poultry has not been considered in the past due to the cheap availability of limestone, the major source of Ca in poultry diets. However, the recent interest in the measurement of digestible P values in feed ingredients (WPSA, 2013) due to increasing price of inorganic phosphates and environmental P pollution necessitates a closer look at digestible Ca. If the poultry industry shifts towards a digestible P system, the close relationship between Ca and P during absorption and post-absorptive utilisation processes requires the development of a digestible Ca system to ensure that the Ca and P requirements of birds are precisely met. It is generally assumed that Ca in feed ingredients is 100% available and this needs to be confirmed. Published data on the digestibility of Ca in feed ingredients for poultry, however, are limited.

12.2 Calcium sources for poultry

Reasonably good data are available on the Ca concentration of inorganic Ca supplements and feed ingredients. The Ca concentration in plant ingredients is low, ranging between 0.02 and 0.06% in cereals and 0.25 and 0.70% in common protein meals. For this reason, inorganic Ca sources (limestone and inorganic phosphates) and/or animal protein meals (meat and bone meal; MBM) are included in poultry diets to satisfy the Ca requirements. In a typical corn-soy broiler starter diet, almost 88% of the Ca originates from inorganic sources.

Commonly used inorganic and organic Ca sources for poultry diets are limestone, dicalcium phosphate (DCP), monocalcium phosphate (MCP), MBM, and oyster shell. These sources are reported to contain an average of 38.0, 22.0, 16.0, 10.3 and 38.0% Ca, respectively (NRC, 1994). Calcium concentrations, however, can vary depending on their origin and the variations are much wider in organic Ca sources. For example, Subalo and Stein (2013) reported a range of 5.1 to 11.0% Ca in eight MBM samples.

12.3 Absorption of calcium

Early studies on the intestinal site of Ca absorption in poultry have yielded conflicting results, largely due to methodological differences. Recent data have conclusively demonstrated that the jejunum is the major site of Ca absorption in poultry and there is essentially no absorption in the ileum (Mutucumarana *et al.*, 2014a). A marked net flow of Ca was observed into the duodenum probably due to the secretion of large amounts of digestive juices from the pancreas and bile. The existence of continuous duodenal-gizzard refluxes in chickens could also account, in part, for the observed net secretion. Digesta, digestive enzymes and bile are known to be shuttled between the gizzard and duodenum to optimise nutrient digestion by prolonging the retention time (Duke, 1982) and this reverse peristalsis can be expected to increase the net concentration of Ca in the duodenum.

12.4 Endogenous calcium losses

Endogenous Ca is secreted into the lumen of the gastrointestinal tract from various sources, including biliary secretions, digestive enzymes, and desquamated mucosa (Davies *et al.*, 2004). Of these, Ca in bile is probably the major contributor where it is mainly found as a conjugate with bile acids (Williamson and Percy-Robb, 1980). As shown in rats, biliary Ca secretions appear dependent on plasma Ca concentrations (Limlomwongse *et al.*, 1988). Calcium is also reported to be secreted into the intestinal tract, primarily in the jejunum and ileum, through a paracellular route when intra-luminal Ca concentration is well below the serum ionised Ca concentration (Favus, 1985). These endogenous secretions, along with dietary Ca, are partially absorbed and the unabsorbed fraction is lost to the animal.

Measurement of the endogenous loss of Ca is necessary for the estimation of true/standardised digestibility of Ca. In several University of Illinois studies, basal endogenous Ca losses in pigs have been quantified following the feeding of Ca-free diets and wide variations have been reported, with values ranging from 123 to 670 mg/kg dry matter intake (Gonzalez-Vega and Stein, 2016). At Massey University, we have used purified diets free of both Ca and P to determine basal endogenous losses of Ca in broilers. Based on nine separate assays, the average (± SE) basal loss was determined to be 108±6 mg/kg dry matter intake (range, 84-127 mg/kg). Interestingly, markedly higher basal ileal endogenous P losses of 354 mg/kg of dry matter in birds fed a Ca- and P-free diet was reported by Mutucumarana and Ravindran (2016).

An issue that needs to be resolved in future studies is whether a Ca-free diet with P or a Ca- and P-free diet should be used in endogenous Ca measurements, because the presence or absence of P will influence both endogenous Ca secretions and reabsorption.

12.5 Calcium bioavailability

The term 'bioavailability' has been used for the measurement of Ca availability of inorganic sources for chickens. In these assays, the value of the test Ca source is determined by slope ratio assay where the birds are fed a Ca-deficient basal diet and test diets with different concentrations of the test supplement. The assay usually is conducted during a 2-3 week period and the response criteria (e.g. weight gain, bone ash, bone Ca content, bone strength, plasma Ca concentration) are compared with those birds fed a standard diet with Ca carbonate assumed to have 100% availability (Coon *et al.*, 2002). These values refer to 'relative bioavailability'; as the term implies, this is a comparative measure and does not reflect the actual amount of Ca absorbed from the test supplement by the bird.

Calcium availability from inorganic Ca supplements is considered to be 100% (Table 12.1). Possibly based on the availability data for inorganic Ca, Ca in feed ingredients is also generally assumed to be highly available. However, no nutrient, inorganic or organic, can be completely digested, absorbed or retained, and Ca is no exception. Some evidence against this popular belief of 100% Ca availability comes from published data for apparent ileal Ca digestibility in practical poultry diets. For example, reported values for the apparent ileal

Table 12.1. *Bioavailability of calcium from different inorganic calcium sources for young chickens – selected examples.*

	Response criteria	Dietary calcium concentration (%)	Relative bioavailability (%)[1]	Reference
Limestone	body ash	0.90	102	Blair *et al.* (1965)
	growth	0.11-0.56	75-96	Reid and Weber (1976)
	bone ash	0.11-0.56	73-109	Reid and Weber (1976)
	bone ash	0.67	77	Augspurger and Baker (2004)
Bone meal	body ash	0.90	109	Blair *et al.* (1965)
Tricalcium phosphate	body ash	0.90	115	Blair *et al.* (1965)
Dicalcium phosphate	body ash	0.90	113	Blair *et al.* (1965)
Oyster shell	growth	0.11-0.64	87	Reid and Weber (1976)
	bone ash	0.11-0.64	108	Reid and Weber (1976)
	bone ash	0.66	95	Augspurger and Baker (2004)
Deflourinated phosphate	bone ash	0.77	88-102	Burnell *et al.* (1990)

[1] Relative to a calcium carbonate standard, which was considered to have 100% availability.

absorption of Ca in corn-soy broiler starter diets, containing no supplemental phytase, are low to moderate ranging between 28 (Ravindran *et al.*, 2006) and 66% (Walk *et al.*, 2012).

12.6 Ileal versus total tract calcium digestibility

In pigs, digestibility of Ca and P in feed ingredients is usually determined over the total tract (Petersen and Stein, 2006; Fang *et al.*, 2007) and this approach is workable because faecal samples can be collected without urine contamination. In poultry, measurements of total tract digestibility are practically unsound because of the collective voiding of urine and faeces. Calculations based on poultry excreta, therefore, reflect retainable, rather than digestible Ca values, of the particular source. For P, it is well documented that the plasma P concentration increases until a physiological threshold is reached. When the dietary non-phytate P exceeds this threshold, the excess P is excreted via urine confounding the data at the excreta level. Studies by Manangi and Coon (2006) suggest that, for broilers, this critical threshold for dietary non-phytate P is between 0.2 to 0.3%. Corresponding data for Ca is not available, but it is likely a physiological Ca threshold similar to that of P may exist and therefore Ca digestibility measurements in poultry must be made at the ileal level.

12.7 Methodologies for calcium digestibility measurement

Currently there is no established method available for the determination of Ca digestibility in poultry. However, three different methods, namely direct, difference, and regression are used for the determination of ileal amino acid digestibility in feedstuffs. These methods can be potentially applied to Ca digestibility assays. Each method has its strengths and limitations, which have been discussed by Ravindran and Bryden (1999) and Lemme *et al.* (2004).

Several studies have been conducted during the recent past to determine the true P digestibility of feedstuffs for broiler chickens using the regression method (Dilger and Adeola, 2006; Liu *et al.*, 2013; Mutucumarana *et al.*, 2014b,c, 2015) and the direct method (Mutucumarana and Ravindran, 2016). In the regression method, diets with graded concentrations of the specific nutrient from the specific assay ingredient are formulated. Theoretically, the digestibility estimates determined by the regression method are automatically corrected for total endogenous losses and represent true digestibility values. In the direct method, test ingredient serve as the sole source of nutrient in the diet. The digestibility values determined using the direct method are apparent values and need to be corrected for endogenous losses.

Calcium digestibility data for poultry do not currently exist. A series of studies in our laboratory were conducted in recent years to determine the true Ca digestibility of common Ca sources for broiler chickens and a brief overview of the findings from six Ca digestibility assays is presented herein.

12.7.1 Digestible calcium in meat and bone meals

Two studies, using different methodologies, were conducted to determine the true ileal Ca digestibility of MBM. In the first study, the regression method was employed (Anwar *et al.*, 2015) to assay three MBM samples. The strong linear relationship observed between dietary Ca concentrations and digesta Ca output indicated that the regression method can be used for the estimation of Ca digestibility of MBM. The true Ca digestibility coefficients of the three MBM samples ranged between 0.46 and 0.60; exact reasons for the observed variability are unclear, but could not be attributed to the contents of ash, Ca and bones or particle size.

In the second study, the direct method (Anwar *et al.*, 2016a) was used to determine the apparent Ca digestibility in four MBM samples and true digestibility values were calculated using basal endogenous losses measured following the feeding of a Ca- and P-free diet. True ileal Ca digestibility coefficients of the four samples varied from 0.41 to 0.56. Three of the MBM samples evaluated in these two studies were the same, thus enabling comparison between the two methods. The digestibility estimates determined by the regression method are automatically corrected for endogenous losses and represent the true digestibility values, while those determined by the direct method are apparent values and need to be corrected for endogenous losses. The advantage of the direct method is that it is less expensive and simple compared with the regression method in part because fewer diets and birds are required. The comparison of true Ca digestibility coefficients of MBM samples determined by the direct and regression methods (Table 12.2) showed that the ranking and trend in variation in digestibility between samples within each method

Table 12.2. *True ileal calcium digestibility coefficients of three meat and bone meal samples as influenced by methodology (Anwar* et al.*, 2016a).*[1]

	True ileal digestibility[2]	
	Direct method	Regression method
MBM-1	0.56[a]	0.60[a]
MBM-2	0.45[bc]	0.46[b]
MBM-3	0.52[ab]	0.50[ab]
Probability, $P\leq$	0.05	0.05

[1] A dietary Ca concentration of 0.83% was used in direct method, while Ca concentrations ranged from 0.14 to 0.95% in the regression method. Data were subjected to two-way analysis of variance. Meat and bone meal (MBM) effect, $P<0.05$; method effect, $P>0.05$; MBM × method, $P>0.05$.
[2] a,b,c values with a different superscript within a column differ significantly ($P<0.05$).

were similar and that there was no difference ($P>0.05$) in digestibility of MBM determined by these two methods.

It must be noted, however, that MBM has a Ca:P ratio of 2:1, indicating a good balance between these inter-dependent minerals and therefore optimal Ca absorption. Given this balance is critical for the absorption of Ca, the applicability of the direct method to common feed ingredients with much lower Ca:P ratios need to be evaluated in future studies. The major finding of these studies was that the Ca in MBM is not highly available, as widely assumed. Industry nutritionists generally assign availability values of 80-100% for Ca in MBM. It must be noted that these values can be even lower in practical diets containing phytate (Selle *et al.*, 2009). A study with pigs has also reported apparent total tract Ca digestibility coefficients ranging between 0.53 and 0.81 in eight MBM sources (Sulabo and Stein, 2013).

12.7.2 Digestible calcium in limestone

As direct and regression methods yielded comparable results for true Ca digestibility for MBM, it was decided to continue the use of direct method in subsequent experiments. Two assays were conducted with limestone.

In the first study, the effects of limestone source and dietary P concentration on the apparent ileal Ca digestibility in three limestone samples of calcitic origin were examined (Anwar *et al.*, 2016b). Two experimental diets, containing 0.9% Ca, were developed from each limestone source with 0 and 0.45% non-phytate P. Apparent ileal Ca digestibility coefficients of Limestone-1, Limestone-2 and Limestone-3 were determined to be 0.58, 0.61 and 0.54, respectively (Table 12.3). Increasing the P concentration from 0 to 0.45% in assay diets increased ($P<0.05$) the average Ca digestibility from 0.56 to 0.60.

Table 12.3. *Effect of limestone source and dietary phosphorus concentration on apparent ileal calcium digestibility coefficients in broiler chickens (Anwar* et al., *2016b).*[1]

Main effects	Calcium digestibility coefficient[2]
Limestone source	
Limestone-1	0.58[ab]
Limestone-2	0.61[a]
Limestone-3	0.54[b]
SEM[3]	0.01
Dietary phosphorus (%)	
0	0.56[b]
0.45	0.60[a]
SEM[3]	0.01
Probabilities, $P\leq$	
Limestone source	0.02
Dietary phosphorus	0.04
Limestone source × Dietary phosphorus	0.16

[1] Each value represents the mean of six replicates (eight birds per replicate). Dietary Ca and P concentration of the diets was 0.9 and 0.45%, respectively.

[2] [a,b,c] Values with a different superscript within a column differ significantly ($P<0.05$).

[3] Pooled standard error of mean.

In a follow-up study, the effect of limestone particle size and Ca:non-phytate P ratio was determined on the true ileal Ca digestibility of limestone for broiler chickens (Anwar *et al.*, in press). A limestone sample was passed through a set of sieves and separated into fine (<0.5 mm) and coarse (1-2 mm) particles. Six experimental diets were developed using each particle size with Ca:non-phytate P ratios of 1.5:1, 2.0:1 and 2.5:1, with ratios being adjusted by manipulating the dietary Ca concentrations. A Ca- and P-free diet was also developed to determine the basal ileal endogenous Ca losses. Increasing Ca:non-phytate P ratios reduced ($P<0.05$) the true Ca digestibility of limestone (Table 12.4). The true Ca digestibility coefficients of limestone with Ca:non-phytate P ratios of 1.5:1, 2.0:1 and 2.5:1 were 0.65, 0.57 and 0.49, respectively. Particle size of limestone had a marked effect ($P<0.05$) on the Ca digestibility, with the digestibility being higher in coarse particles.

12.7.3 Digestible calcium in limestone and oyster shell

Oyster shell, a marine source, is another source of Ca used in poultry diets. Being an organic source, Ca digestibility of this source may be different from that of calcitic limestone. A recent study (Anwar *et al.*, unpublished) was conducted to determine the effect of Ca source (limestone vs oyster shell) and particle size (<0.5 and 1-2 mm) on true Ca digestibility for broiler chickens. The true ileal Ca digestibility coefficient of limestone and oyster shell was found to be similar. Increasing the particle size from <0.5 mm to 1-2 mm increased ($P<0.05$) the Ca digestibility of both Ca sources. The Ca digestibility

Table 12.4. *Effect of calcium to non-phytate phosphorus ratio and particle size of limestone on apparent and true calcium digestibility coefficients in broiler chickens (Anwar* et al., *in press).*[1]

Main effects	Apparent ileal calcium digestibility[2]	True ileal calcium digestibility[2,3]
Ca:non-phytate P ratio		
1.5:1	0.63[a]	0.65[a]
2.0:1	0.56[b]	0.57[b]
2.5:1	0.48[c]	0.49[c]
SEM[4]	0.01	0.01
Particle size (mm)		
<0.5	0.42[b]	0.43[b]
1.0-2.0	0.70[a]	0.71[a]
SEM[4]	0.01	0.01
Probabilities, $P\leq$		
Ca:non-phytate P ratio	<0.001	<0.001
Particle size	<0.001	<0.001
Ca:non-phytate P ratio × Particle size	0.15	0.14

[1] Each treatment value represents the mean of six replicates (eight birds per replicate). Dietary Ca concentration of diets with Ca:non-phytate P ratio of 1.5:1, 2.0:1 and 2.5:1 was 0.68,0.90 and 1.13%, respectively. Dietary P concentration was maintained at 0.45% in all diets.

[2] a,b,c values with a different superscript within a column differ significantly ($P<0.05$).

[3] Ileal endogenous calcium losses were determined to be (mean ± SE) 127±12 mg/kg of dry matter intake and were used to determine the true calcium digestibility.

[4] Pooled standard error of mean.

estimates of fine and coarse particles of limestone and oyster shell were 0.38 and 0.62, and 0.33 and 0.56, respectively.

12.7.4 Digestible calcium in other calcium supplements and feedstuffs

Dicalcium phosphate and MCP are also commonly used as Ca sources in poultry diets. In addition, ingredients such as fish meal, poultry by-product meal and canola meal can contribute reasonable amounts of dietary Ca when included in feed formulations. The true Ca digestibility of these feed ingredients was assayed in the most recent study (Anwar *et al.*, unpublished) using direct method. The true Ca digestibility coefficients for DCP, MCP, poultry by-product meal, fish meal and canola meal were determined to be 0.28, 0.33, 0.29, 0.24 and 0.31, respectively. These values are much lower than those reported in pigs for DCP, MCP and fish meal by Gonzalez-Vega *et al.* (2015a,b). These pig studies also used the direct method, but the diets were maize-based instead of maize starch used in our assays. It is likely, however, that the high dietary P in our assay diets, which was unavoidable, due to the high P content relative to Ca in these ingredients, and the resultant Ca:P imbalance was responsible for the observed low digestibility values. In this study, the Ca and total P concentrations in assay diets containing DCP, MCP, poultry by-product meal, fish meal and canola meal were 0.90 and 0.83; 0.90 and 1.29; 0.90 and 0.60; 0.63 and 0.63, and 0.57 and 0.63%, respectively.

12.8 Challenges for the future

Some challenges that need to be resolved in future studies are briefly discussed below. Several of these issues have been highlighted previously (Adedokun and Adeola, 2013; Proszkowiec-Weglarz and Angel, 2013).

12.8.1 Methodology

Further studies comparing the different methodologies (regression, direct, and substitution) are warranted, before agreement on a consensus methodology is reached. In the direct method, a 2:1 Ca:P ratio was maintained in the assay diets used for limestone and MBM. But this is not possible in assays for inorganic phosphates (MCP and DCP) and common protein meals with the high P concentrations (relative to Ca), resulting in much narrower ratios and possibly leading to the observed low Ca digestibility estimates. Future studies are needed to understand the reasons for the low digestibility values for these ingredients observed with the direct method. In particular, the effects of dietary P concentrations and Ca:P ratios on Ca digestibility will be pertinent.

12.8.2 Undigested/ unabsorbed calcium

Our data, presented above, show that almost half the Ca in limestone and meat and bone meal is unabsorbed and excreted. In this context, future research should explore the reasons for the poor absorption and how the absorption can be improved (Dudley-Cash, 2016).

12.8.3 Supplemental phytase

Given that the use of microbial phytase in poultry diets is routine nowadays, generation of data on ileal Ca and P digestibility of organic sources with phytase supplementation is needed to be relevant in practical situations. This consideration is particularly relevant given the differential effects of phytase on the release of Ca and P (Walk, 2016).

12.8.4 Age effect

The same nutrient digestibility values are used in feed formulations for broilers of different ages and for broilers and laying hens. Comparative values between bird classes are limited and the assumption of similar digestibility has never been validated. Some evidence indicates that Ca digestibility is reduced with age of broilers (Angel *et al.*, 2013). As Ca metabolism is influenced by physiological needs (egg production, growth rate and skeletal growth) and has a very tight homeostatic control, such a validation is urgent.

12.8.5 Adaptation length

Since animals have a very tight homeostatic regulation of Ca, a time course of changes in Ca digestibility must be established. Angel *et al.* (2013), using the direct method, determined the apparent Ca digestibility of four ingredients from 8 to 96 hours after the introduction of assay diets and reported that, based on changes in Ca digestibility, the best time to determine digestibility was around 32-40 h. An assay period of 5 days has been suggested by WPSA (2013) for P utilisation measurements. According to Tamim *et al.* (2004), a shorter duration minimises the physiological adaptation of digestive tract to P-deficient diets. It is therefore important that a standard dietary adaptation length should be established to reduce variability associated with the measurement of Ca digestibility of ingredients.

12.8.6 Vitamin D_3

The effect of vitamin D_3 (1, 25 dihydroxy D_3) on the absorption of Ca from the intestine and its homeostasis in the body is well known (Veum, 2010). In our assay diets, vitamin D_3 was supplied as 4,000 International Units cholecalciferol per kg diet. Although this supplementation level is reported to be sufficient to support live performance and bone mineralisation (Fritts and Waldroup, 2003), adequacy of this inclusion level for maximum Ca absorption has never been tested.

12.8.7 Particle size

Physical form of Ca supplements is important to allow for maximum utilisation and this has been demonstrated in our studies with limestone and oyster shell. However, because broiler diets are pelleted, particle size of limestone may not be directly relevant to broiler feeding. A practical solution may be to feed these Ca sources as large particles separate from the rest of the broiler feed (Wilkinson *et al.*, 2013; Abdollahi *et al.*, 2016).

12.8.8 Mash vs pelleted diets

The use of mash diets in the digestibility assay may also be criticised. Use of semi-purified mash diets is standard practice to determine the ileal digestibility of nutrients in feed ingredients – whether of amino acids, P or Ca. In addition, semi-purified diets, such as those used in the current work, are difficult to pellet and are easily broken down. It is recognised, however, since practical broiler diets are fed as pellets, it will be of interest to investigate the effect of pellet vs mash diets on ileal Ca and P digestibility in future experiments as published data on the effect of feed form on the digestibility of these minerals are contradictory (Table 12.5).

12.8.9 Applicability to practical diets

The results generated using purified or semi-purified diets may not be applicable to practical-type diets. There are number of complex and interacting factors in practical diets that influence Ca digestibility, the most important of which is the phytate. It noteworthy,

Table 12.5. *Effect of feed form on apparent ileal calcium digestibility in broiler chickens.*

	Dietary calcium (%)	Dietary phosphorus (%)	Apparent ileal calcium digestibility coefficient	
			Mash	Pellet
Abdollahi *et al.* (2013)[1]	1.0	0.52	0.42	0.50
Abdollahi *et al.* (2013)[2]	1.0	0.52	0.37	0.29
Abdollahi *et al.* (2014)[3]	1.0	0.53	0.49	0.36
Naderinejad *et al.* (2016)[1]	1.0	0.52	0.53-0.57	0.33-0.45

[1] Maize-based diets.
[2] Wheat-based diets.
[3] Sorghum-based diets.

however, the use of semi-purified diets to determine the amino acid digestibility and the use of these values in practical feed formulations had been a great success (Dudley-Cash, 2016).

12.8.10 Calcium analysis

A major problem often faced in Ca research was the notable differences between analysed and calculated dietary Ca concentrations, especially at lower dietary Ca concentrations. Given that over 85% of dietary Ca is supplied by Ca sources, analysing them prior to formulation should resolve this problem, but our experience shows that is not always the case.

12.9 Digestible calcium requirements

Calcium requirements for broiler chickens have been studied extensively over the past 85 years (Driver *et al.*, 2005; Angel, 2011). Currently Ca requirements are considered on a total basis and the available data on total Ca requirements are far from being conclusive. There is diversity in the recommended Ca requirement of broilers of different ages, due to differences *inter alia* in response criteria, methodology, strain, diet type, Ca and P sources, Ca:P ratio, and non-phytate P concentrations. The only agreements in these studies are that that Ca requirement decreases with age and higher dietary Ca is required for maximum bone mineralisation compared to growth.

Validation of the ileal assay methodology will ultimately lead to a re-examination of the requirements for Ca and P (Dudley-Cash, 2016). It is also readily evident that the requirement for Ca is largely dependent upon the dietary P concentration and 'ideal' Ca:P ratio becomes critical. The requirement values for digestible Ca, therefore, can be finalised only after the digestible P requirement values are established and appropriate methodologies are agreed for the determination of Ca digestibility. Ideally, the digestible

Ca requirements would have to be determined in studies where ingredients used have been previously assayed for Ca digestibility (Angel, 2011).

12.10 Conclusions

The genesis of the studies reported in this paper was the necessity to consider an appropriate ratio between digestible Ca and digestible P for optimum absorption and utilisation of both minerals. Currently, except for that generated from the studies mentioned above, there is no published data available for the ileal digestibility of Ca in ingredients for poultry. The first step in this exercise is to develop an acceptable method that can be used across laboratories to assay large number of ingredient samples to generate matrix values. It is evident that such a development of database for digestible P and Ca will require considerable investment in terms of research and scientific collaboration, and will take time. Experience in Ca digestibility research presented in this overview will hopefully be of interest and useful in future discussions.

1. Ileal Ca digestibility in limestone and meat and bone meal for poultry is not high and much lower than the bioavailability values reported in the past.
2. Particle size of limestone and oyster shell markedly influences the Ca digestibility.
3. The need to develop and agree on an acceptable method of ileal Ca digestibility assay to be used across laboratories is highlighted.

References

Abdollahi, M.R., Ravindran, V. and Svihus, B., 2014. Influence of feed form on performance, ileal nutrient digestibility and energy utilisation in broiler starters fed sorghum-based diets. Livestock Production Science 165:80-86.

Abdollahi, M.R., Ravindran, V. and Svihus, B., 2013. Influence of grain type and feed form on performance apparent Metabolisable energy and ileal digestibility of nitrogen, starch, fat, calcium and phosphorus in broiler chickens. Animal Feed Science and Technology 186:193-203.

Abdollahi, M.R., Duangnumsawang, Y., Kwakkel, R.P., Steenfeldt, S., Bootwalla, S.M. and Ravindran, V. 2016. Investigation of the interaction between separate calcium feeding and phytase supplementation on the growth performance, calcium intake, nutrient digestibility and energy utilisation in broiler starters. Animal Feed Science and Technology 219:48-58.

Adedokun, S.A. and Adeola, O. 2013. Calcium and phosphorus digestibility: metabolic limits. Journal of Applied Poultry Research 22:600-608.

Angel, R., 2011. Calcium and phosphorus requirements in broilers and laying hens. Australian Poultry Science Symposium 22:32-48.

Angel, R., Proszkowiec-Weglarz, M., Jimenez-Moreno, E., Kim, S.-W. and Plumstead, P., 2013. Impact of time and dietary calcium and phosphorus deficiencies on their digestibilities in single ingredients. Proceedings European Symposium of Poultry Nutrition, Potsdam, Germany.

Anwar, M.N., Ravindran, V., Morel, P.C.H., Ravindran, G. and Cowieson, A.J., 2015. Measurement of true ileal calcium digestibility in meat and bone meal for broiler chickens. Animal Feed Science and Technology 206:100-107.

Anwar, M.N., Ravindran, V., Morel, P.C.H., Ravindran, G. and Cowieson, A.J., 2016a. Measurement of true ileal calcium digestibility in meat and bone meal for broiler chickens using the direct method. Poultry Science 95:70-76.

Anwar, M.N., Ravindran, V., Morel, P.C.H., Ravindran, G. and Cowieson, A.J., 2016b. Apparent ileal digestibility of calcium in limestone for broiler chickens. Animal Feed Science and Technology 213:142-147.

Anwar, M.N., Ravindran, V., Morel, P.C.H., Ravindran, G. and Cowieson, A.J., in press. Effect of limestone particle size and calcium to non-phytate phosphorus ratio on true ileal calcium digestibility of limestone for broiler chickens. British Poultry Science. DOI: http://doi.org/10.1080/00071668.2016.1196341.

Augspurger, N.R. and Baker, D.H., 2004. Phytase improves dietary calcium utilization in chicks, and oyster shell, carbonate, citrate, and citrate-malate forms of calcium are equally bioavailable. Nutrition Research, 24:293-301.

Ballam, G.C., Nelson, T.S. and Kirby, L.K., 1984. Effect of fiber and phytate source and of calcium and phosphorus level on phytate hydrolysis in the chick. Poultry Science 63:333-338.

Blair, R., English, P. and Michie, W., 1965. Effect of calcium source on calcium retention in the young chick. British Poultry Science 44:355-356.

Burnell, T.W., Cromwell, G.L. and Stahly, T.S., 1990. Effects of particle size on the biological availability of calcium and phosphorus in defluorinated phosphate for chicks. Poultry Science 69:1110-1117.

Coon, C., Leske, K. and Seo, S., 2002. The availability of calcium and phosphorus in feedstuffs. In: McNab, J.M. and Boorman, N. (eds.) Poultry feedstuffs: supply, composition, and nutritive value. CABI Publishing, New York, NY, USA, pp. 151-186.

Davies, K.M., Rafferty, K. and Heaney, R.P., 2004. Determinants of endogenous calcium entry into the gut. American Journal of Clinical Nutrition 80:919-923.

Dilger, R.N. and Adeola, O., 2006. Estimation of true phosphorus digestibility and endogenous phosphorus loss in growing chicks fed conventional and low-phytate soybean meals. Poultry Science 85:661-668.

Driver, J., Pesti, G., Bakalli, R. and Edwards, H., 2005. Calcium requirements of the modern broiler chicken as influenced by dietary protein and age. Poultry Science 84:1629-1639.

Duke G.E., 1982. Gastrointestinal motility and its regulation. Poultry Science 61:1245-1256.

Dudley-Cash, W.A., 2016. Digestible phosphorus, calcium in MBM. Feedstuffs (USA) 88(06), June 6 2016.

Fang, R.J., Yin, Y.L., Wang, K.N., He, J.H., Chen, Q.H., Li, T.J., Fan, M.Z. and Wu, G., 2007. Comparision of the regression analysis technique and the substitution method for the determination of true phosphorus digestibility and faecal endogenous phosphorus losses associated with feed ingredients for growing pigs. Livestock Science 109:251-254.

Favus, M.J., 1985. Factors that influence absorption and secretion of calcium in the small intestine and colon. American Journal of Physiology (Gastrointestinal and liver Physiology) 248:G147-G157.

Fritts, C.A. and Waldroup, P.W., 2003. Effect of source and level of vitamin D on live performance and bone development in growing broilers. Journal of Applied Poultry Research 12:45-52

González-Vega, J.C., Walk, C.L. and Stein, H.H., 2015a. Effects of microbial phytase on apparent and standardized total tract digestibility of calcium in calcium supplements fed to growing pigs. Journal of Animal Science 93:2255-2264.

González-Vega, J.C., Walk, C.L. and Stein, H.H., 2015b. Effect of phytate, microbial phytase, fiber, and soybean oil on calculated values for apparent and standardized total tract digestibility of calcium and apparent total tract digestibility of phosphorus in fish meal fed to growing pigs. Journal of Animal Science 93:4808-4818.

Gonzalez-Vega, J.C. and Stein, H.H., 2016. Digestibility of calcium in feed ingredients and requirements of digestible calcium for growing pigs. Animal Production Science 56:1339-1344.

Lemme, A., Ravindran, V. and Bryden, W.L., 2004. Ileal digestibility of amino acids in feed ingredients for broilers. World's Poultry Science Journal 60:423-438.

Limlomwongse, L., Deachapunya, C. and Krishnamra, N., 1988. Biliary calcium and bile acid secretion in intact and TPTX rats with varying plasma calcium concentration. Digestive Diseases and Sciences 33:685-691.

Liu, J., Chen, D. and Adeola, O., 2013. Phosphorus digestibility response of broiler chickens to dietary calcium-to-phosphorus ratios. Poultry Science 92:1572-1578.

Manangi, M.K. and Coon, C.N., 2006. Phosphorus utilization and environment concerns. In: Zimmermann, N.G. (ed.) Proceedings of the 4th Mid-Atlantic Nutrition Conference University of Maryland, College Park, MD, USA.

Mutucumarana, R.K. and Ravindran, V., 2016. Measurement of true ileal phosphorus digestibility in meat and bone meal for broiler chickens using the direct method. Animal Feed Science and Technology 219: 249-256.

Mutucumarana, R.K., Ravindran, V., Ravindran, G. and Cowieson, A.J., 2014a. Influence of dietary calcium concentration on the digestion of nutrients along the intestinal tract of broiler chickens. Journal of Poultry Science 51:392-401.

Mutucumarana, R.K., Ravindran, V., Ravindran, G. and Cowieson, A.J., 2014b. Measurement of true ileal digestibility and total tract retention of phosphorus in corn and canola meal for broiler chickens. Poultry Science 93:412-419.

Mutucumarana, R.K., Ravindran, V., Ravindran, G. and Cowieson, A.J., 2014c. Measurement of true ileal digestibility of phosphorus in some feed ingredients for broiler chickens. Journal of Animal Science 92:5520-5529.

Mutucumarana, R.K., Ravindran, V., Ravindran, G. and Cowieson, A.J., 2015. Measurement of true ileal phosphorus digestibility in maize and soybean meal for broiler chickens: Comparison of two methodologies. Animal Feed Science and Technology 206:76-80.

Naderinejad, S., Zaefarian, F., Abdollahi, M.R., Hassanabadi, A., Kermanshahi, H. and Ravindran, V. 2016. Influence of feed form and particle size on performance, nutrient utilisation and gastrointestinal tract development and morphometry in broiler starters fed maize-based diets. Animal Feed Science and Technology 215:92-94.

NRC, 1994. Nutrient requirements of poultry (9th rev. ed.). National Research Council, National Academic Press, Washington, DC, USA.

Petersen, G.I. and Stein, H.H., 2006. Novel procedure for estimating endogenous losses and measurement of apparent and true digestibility of phosphorus by growing pigs. Journal of Animal Science 84:2126-2132.

Proszkowiec-Weglarz, M. and Angel, R. 2013. Calcium and phosphorus metabolism in broilers: Effect of homeostatic mechanism on calcium and phosphorus digestibility. Journal of Applied Poultry Research 22:609-627.

Ravindran, V. and Bryden, W.L., 1999. Amino acid availability in poultry – *in vitro* and *in vivo* measurements. Australian Journal of Agricultural Research 50:889-908.

Ravindran, V., Morel, P.C.H., Partridge, G.G., Hruby, M. and Sands, J.S., 2006. Influence of an *Escherichia* coli-derived phytase on nutrient utilization in broiler starters fed diets containing varying concentrations of phytic acid. Poultry Science 85:82-89.

Reid, B.L. and Weber, C.W., 1976. Calcium availability and trace mineral composition of feed grade calcium supplements. Poultry Science 55:600-605.

Selle, P.H., Cowieson, A.J. and Ravindran, V., 2009. Consequences of calcium interactions with phytate and phytase for poultry and pigs. Livestock Science 124:126-141.

Sulabo, R. and Stein, H.H., 2013. Digestibility of phosphorus and calcium in meat and bone meal fed to growing pigs. Journal of Animal Science 91:1285-1294.

Suttle N.F., 2010. Mineral nutrition of livestock. 4th ed. CABI Publishing, CAB International, Wallingford, UK.

Tamim, N.M. and Angel, R., 2003. Phytate phosphorus hydrolysis as influenced by dietary calcium and micro-mineral source in broiler diets. Journal of Agricultural and Food Chemistry 51:4687-4693.

Tamim, N.M., Angel, R. and Christman, M., 2004. Influence of dietary calcium and phytase on phytate phosphorus hydrolysis in broiler chickens. Poultry Science 83:1358-1367.

Veum, T.L., 2010. Phosphorus and calcium nutrition and metabolism. In: Vitti, D.M.S.S. and Kebreab, E. (eds.) Phosphorus and calcium utilization and requirements in farm animals. CAB International, Oxfordshire, UK. pp. 94-112.

Walk C.L., 2016. The influence of calcium on phytase efficacy in non-ruminant animals. Animal Production Science 56:1345-1349.

Walk C.L., Addo-Chidie E., Bedford M. and Adeola O., 2012. Evaluation of a highly soluble calcium source and phytase in the diets of broiler chickens. Poultry Science 91:2255-2263.

Wilkinson, S.J., Selle, P.H., Bedford M.R. and Cowieson, A.J., 2013. Separate feeding of calcium improves performance and ileal nutrient digestibility in broiler chicks. Animal Production Science 54:172-178.

Williamson, B. and Percy-Robb, I., 1980. Contribution of biliary lipids to calcium binding in bile. Gastroenterology, 78:696-702.

WPSA., 2013. Determination of phosphorus availability in poultry. Working Group No 2 (Nutrition), World's Poultry Science Journal 69:687-698.

Chapter 13
Establishing a digestible calcium requirement for pigs

H.H. Stein[1], L.A. Merriman[1] and J.C. González-Vega[2]*
[1]University of Illinois at Urbana Champaign, Department of Animal Science, Monogastric Nutrition Laboratory, 1207 West Gregory Dr., Urbana, IL 61801, USA; [2]Evonik Industries AG, Rodenbacher Chaussee 4, 63457 Hanau, Germany; hstein@illinois.edu

Abstract

The standardized total tract digestibility (STTD) of calcium (Ca) in most feed ingredients used in diets fed to pigs has been determined, and the effect of microbial phytase on STTD of Ca has also been determined. This has made it possible to formulate diets based on values for STTD of Ca and attempts to determine requirements for STTD of Ca for 11 to 25, 25 to 50, and 100 to 130 kg pigs have been made. Results of these experiments have indicated that STTD Ca provided in excess of the requirement is detrimental to pig growth performance. This is particularly true if dietary STTD phosphorus (P) is marginal, whereas the negative effects of excess Ca is mitigated if STTD P is provided in excess of the requirement. As a consequence, effects of microbial phytase on the STTD of Ca need to be accounted for in diet formulation. Based on currently available information, it is recommended that diets for pigs less than 50 kg are formulated to have a STTD Ca:STTD P ratio between 1:1 and 1.35:1, whereas diets for pigs greater than 100 kg should have a STTD Ca:STTD P ratio that does not exceed 1.1:1.

Keywords: digestibility, minerals, phosphorus, pigs

13.1 Introduction

Historically, requirements for calcium (Ca) in diets fed to pigs have been expressed on the basis of total Ca and all feed requirement tables include values for the requirement for total Ca by different categories of pigs. This is in contrast to values for phosphorus (P) where requirements are usually expressed as values for either apparent total tract digestible (ATTD) P or standardized total tract digestible (STTD) P. In the current version of 'Nutrient Requirements of Swine' requirements for total Ca were calculated by multiplying requirements for STTD P by 2.15 (NRC, 2012). However, the committee that developed 'Nutrient Requirements of Swine' acknowledged that this approach was used simply because no data for the digestibility of Ca in feed ingredients were available and the committee specifically stated that 'A preferred ratio would have been a ratio between digestible Ca and digestible P, but again, because of lack of data the ratios between total Ca and STTD P are used' (NRC, 2012). Thus, the lack of data for digestibility of Ca in feed ingredients prevented the committee from estimating requirements for STTD of Ca.

DOI 10.3920/978-90-8686-836-0_13, © Wageningen Academic Publishers 2016

However, during the last few years, values for the digestibility of Ca in most Ca containing feed ingredients have been determined, and it has, therefore, been possible to initiate work to estimate requirements for digestible Ca in diets fed to different categories of pigs. It is the objective of the current contribution to present currently available data for digestibility of Ca in feed ingredients and preliminary data for the requirement for STTD Ca by growing-finishing pigs.

Nutrient requirements are determined by factorial calculations or empirical measurements. Because the factorial technique accounts for several factors such as availability, obligatory losses of the nutrient in the body, and retention of nutrients, it is believed that requirements obtained using this approach is more precise (Weremko *et al.*, 1997). However, empirical measurements, which use one or several response criteria to determine the nutrient requirement, have been most commonly used (NRC, 2012; Weremko *et al.*, 1997). However, factors such as sex and age may influence the requirement values (NRC, 2012).

Muscle and bones grow independently, therefore, deposition of Ca and P in lean tissue and skeletal tissue are not directly proportional (Crenshaw, 2001). Thus, the ratio between Ca and P in the body ranges between 1.20:1 and 1.60:1 for a 25 kg pig, and between 1.25:1 and 1.70:1 for a 50 kg pig (Hendriks and Moughan, 1993; Mahan and Shields, 1998; Pettey *et al.*, 2015; Rymarz *et al.*, 1982; Wiseman *et al.*, 2009). However, because 96 to 99% of Ca is believed to be present in skeletal tissue (Crenshaw, 2001), it is expected that requirements of STTD Ca to maximize bone ash are close to requirements to maximize Ca retention in the body.

13.2 Digestibility of calcium in feed ingredients

Preliminary work to determine the digestibility of Ca in feed ingredients fed to pigs indicated that the basal endogenous loss of Ca from pigs is between 150 and 400 mg per kg dry matter intake (Gonzalez-Vega *et al.*, 2013). As a consequence, values for ATTD of Ca are influenced by the inclusion level of Ca in the diets because the endogenous Ca contributes a greater proportion of Ca to the faecal output of Ca at low levels of Ca intake compared with greater levels of intake. Values for the ATTD of Ca are, therefore, not additive in mixed diets and as a consequence, digestibility values for Ca need to be corrected for the basal endogenous losses and expressed as values for the STTD of Ca.

A number of experiments have been conducted in recent years to determine the STTD of Ca in feed ingredients fed to pigs (Table 13.1). In general, the greatest STTD values have been obtained in feed ingredients of animal origin whereas the STTD of Ca in most inorganic sources of Ca has been observed to be less than in most animal proteins. Cereal grains and most cereal co-products contain very little or no Ca, whereas oilseed meals provide some Ca to the diets with canola meal having the greatest concentration of Ca among oilseed meals. However, as is the case for P, a proportion of the Ca in oilseed meals is bound to phytate with a reduced digestibility as a consequence. Therefore, inclusion of microbial phytase to diets that contain oilseed meals will result in an increased STTD of Ca (Gonzalez-Vega *et al.*, 2013). It has, however, been demonstrated that the increase

Table 13.1. *Standardized total tract digestibility (STTD) of calcium (Ca) in feed ingredients (%).*[1]

Ingredient	STTD of Ca without phytase	STTD of Ca with phytase
Mineral supplements		
Monocalcium phosphate[1]	77	80
Dicalcium phosphate[1]	73	75
Calcium carbonate[1]	64	71
Plant feed ingredients		
Canola meal[2]	42	-
Soybean meal[2]	78	-
Animal feed ingredients		
Meat and bone meal[3]	77	82
Meat meal[3]	77	86
Fish meal[4]	76	87
Poultry meal[3]	82	76
Poultry by product meal[3]	88	87
Skim milk powder[2]	93	97
Whey powder[2]	96	93
Whey permeate[2]	58	71

[1] Gonzalez-Vega *et al.*, 2015a.
[2] Unpublished data from the University of Illinois.
[3] Merriman *et al.*, 2016b.
[4] Gonzalez-Vega *et al.*, 2015b.

in digestibility of Ca that is observed in response to microbial phytase in diets based on corn and soybean meal is greater than what can be explained by the release of Ca bound to phytate in soybean meal (Almeida and Stein, 2013), and it was, therefore, hypothesized that other dietary Ca from mineral supplements may be bound to phytate in corn-soybean meal diets. Indeed, working with corn-based diets, it was subsequently demonstrated that the STTD of Ca in calcium carbonate, but not in monocalcium phosphate or dicalcium phosphate, is increased if microbial phytase is added to the diet (Gonzalez-Vega *et al.*, 2015a). Likewise, the STTD of Ca in fish meal included in a corn-based diet is increased by microbial phytase (Gonzalez-Vega *et al.*, 2015b), but for other animal proteins such as meat and bone meals and poultry meals, effects of microbial phytase have been less clear (Merriman *et al.*, 2016b). The STTD of Ca in whey permeate is also increased by microbial phytase, but that is not the case for skim milk powder and whey powder (unpublished data). There is, however, no impact of particle size of calcium carbonate on the STTD of Ca, and growth performance of pigs is not affected by the particle size of the calcium carbonate in the diets (Merriman and Stein, in press).

The reason the STTD of Ca in calcium carbonate and some animal proteins is increased if microbial phytase is used is most likely that the phytate from the plant ingredients may chelate the Ca from non-plant feed ingredients in the stomach of the pigs, which prevents absorption, but if microbial phytase reduces the chelating abilities of phytate, less Ca will be bound to phytate, which will result in increased digestibility of Ca. However, because excess dietary Ca will reduce the digestibility of P and thereby increase P excretion in the faeces (Stein *et al.*, 2011) and reduce growth performance of pigs (Gonzalez-Vega *et al.*, in press, 2016; Merriman *et al.*, 2016a), it is important to take the increased digestibility of Ca that is a result of microbial phytase into account in the formulation of diets fed to growing pigs.

13.3 Calcium absorption

In pigs, Ca may be absorbed via passive or via active transport (Bronner, 2003). Active absorption is the primary route if pigs are supplied low levels of dietary Ca, whereas passive absorption is the primary route if dietary Ca supply is at or above the requirement. However, combined, the two sources of absorption results in a near constant rate of absorption, regardless of the concentration of Ca in the diet, and there is only limited regulation of Ca absorption at the intestinal level (Stein *et al.*, 2011). In the kidneys, Ca can also be absorbed into the bloodstream form the distal tubule, but in this case, only active transport is used, which means that Ca is absorbed only if there is a low concentration of plasma Ca. As a consequence, if plasma concentrations of Ca are adequate, less Ca will be absorbed from the kidney tubules, and more will be excreted in the urine. Thus the main regulatory site of Ca homeostasis in pigs is the kidneys (Gonzalez-Vega *et al.*, in press).

13.4 Requirements for digestible calcium by growing pigs

Three experiments were conducted to determine the requirements for digestible Ca by growing pigs. In the first experiment, six diets that contained 0.36% STTD P and six levels of digestible Ca from 0.32 to 0.72% were used (Gonzalez-Vega *et al.*, in press). Two additional diets containing 0.72% Ca and either 0.33 or 0.40% STTD P were also formulated. The eight diets were fed for 22 days to pigs that were 11.39±1.21 kg at the start of the experiment with four pigs per pen and eight replicate pens per treatment. Individual pig weights were recorded at the beginning and at the conclusion of the experiment and on the last day of the experiment, one pig per pen was sacrificed and the right femur was removed. Data for average daily gain (ADG), average daily feed intake (ADFI) and average gain to feed (G:F) ratios were calculated at the end of the experiment and the femurs from the sacrificed pigs were ashed and analysed for Ca and P. Diets were also fed to 80 pigs that were placed individually in metabolism crates with 10 replicate pigs per diet. Faeces and urine were collected from all pigs and balances for Ca and P were calculated.

In the second experiment, a total of 20 diets were used. Diets contained 0.15, 0.31, 0.39, or 0.47% STTD P and 0.13, 0.27, 0.42, 0.57, or 0.72% STTD Ca. Thus, diets were formulated in a 4×5 factorial design. A total of 240 pigs (initial BW: 24.70±1.27 kg) were allotted to the 20 diets with two pigs per pen and six pen replicates per diet (Gonzalez-Vega *et al.*, 2016). Pigs were fed experimental diets for 28 days and on the last day of the experiment, one pig per pen was sacrificed and the right femur was removed. Diets were also fed to 120 pigs that were placed individually in metabolism crates. All measurements in this experiment were similar to those in Exp. 1.

In the third Experiment, 90 growing pigs (initial BW: 99.89±3.34 kg) were housed individually and randomly allotted to 15 diets with six pigs per diet (Merriman *et al.*, 2016a). Experimental diets were formulated as a 3×5 factorial with three dietary concentrations of STTD P (0.11, 0.21, or 0.31%) and five concentrations of STTD Ca (0.08, 0.18, 0.29, 0.38, or 0.49%). Pigs were fed experimental diets for 28 days. Initial and final BW were recorded and daily feed allotments were recorded as well for calculation of ADG, ADFI, and average G:F for each treatment group. All pigs were sacrificed at the conclusion of the experiment and the right femur was collected from each pig and bone ash, bone Ca and bone P were measured.

Results from Exp. 1 indicated that there were no negative effects of reducing the concentration of STTD Ca in the diets, but adding more than 0.54 or 0.50% STTD Ca to the diets resulted in a negative response to ADG and G:F, respectively (Figures 13.1 and 13.2). In contrast, bone ash increased as the concentration of STTD Ca increased in the diet and the optimum amount of bone ash was achieved if STTD Ca was 0.48% (Figure 13.3). Concentrations of bone P and bone Ca were maximized at dietary concentrations of STTD Ca of 0.56 and 0.50%, respectively. Likewise, the retention of Ca and P in the body was maximized at concentrations of STTD Ca of 0.60 and 0.49%, respectively. Results of this experiment indicated that under the condition of the experiment, there were no negative effects of reducing dietary Ca to well below the expected requirement, but adding excess Ca to the diets is detrimental to ADG and G:F. Because bone ash and bone Ca were maximized at 0.48 to 0.50% STTD Ca in the diets, and because ADG and G:F declined

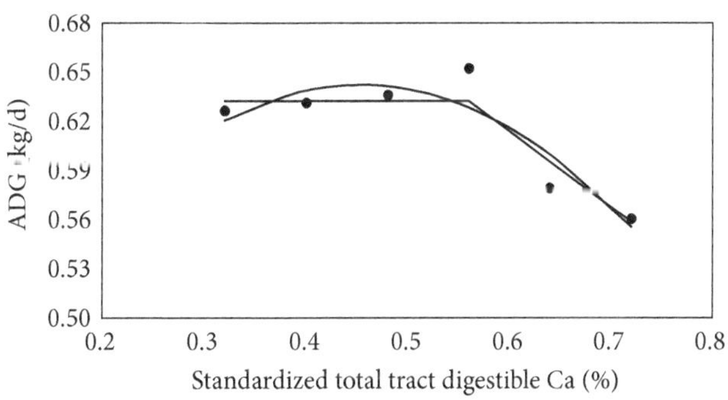

Figure 13.1. *Fitted broken line of ADG (kg/d, d 1 to 22) as a function of standardized total tract digestible Ca (STTD Ca; Exp. 1). The mean of each diet (•) represents the mean of 8 replicated pens per diet. The breakpoint was 0.56±0.10% STTD Ca if the broken line analysis was used (plateau = 0.63 kg/d ADG). The maximum ADG determined from the quadratic analysis was at 0.45±0.04% STTD Ca. The intersections between the broken line and quadratic analysis were 0.36 and 0.54% STTD Ca (adapted from Gonzalez-Vega* et al., *in press).*

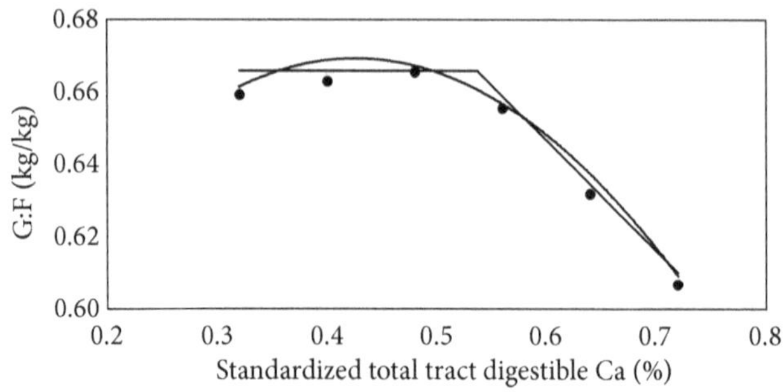

Figure 13.2. *Fitted broken line of G:F (kg/kg, d 1 to 22) as a function of standardized total tract digestible Ca (STTD Ca; Exp. 1). The mean of each diet (•) represents the mean of 8 replicated pens per diet. The break point for G:F was at 0.54±0.04% STTD Ca (plateau = 0.67 kg/kg G:F). The maximum G:F determined from the quadratic analysis was at 0.43±0.04% STTD Ca. The intersections between the broken line and quadratic analysis were 0.35 and 0.50% STTD Ca (adapted from Gonzalez-Vega* et al., *in press).*

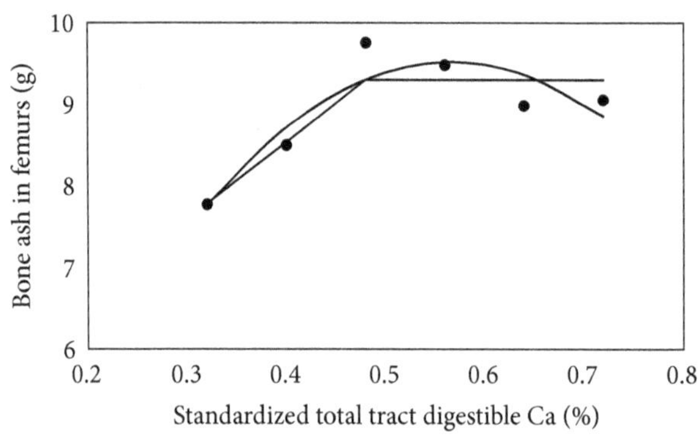

Figure 13.3. *Fitted broken line of bone ash in femurs (g) as a function of standardized total tract digestible Ca (STTD Ca; Exp. 1). The mean of each diet (•) represents the mean of 8 replicated pens per diet. The break point for g bone ash was at 0.48±0.11% STTD Ca if the broken line analysis was used (plateau = 9.30 g bone ash). The maximum concentration of bone ash determined from the quadratic analysis was 0.57±0.03% STTD Ca. The intersections between the broken line and quadratic analysis were 0.48 and 0.66% STTD Ca (adapted from Gonzalez-Vega* et al., *in press).*

at greater dietary concentrations of Ca, these levels should be considered the maximum levels of Ca that can be used in diets fed to pigs from 11 to 25 kg. These levels correspond to a ratio between STTD Ca and STTD P of approximately 1.35:1 and it appears that the best growth performance and bone ash concentrations in 11 to 25 kg pigs are obtained if the ratio between STTD Ca and STTD P is between 1:1 and 1.35:1, with the latter value being the absolute maximum that should be used.

Results of Exp. 2 confirmed that excess dietary Ca is detrimental to pig growth performance. This is particularly the case if dietary P is marginal or below the requirement (Table 13.2). However, results also confirmed that low dietary concentrations of Ca do not reduce growth performance of pigs. In contrast, bone ash, bone Ca, and bone P, and retention of Ca and P in the body are improved if diet concentrations of STTD Ca and STTD P are increased. However, it was observed that the concentration of Ca that is needed to maximize Ca retention is much greater than the concentration needed to maximize Ca in bone ash, which contradicts the expectation that 96 to 99% of all Ca in the body is stored in bones (Crenshaw, 2001). Overall, results of experiment 2 indicated that ADG, G:F, and bone ash were optimized if diets were formulated to a STTD Ca to STTD P ratio between 1.16:1 and 1.43:1. Formulating diets at a greater ratio between STTD Ca and STTD P will result in reduced growth performance of pigs.

Results of Exp. 3 confirmed the negative effects of excess Ca in the diets and indicated that both ADFI and ADG is reduced as dietary Ca increases, whereas dietary STTD P concentrations had no impact on ADG and ADFI (Table 13.3). However, there were no consistent impacts of dietary Ca or P on G:F in this experiment. The model to describe maximum ADG was negatively impacted by STTD Ca in the diet, but not by STTD P. However, bone ash linearly increased as STTD Ca or STTD P in the diets increased and the model to describe the response on bone ash was impacted by STTD Ca as well as STTD P (Figure 13.4).

Table 13.2. *Predicted values for average daily gain (ADG), gain to feed (G:F), bone ash, and calcium (Ca) retention of pigs fed diets containing 0.15, 0.31, 0.39, or 0.47% standardized total tract digestible (STTD) P and 0.13, 0.27, 0.42, 0.57, or 0.72% STTD Ca, Exp. 2 (adapted from Gonzalez-Vega* et al., *2016).*[1]

Item	Dietary STTD P			
	0.15	0.31	0.39	0.47
ADG (kg)	0.76 (0.12)	0.87 (0.36)	0.90 (0.47)	0.92 (0.59)
G:F	0.43 (0.09)	0.46 (0.38)	0.48 (0.52)	0.50 (0.67)
Bone ash (g)	14.3 (0.43)	21.3 (0.56)	23.2 (0.63)	24.0 (0.69)
Ca retention (g/d)	1.8 (0.76)	7.1 (0.93)	8.4 (1.02)	9.9 (1.11)

[1] The first value in each cell is the maximum response obtained at the STTD P concentration indicated, and the value in parenthesis is the % STTD Ca that will result in this response.

Table 13.3. *Growth performance of pigs fed experimental diets with varying concentrations of standardized total tract digestible (STTD) Ca and P for 28 d (adapted from Merriman* et al., *2016a).*

Item	Total Ca (STTD Ca) (%)				
STTD Ca (%)	0.12 (0.08)	0.29 (0.18)	0.46 (0.29)	0.61 (0.38)	0.78 (0.49)
Initial BW, kg					
0.11% STTD P	98.83	100.28	101.75	97.62	99.23
0.21% STTD P	99.42	98.38	101.07	98.47	100.45
0.31% STTD P	101.13	99.15	99.25	102.60	100.72
ADG, kg[1,2]					
0.11% STTD P	1.21	1.14	1.17	0.89	0.83
0.21% STTD P	1.16	1.20	1.17	1.15	0.96
0.31% STTD P	1.11	1.10	1.08	1.00	1.10
ADFI, kg[3,4]					
0.11% STTD P	3.70	3.29	3.27	3.16	2.88
0.21% STTD P	3.72	3.58	3.31	3.32	3.16
0.31% STTD P	3.46	3.32	3.27	3.16	3.16
G:F, d1-28[5,6]					
0.11% STTD P	0.33	0.35	0.37	0.28	0.29
0.21% STTD P	0.32	0.34	0.36	0.35	0.31
0.31% STTD P	0.33	0.33	0.34	0.32	0.35
Final BW, kg[7,8]					
0.11% STTD P	132.68	132.33	134.48	122.42	122.52
0.21% STTD P	132.02	131.85	133.75	130.77	127.32
0.31% STTD P	132.33	129.90	129.62	130.57	131.47

[1] Results indicated that ADG from d 1 to 28 at different combinations of STTD Ca and STTD P can be described by the following model: 1.2141 – 0.6230 × STTD Ca (P=0.008).

[2] Standard error of the within treatment least squares means = 0.09.

[3] Results indicated that ADFI from d 1 to 28 at different concentrations of STTD Ca can be described by the following model: 3.6782 – 1.2722 × STTD Ca (P=0.001).

[4] Standard error of the within treatment least squares means = 0.23.

[5] Results indicated that G:F from d 1 to 28 could not be predicted using STTD Ca or STTD P.

[6] Standard error of the within treatment least squares means = 0.02.

[7] Results indicated that final BW at different combinations of STTD Ca and STTD P can be described by the following model: 140.4729 – 42.9212 × STTD Ca – 30.3919 × STTD P + 140.2884 STTD Ca × STTD P (P=0.006).

[8] Standard error of the within treatment least squares means = 2.87.

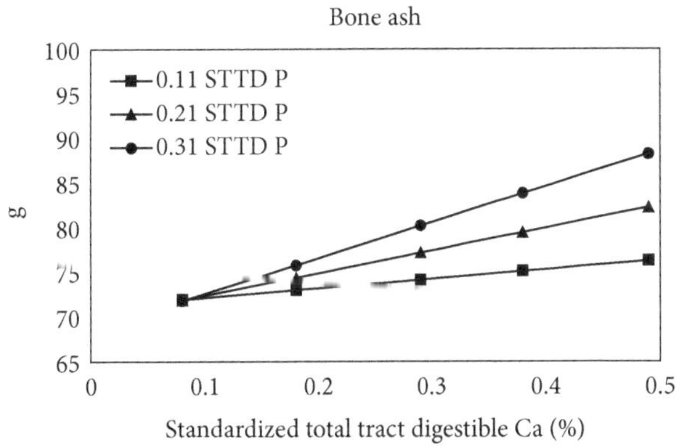

Figure 13.4. *Predicted values, based on the interaction between STTD Ca and STTD P (P=0.049), for bone ash (g) in pigs fed diets containing from 0.08 to 0.49% standardized total tract digestible (STTD) Ca and from 0.11 to 0.31% STTD P. All responses were linear, therefore, no maximum values were estimated (adapted from Merriman* et al., *2016a).*

13.5 Conclusions

1. Results of three experiments with 11 to 25 kg pigs, 25 to 50 kg pigs, and 100 to 130 kg pigs indicate that during short term experiments, it is not possible to reduce growth performance of pigs by reducing the concentration of STTD Ca in the diets.
2. In contrast, inclusion of excess STTD Ca in the diet will reduce ADG and G:F and this effect is more pronounced if dietary STTD P is marginal or below the requirement.
3. It is also clear that the amount of STTD Ca needed to maximize bone ash, bone Ca, bone P, or retention of Ca and P in the body is much greater than the amount needed to maximize growth performance.
4. Based on these results it is recommended that the STTD Ca to STTD P ratio for pigs less than 50 kg should be between 1:1 and 1.35:1, whereas diets fed to pigs above 100 kg should have a STTD Ca to STTD P ratio that does not exceed 1.1:1.
5. Because the STTD of Ca is increased in diets containing microbial phytase, it is important that the effect of phytase on STTD of Ca is taken into account in diet formulation.

References

Almeida, F.N. and Stein, H.H., 2013. Effects of a novel phytase on the digestibility of phosphorus in corn-soybean meal diets fed to weanling and growing pigs. Journal of Animal Science and Biotechnology 4:8.

Bronner, F., 2003. Mechanisms of intestinal calcium absorption. Journal of Cellular Biochemistry 88:387-393.

Crenshaw, T.D., 2001. Calcium, phosphorus, vitamin D, and vitamin K in swine nutrition. In: Lewis, A.J. and Southern, L.L. (eds.) Swine nutrition. 2nd ed. CRC Press, Boca Raton, FL, USA, pp. 187-212.

Gonzalez-Vega, J.C., Liu, Y., McCann, J.C., Walk, C.L., Loor, J.J. and Stein, H.H., in press. Requirement for digestible calcium by 11 to 25 kg pigs as determined by growth performance, bone ash concentration, calcium and phosphorus balances, and expression of genes involved in transport of calcium in intestinal and kidney cells. Journal of Animal Science. DOI: http://doi.org/10.2527/jas.2016-0444.

González-Vega, J.C., Walk, C.L. and Stein, H.H., 2015a. Effects of microbial phytase on apparent and standardized total tract digestibility of calcium in calcium supplements fed to growing pigs. Journal of Animal Science 93:2255-2264.

González-Vega, J.C., Walk, C.L. and Stein, H.H., 2015b. Effect of phytate, microbial phytase, fiber, and soybean oil on calculated values for apparent and standardized total tract digestibility of calcium and apparent total tract digestibility of phosphorus in fish meal fed to growing pigs. Journal of Animal Science 93:4808-4818.

González-Vega, J.C., Walk, C.L. and Stein, H.H., 2016. Digestible calcium requirements for 25 to 50 kg pigs. Journal of Animal Science 94(Suppl. 2):101 (Abstr.)

González-Vega, J.C., Walk, C.L., Liu, Y. and Stein, H.H., 2013. Determination of endogenous intestinal losses of calcium and true total tract digestibility of calcium in canola meal fed to growing pigs. Journal of Animal Science 91:4807-4816.

Hendriks, W.H. and Moughan, P.J., 1993. Whole-body mineral composition of entire male and female pigs depositing protein at maximal rates. Livestock Production Science 33:161-170.

Mahan, D.C. and Shields, R.G. Jr., 1998. Macro- and micromineral composition of pigs from birth to 145 kilograms of body weight. Journal of Animal Science 76:506-512.

Merriman, L.A. and Stein, H.H., in press. Particle size of calcium carbonate does not affect apparent and standardized total tract digestibility of calcium, retention of calcium, or growth performance of growing pigs. Journal of Animal Science. DOI: http://doi.org/10.2527/jas.2015-0252.

Merriman, L.A., Walk, C.L. and Stein, H.H., 2016b. The effect of microbial phytase on the apparent and standardized total tract digestibility of calcium in feed ingredients of animal origin. Journal of Animal Science 94(Suppl. 2):110 (Abstr.)

Merriman, L.A., Walk, C.L., Parsons, C.M. and Stein, H.H., 2016a. Digestible calcium requirement for 100 to 130 kg pigs. Journal of Animal Science 94(E-Suppl. 5):458 (Abstr.)

NRC., 2012. Nutrient Requirements of Swine. 11th rev. ed. National Academy Press, Washington, DC, USA.

Pettey, L.A., Cromwell, G.L., Jang, Y.D. and Lindemann, M.D., 2015. Estimation of calcium and phosphorus content in growing and finishing pigs: Whole empty body components and relative accretion rates. Journal of Animal Science 93:158-167.

Rymarz, A., Fandrejewski, H. and Kielanowski, J., 1982. Content and retention of calcium, phosphorus, potassium and sodium in the bodies of growing gilts. Livestock Production Science 9:399-407.

Stein, H.H., Adeola, O., Cromwell, G.L, Kim, S.W., Mahan, D.C. and Miller, P.S., 2011. Concentration of dietary calcium supplied by calcium carbonate does not affect the apparent total tract digestibility of calcium, but reduces digestibility of phosphorus by growing pigs. Journal of Animal Science 89:2139-2144.

Weremko, D., Fandrejewski, H. and Zebrowska, T., 1997. Bioavailability of phosphorus in feeds of plant origin for pigs – Review. Asian Australian Journal of Animal Science 10:551-566.

Wiseman, T.G., Mahan, D.C. and St-Pierre, N.R., 2009. Mineral composition of two genetic lines of barrows and gilts from twenty to one hundred twenty-five kilograms of body weight. Journal of Animal Science 87:2306-2314.

Chapter 14
Calcium transporters and gene expression and absorption of calcium in pigs

J.C. González-Vega[1] and H.H. Stein[2]*
[1]Evonik Industries AG, Rodenbacher Chaussee 4, 63457 Hanau, Germany;
[2]University of Illinois at Urbana-Champaign, Department of Animal Science, Monogastric Nutrition Laboratory, 1207 West Gregory Dr., Urbana, IL 61801, USA; hstein@illinois.edu

Abstract

Calcium (Ca) absorption occurs by two mechanisms, nonsaturable paracellular absorption and saturable transcellular absorption. The activity of these mechanisms depends on the concentration of Ca in the diet. At high dietary Ca concentration, the nonsaturable mechanism is more active, whereas at low dietary Ca concentration, the saturable mechanism is more involved. The saturable mechanism is vitamin-D dependent, because the expression of genes related to the transcellular transport may be influenced by the active form of vitamin D. Increasing dietary Ca decreases the mRNA expression of transient receptor potential cation channel, subfamily V, member 6 (TRPV6), which is used for absorption of Ca into the enterocytes in the jejunum. However, this does not result in less absorption of Ca, because as dietary Ca increases, an increased proportion of Ca is absorbed via the paracellular route in the small intestine and the total percentage of Ca that is absorbed is almost constant regardless of dietary Ca concentration. High concentrations of dietary Ca will also reduce the mRNA expression in the kidneys of TRPV6, TRPV5, S100 calcium binding protein G, and calbindin 1, but because there is no paracellular reabsorption of Ca in the kidneys excretion of Ca in the urine is increased as dietary Ca concentration increases. As a consequence, body Ca concentrations are primarily regulated in the kidneys.

Keywords: calbindin, calcium channels, paracellular absorption, transcellular absorption, vitamin D receptor

14.1 Introduction

Calcium (Ca) is important for bone health, muscle contraction, transmission of nerve pulses, and other important physiological functions (Crenshaw, 2001; Ewing and Charlton, 2007; Vitti *et al.*, 2010). Calcium is mainly absorbed in the small intestine by two mechanisms: paracellular absorption and transcellular absorption (Bronner, 1987). Paracellular absorption of Ca is nonsaturable and does not require energy, whereas transcellular absorption of Ca is a saturable process that not only requires energy, but also needs Ca channels and Ca transporters (Christakos, 2012).

DOI 10.3920/978-90-8686-836-0_14, © Wageningen Academic Publishers 2016

Vitamin D may influence the expression of genes involved in transcellular transport of Ca, which is mainly active if dietary Ca is relatively low (Kutuzova and DeLuca, 2004). Extensive research has been conducted in broiler chickens, rats, and mice to evaluate the effect of dietary Ca concentrations on expression of genes involved in transcellular transport of Ca in the small intestine and kidneys (Armbrecht *et al.*, 1980, 2003; Healy *et al.*, 2005; Hurwitz *et al.*, 1995; Ko *et al.*, 2009; Rosenberg *et al.*, 1986; Van de Graaf *et al.*, 2004). However, for pigs, only limited data have been reported.

14.2 Calcium absorption

Calcium absorption in pigs is similar to that in other species with the active energy-requiring transcellular route being the primary form of absorption at low concentrations of dietary Ca, whereas the paracellular route is used more extensively at greater dietary concentrations of Ca (Bronner, 2003). When combined, the two routes of absorption result in a near constant rate of absorption of Ca from the intestines regardless of the concentration of Ca in the diet (Stein *et al.*, 2011). However, increased dietary Ca has a negative impact on the digestibility and absorption of phosphorus (Stein *et al.*, 2011), which may be a result of formation of Ca-phosphorus complexes or possibly Ca-phosphorus-phytate complexes in the intestinal tract.

In the kidneys, Ca may also be absorbed into the bloodstream from the distal tubule, but in this case, only active transport is used, which means that Ca is absorbed only if there is a low concentration of plasma Ca. As a consequence, if plasma concentrations of Ca are adequate, less Ca will be absorbed from the kidney tubules, and more will be excreted in the urine (Blaine *et al.*, 2015).

14.2.1 Active absorption of calcium

Active absorption of Ca is also referred to as transcellular absorption because Ca is absorbed via active absorption passes through the cell. Active transport of Ca occurs primarily in the proximal small intestine (duodenum and proximal jejunum) and is a saturable process requiring energy and a Ca-binding protein for transport (Bouillon *et al.*, 2003; Gropper *et al.*, 2009). The low pH in the stomach causes dietary Ca to solubilize and Ca enters the lumen of the small intestine in the form of Ca salts. The free Ca enters the enterocyte by Ca channels located in the brush border membrane (Bronner, 2003; Gonzalez-Vega *et al.*, in press). The Ca channels are referred to as transient reception potential vanilloids (TRPV). In the small intestine TRPV6 is present, whereas TRPV5 is located in the kidneys (Christakos, 2012). Upon absorption into the enterocytes, Ca binds to the Ca-binding proteins (CaBP) which are upregulated or down regulated by Vitamin D. The Ca-binding proteins, also known as calbindin, move Ca across the enterocyte (Gropper *et al.*, 2009). The CaBP9k is mainly found in the intestines and CaBP28k in kidneys (Bouillon *et al.*, 2003). In mice, it has been demonstrated that Ca exits the cell on the basolateral membrane by active transport using a Ca-sodium (Na) pump or an ATPase (Bronner, 2003), an enzyme that releases ATP and allows Ca to exit from the cell (Gropper *et al.*, 2009), and it is believed that a similar mechanism is present in most other species (Bouillon *et al.*, 2003).

14.2.2 Passive absorption of calcium

Passive absorption is also referred to as passive diffusion or paracellular absorption. This process occurs primarily in the jejunum and ileum in the small intestine and is dependent on the concentration of Ca in the intestinal lumen and the electrochemical gradient across the epithelium (Gropper *et al.*, 2009). It is imperative that this pathway is regulated to allow selective permeability. Using the passive transport pathway, Ca is passively moved from the lumen of the small intestine through the tight junctions between the enterocytes (Gropper *et al.*, 2009). Passive absorption is increased with increased solubility of Ca in the distal small intestine and increased by the amount of time the chyme resides in these segments of the intestine (Gropper *et al.*, 2009).

14.3 Site of calcium absorption

Although Ca source and type of diet may influence the place of Ca absorption (González-Vega *et al.*, 2014; Partridge, 1978), Ca absorption mainly takes place in the small intestine (Liu *et al.*, 2000; Partridge, 1978), mostly in the duodenum at low Ca intake using active absorption, but at greater Ca intake, absorption occurs primarily via the paracellular route in the jejunum and ileum. There are however, differences among Ca sources in the exact place in the small intestine where Ca is absorbed, presumably due to differences in solubility, and for some sources of Ca, the main site of absorption appears to be the jejunum or ileum (Gonzalez-Vega *et al.*, 2014). Results of some experiments also indicated that absorption of Ca may take place in the colon (Liu *et al.*, 2000). However, results of recent studies have indicated that no absorption of Ca takes place in the large intestine (Bohlke *et al.*, 2005; González-Vega *et al.*, 2014).

14.4 Endocrine regulation of calcium homeostasis

The three primary hormones that are responsible for maintaining Ca homeostasis are parathyroid hormone (PTH), vitamin D, and calcitonin (Crenshaw, 2001; Eklou-Kalonji *et al.*, 1999; Gropper *et al.*, 2009). The PTH is produced and secreted by the chief cells in the parathyroid gland and secretion of PTH results in an increase in serum Ca levels, but a reduction in phosphorus levels. The primary regulation of PTH is a conformational change in the Ca sensing receptor, which indirectly inhibits PTH secretion from the parathyroid gland (Gropper *et al.*, 2009). Vitamin D in the active form is referred to as calcitriol. The overall effect of calcitriol is to increase levels of Ca in plasma. Calcitriol is synthesized in the kidneys from 25-OH vitamin D using the enzyme 1-hydroxylase and this process is stimulated by PTH. Calcitriol causes reabsorption of Ca in the kidneys using CaBP28k, and in the intestines calcitriol increases CaBP9k synthesis. Both CaBP28k and CaBP9k are used to transport Ca across the epithelial cells from the luminal site to the basolateral site and these proteins, therefore, are needed for the active absorption process of Ca.

Calcitonin is synthesized by the parafollicular cells in the thyroid gland and has the primary function of reducing serum Ca concentrations. This is primarily accomplished by inhibiting the action of osteoclasts in the bone and increasing Ca excretion in the urine. Calcitonin also inhibits bone resorption by osteoblasts (Gropper *et al.*, 2009). The primary function of calcitonin is to maintain the skeleton when Ca demands increase such as in growth and lactation (Wimalawansa, 1996).

14.5 Gene expression of calcium transporters

The release of 1-hydroxylase in the kidneys at low plasma Ca concentrations results in synthesis of 1,25-dihydroxycholecalciferol (1,25-$(OH)_2D_3$), which is the active form of vitamin D (Crenshaw *et al.*, 2011; Hurwitz, 1996). Thus, the energy-requiring transcellular absorption of Ca is vitamin D-dependent (Healy *et al.*, 2005) and 1,25-$(OH)_2D_3$ increases the expression of CaBP9k and TRPV6 (Christakos, 2012). Expression of the Ca-ATPase that is needed to transport Ca out of the enterocyte via the basolateral membrane is also up-regulated by 1,25-$(OH)_2D_3$ (Kutuzova and DeLuca, 2004). Combined, the action of 1,25-$(OH)_2D_3$ results in increased transport of Ca from the small intestinal lumen into the enterocyte, transport inside the enterocyte to the basolateral membrane, and transport out of the enterocyte over the basolateral membrane. In addition, 1,25-$(OH)_2D_3$ increases expression of TRPV5 and CaBP28k in the kidneys which results in increased reabsorption of Ca from the distal kidney tubules to plasma (Bouillon *et al.*, 2003).

In broiler chickens, rats, and mice, the expression of TRPV6, TRPV5, the intestinal binding protein S100G, and/or the ATPase PMCA1 was increased in the small intestine and/or in the kidneys at low dietary Ca concentrations (Armbrecht *et al.*, 1980, 2003; Hurwitz *et al.*, 1995; Ko *et al.*, 2009; Rosenberg *et al.*, 1986; Van de Graaf *et al.*, 2004). In laying hens, expression of intestinal calbindin and vitamin D receptor (VDR) increased at low concentrations of non-phytate phosphorus, whereas at high concentration of non-phytate phosphorus, the mRNA expression of calbindin and VDR decreased (Nie *et al.*, 2013).

However, limited studies with pigs have been conducted to evaluate the effect of increasing concentrations of Ca on the gene expression of transporters of nutrients. In the jejunum, the concentration of Ca influenced the expression of some genes involved in nutrient absorption such as zonula occludens 1, occludin, toll-like receptor 2, and colonic GLUT2 (Metzler-Zebeli *et al.*, 2015). An experiment was also conducted to determine the effect of dietary Ca on the expression of genes involved in the transcellular transport of Ca in the jejunum and kidneys in pigs (González-Vega *et al.*, in press). Six diets were formulated to contain 0.36% standardized total tract digestible (STTD) phosphorus and 0.32, 0.40, 0.48, 0.56, 0.64, or 0.72% STTD Ca, by including increasing quantities of Ca carbonate in the diets at the expense of cornstarch. The three diets with the least concentrations of Ca contained less Ca than the requirement according to NRC (2012) and the three diets with the greatest concentrations of Ca met or exceeded the Ca requirement (NRC, 2012).

Two internal control genes, glyceraldehyde 3-phosphate dehydrogenase (GAPDH) and hydroxymethylbilane synthase (HMBS) were used to normalize the expression of tested genes (Vigors *et al.*, 2014). The selected tested genes involved in the active transcellular transport of Ca were S100 calcium binding protein G (S100G), TRPV6, PMCA1, and VDR in jejunum samples; and S100G, calbindin 1, 28kDa (CALB1), TRPV6, TRPV5, ATP2B1, and VDR in kidney samples.

Results from this experiment indicated that the mRNA expression of TRPV6 and S100G in the jejunum (Figure 14.1) and the mRNA expression of TRPV5, TRPV6, S100G, and CALB1 in the kidneys (Figure 14.2) were down-regulated at high concentrations of dietary Ca. These results indicated that dietary Ca concentrations regulated the mRNA expression of Ca binding proteins and Ca channel proteins in the jejunum and the kidneys. These results are in close agreement with data obtained in mice, rats, and broiler chickens (Armbrecht *et al.*, 1980, 2003; Hurwitz *et al.*, 1995; Ko *et al.*, 2009; Rosenberg *et al.*, 1986; Van de Graaf *et al.*, 2004). In mice, high concentrations of dietary Ca increased mRNA expression of VDR in the kidney, but not in the duodenum (Healy *et al.*, 2005), which was in agreement with results observed by González-Vega *et al.* (in press). Thus, results of this experiment indicate that at high dietary concentrations of Ca, mRNA expression of Ca transporters involved in the transcellular route were down-regulated in the jejunum, however, most Ca may have been absorbed using the paracellular route. As a consequence, Ca that was not needed for bone tissue synthesis or was not retained in the body was subsequently excreted in the urine. Therefore, vitamin D-dependent transport may play a role in increasing Ca absorption from the small intestine and in Ca reabsorption in the

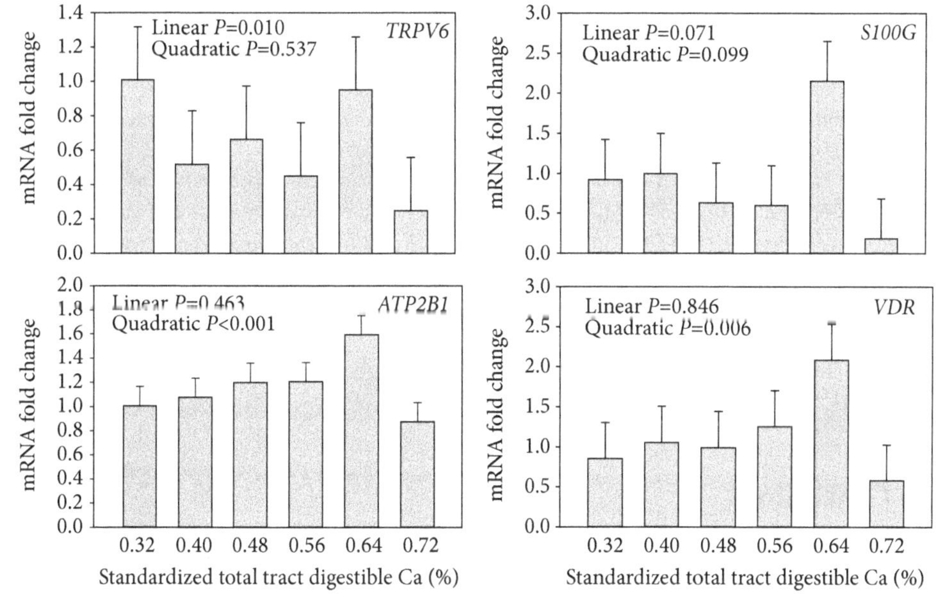

Figure 14.1. *Expression of genes related to transcellular transportation of Ca in jejunum of pigs fed 0.32, 0.40, 0.48, 0.56, 0.64, or 0.72% STTD Ca and 0.36% STTD P. The* P-*values for linear and quadratic effect of increasing concentration of STTD Ca, and means for each diet and standard errors (vertical bars) are indicated.*

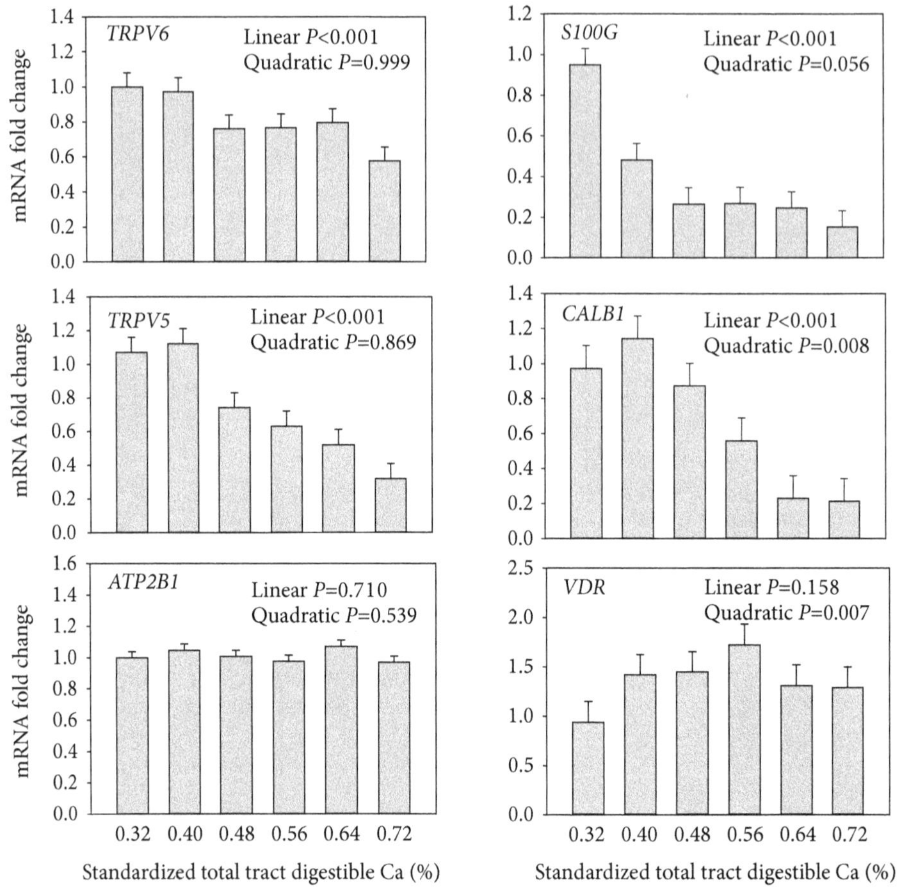

Figure 14.2. *Expression of genes related to transcellular transportation of Ca in kidney of pigs fed 0.32, 0.40, 0.48, 0.56, 0.64, or 0.72% STTD Ca and 0.36% STTD P. The* P*-values for linear and quadratic effect of increasing concentration of STTD Ca and means for each diet and standard errors (vertical bars) are indicated.*

kidneys at low concentrations of dietary Ca, but at high Ca concentrations, homeostasis is regulated mainly at the renal level.

14.6 Conclusions

1. Calcium is absorbed by paracellular and transcellular routes.
2. Expression of genes related to Ca transcellular transport in the jejunum and kidneys of pigs decrease if concentrations of dietary Ca increase. However, although absorption of Ca by the transcellular route decreases, more Ca is absorbed via the paracellular route if dietary Ca concentration increases, and there is, therefore, very limited regulation of Ca absorption from the intestinal tract.

3. However, expression of genes for proteins involved in reabsorption of Ca in the kidneys is upregulated if dietary Ca is low with a subsequent reduction in urinary excretion of Ca. In contrast, if dietary Ca concentrations are above the requirement, expression of genes for kidney proteins involved in reabsorption of Ca are downregulated with a subsequent increase in urinary excretion of Ca.
4. Thus the kidneys are the main regulators of body Ca concentrations in pigs.

References

Armbrecht, H.J., Boltz, M.A. and Bruns, M.E.H., 2003. Effect of age and dietary calcium on intestinal calbindin D-9k expression in the rat. Archives of Biochemistry and Biophysics 420:194-200.

Armbrecht, H.J., Zenser, T.V., Gross, C.J. and Davis, B.B., 1980. Adaptation to dietary calcium and phosphorus restriction changes with age in the rat. American Journal of Physiology 239:E322-327.

Blaine, J., Chonchol, M. and Levi, M., 2015. Renal control of calcium, phosphorus, and magnesium homeostasis. Clinical Journal of American Society of Nephrology 10:1257-1252.

Bohlke, R.A., Thaler, R.C. and Stein, H. H., 2005. Calcium, phosphorus, and amino acid digestibility in low-phytate corn, normal corn, and soybean meal by growing pigs. Journal of Animal Science 83:2396-2403.

Bouillon, R., Van Cromphaut, S. and Carmeliet, G., 2003. Intestinal calcium absorption: Molecular vitamin D mediated mechanisms. Journal of Cellular Biochemistry 88:332-339.

Bronner, F. 2003. Mechanisms of intestinal calcium absorption. Journal of Cell Biochemistry 88:387-393.

Bronner, F., 1987. Intestinal calcium absorption: Mechanisms and implications. Journal of Nutrition 117:1347-1352.

Christakos, S., 2012. Recent advances in our understanding of 1,25-dihydroxyvitamin D3 regulation of intestinal calcium absorption. Archives of Biochemistry and Biophysics 523:73-76.

Crenshaw, T.D., 2001. Calcium, phosphorus, vitamin D, and vitamin K in swine nutrition. In: Lewis, A.J. and Southern, L.L. (eds.) Swine nutrition. 2nd ed. CRC Press, Boca Raton, FL, USA, pp. 187-212.

Crenshaw, T.D., Rortvedt, L.A. and Hassen, Z., 2011. Triennial growth symposium: a novel pathway for vitamin D-mediated phosphate homeostasis: implications for skeleton growth and mineralization. Journal of Animal Science 89:1957-1964.

Eklou Kalonji, E., Zerath, E., Colin, C., Lacroix, C., Holy, X., Denis, I. and Pointllart, A., 1999. Calcium-regulating hormones, bone mineral content, breaking load and trabecular remodeling are altered in growing pigs fed calcium-deficient diets. Journal of Nutrition 129:188-193.

Ewing W. and Charlton, S.J., 2007. The minerals directory. 2nd ed. Context Products Ltd; Leicestershire, United Kingdom: Calcium; pp. 5a-5f, 19a.

González-Vega, J.C., Liu, Y., McCann, J.C., Walk, C.L., Loor, J.J. and Stein, H.H., in press. Requirement for digestible calcium by 11 to 25 kg pigs as determined by growth performance, bone ash concentration, calcium and phosphorus balances, and expression of genes involved in transport of calcium in intestinal and kidney cells. Journal of Animal Science. DOI: http://doi.org/10.2527/jas.2016-0444.

González-Vega, J.C., Walk, C.L., Liu, Y. and Stein, H.H., 2014. The site of net absorption of Ca from the intestinal tract of growing pigs and effect of phytic acid, Ca level and Ca source on Ca digestibility. Archives of Animal Nutrition 68:126-142.

Gropper, S.S., Smith, J.L. and Groff, J.L., 2009. Macrominerals. In: Advanced nutrition and human metabolism. 5th ed. Wadsworth, Cengage Learning, Belmont, CA, USA, pp 429-457.

Healy, K.D., Frahm, M.A. and DeLuca, H.F., 2005. 1,25-Dihydroxyvitamin D3 up-regulates the renal vitamin D receptor through indirect gene activation and receptor stabilization. Archives of Biochemistry and Biophysics 433:466-473.

Hurwitz, S., 1996. Homeostatic control of plasma calcium concentration. Critical Reviews in Biochemistry and Molecular Biology 31:41-100.

Hurwitz, S., Plavnik, I., Shapiro, A., Wax, E., Talpaz, H. and Bar, A., 1995. Calcium metabolism and requirements of chickens are affected by growth. Journal of Nutrition 125:2679-2686.

Ko, S., Lee, G., Vo, T.T.B., Jung, E., Choi, K., Cheung, K., Kim, J.W., Park, J., Oh, G.T. and Jeung, E., 2009. Dietary Ca and 1,25-dihydroxyvitamin D3 regulate transcription of calcium transporter genes in calbindin-D9k knockout mice. Journal of Reproduction and Development 55:137-142.

Kutuzova, G.D. and DeLuca, H.F., 2004. Gene expression profiles in rat intestine identify pathways for 1,25-dihydroxyvitamin D3 stimulated calcium absorption and clarify its immunomodulatory properties. Archives of Biochemistry and Biophysics 432:152-166.

Liu, J., Bollinger, D.W., Ledoux, D.R. and Veum, T.L., 2000. Effects of dietary calcium:phosphorus ratios on apparent absorption of calcium and phosphorus in the small intestine, cecum, and colon of pigs. Journal of Animal Science 78:106-109.

Metzler-Zebeli, B.U., Mann, E., Ertl, R., Schmitz-Esser, S., Wagner, M., Klein, D., Ritzmann, M. and Zebeli, Q., 2015. Dietary calcium concentration and cereals differentially affect mineral balance and tight junction proteins expression in jejunum of weaned pigs. British Journal of Nutrition 113:1019-1031.

Nie, W., Yang, Y., Yuan, J., Wang, Z. and Guo, Y., 2013. Effect of dietary nonphytate phosphorus on laying performance and small intestinal epithelial phosphate transporter expression in Dwarf pink-shell laying hens. Journal of Animal Science and Biotechnology 4:34.

NRC, 2012. Nutrient requirements of swine. 11th rev. ed. National Academic Press, Washington, DC, USA.

Partridge, I.G., 1978. Studies on digestion and absorption in the intestines of growing pigs: 3. Net movements of mineral nutrients in the digestive tract. British Journal of Nutrition 39:527-537.

Rosenberg, J., Hurwitz, S. and Bar. A., 1986. Regulation of kidney calcium-binding protein in the bird (*Gallus Domesticus*). Comparative Biochemistry and Physiology 83A:277-281.

Stein, H.H., Adeola, O., Cromwell, G.L., Kim, S.W., Mahan, D.C. and Miller, P.S., 2011. Concentration of dietary calcium supplied by calcium carbonate does not affect the apparent total tract digestibility of calcium, but reduces digestibility of phosphorus by growing pigs. Journal of Animal Science 89:2139-2144.

Van de Graaf, S.F.J., Boullart, I., Hoenderop, J.G.J. and Bindels, R.J.M., 2004. Regulation of the epithelial Ca2+ channels TRPV5 and TRPV6 by 1α,25-dihydroxy vitamin D3 and dietary Ca2+. Journal of Steroid Biochemistry 89-90:303-308.

Vigors, S., Sweeney, T., O'Shea, C.J., Browne, J.A. and O'Doherty, J.V., 2014. Improvements in growth performance, bone mineral status and nutrient digestibility in pigs following the dietary inclusion of phytase are accompanied by modifications in intestinal nutrient transporter gene expression. British Journal of Nutrition 112:688-697.

Vitti, D.M.S.S., Da Silva Filho, J.C., Louvandini, H., Dias, R.S., Bueno I.C.S. and Kebreab, E., 2010. Phosphorus and calcium utilization in ruminants using isotope dilution technique. In: Vitti, D.M.S.S. and Kebreab, E. (eds.) Phosphorus and calcium utilization and requirements in farm animals. CAB International, London, UK. p. 45-67.

Wimalawansa, S.J., 1996. Calcitonin gene-related peptide and its receptors: molecular genetics, physiology, pathophysiology, and therapeutic potentials. Endocrine Reviews 17:533-585.

Chapter 15
The role of phytate in formulating diets for trace minerals

P. Schlegel[1], C. Jondreville[2] and Y. Nys[3]*
[1]Agroscope, Rte de la Tioleyre 4, 1725 Posieux, Switzerland; [2]AFPA, INRA, Université de Lorraine, 54500 Vandoeuvre-les-Nancy, France; [3]URA, INRA, 37380 Nouzilly, France; patrick.schlegel@agroscope.admin.ch

Abstract

Phytate is the principal storage form of phosphorus (P) in plants and is identified as the major antagonist for the bioavailability of essential trace minerals in monogastrics. The negative impact of phytate needs to be quantified to reduce safety margins and limit environmental excretion of dietary trace minerals, especially for copper (Cu) and zinc (Zn), which are toxic for plants when accumulated in soils. In semi-synthetic diets fed to rats, poultry or pigs, the addition of sodium-phytate reduces the bioavailability of iron (Fe) and Zn and could also affect Cu and manganese (Mn) bioavailability. In cereal-based diets for pigs and chickens without trace mineral supplementation, plant based phytate reduces Zn bioavailability and possibly that of Fe and Mn, but not Cu. This as Cu is the only of these minerals not found bond to plant based phytic acid. In contrary to trace minerals from plant origin, the supplementation of Zn, and eventually that of other trace minerals, is not affected by plant based phytate. Hydrolysing phytate thus releases Fe, Mn and Zn from plant origin with different efficiencies depending on the specie. Microbial phytase releases only an equivalency of 5-10 mg Zn (as sulphate)/500 FTU in chickens. In pigs, the equivalency depends on the presence of non-hydrolysed phytic P. Thus, microbial phytase liberates an equivalency of 17, 34 and 48 mg Zn (as sulphate) with 250, 500 and 750 FTU/kg diet, respectively in a corn-soybean meal based diet, but these equivalencies are lower in diets containing less phytate. The phytate to Zn antagonism has been quantified in pig and chicken diets and can be implemented into Zn recommendations. The phytate to Cu antagonism is considered as inexistent in practical diets. The phytate to Fe and Mn antagonisms require further research in practical diets to merge recommendations.

Keywords: phytase, zinc, copper, pig, poultry, antagonism

15.1 Introduction

Trace minerals, such as copper (Cu), iron (Fe), manganese (Mn) and zinc (Zn) are essential nutrients for livestock. Their requirements are determined, for a given animal category, by using the empirical method. This method evaluates the necessary mineral concentration to maximize or minimize one or more tissues or performance parameters when supplying animals with increasing dietary levels of a trace element (dose-response). Feedstuffs mostly do not contain a sufficient trace mineral concentration (Table 15.1), so

DOI 10.3920/978-90-8686-836-0_15, © Wageningen Academic Publishers 2016

Table 15.1. *Copper, iron, manganese, zinc (mg/kg) concentrations in plant based feedstuffs collected on the Swiss market (Schlegel, unpublished) and phytic P (g/kg) concentration according to EFSA, 2014.*

Feedstuff	No.	Phytic P	Cu		Fe		Mn		Zn	
		Mean	Mean	SD	Mean	SD	Mean	SD	Mean	SD
Barley	51	2.2	5.2	0.9	36	5	14.9	1.9	25.6	3.1
Wheat	54	2.4	3.8	0.9	40	10	31.9	6.4	26.6	4.1
Triticale	25	2.8	5.3	0.7	44	30	30.4	7.9	32.1	3.2
Oats	20	2.3	3.7	0.5	89	81	41.7	10.5	26.9	3.9
Maize	46	1.9	1.8	0.8	19	2	4.0	1.7	18.2	1.2
Millet	13	1.9	8.9	1.6	48	6	11.8	3.6	30.9	2.4
Sorgum	19	1.8	4.9	1.6	35	8	15.1	3.9	20.3	3.9
Oat hulls	7		1.1	0.8	23	7	25.1	3.8	10.8	2.0
Barley offal	5		6.7	0.6	79	9	38.0	5.3	48.1	7.9
Oat offal	9		2.8	1.4	65	15	58.0	17.9	29.8	10.1
Wheat middlings	8	6.7	14.5	2.5	132	9	151.5	13.9	122.8	15.5
Wheat feedmeal	6	3.7	6.9	2.2	52	13	35.2	12.8	40.0	11.2
Oat feedmeal	6		5.6	1.2	62	5	55.8	9.9	37.6	4.3
Wheat bran	11	9.7	14.4	2.0	152	23	148.5	18.8	103.6	12.7
Wheat starch	10	0.0	2.3	2.1	7	4	4.8	3.4	4.6	7.0
Maize gluten	10	4.7	12.6	3.6	277	131	4.7	1.6	23.2	7.2
Distiller dried grain	8		8.5	2.9	97	9	52.0	5.1	61.5	5.0
Brewery by-product	7		19.9	2.3	240	33	45.8	2.2	93.3	3.1
Horse beans	5	3.1	15.1	1.2	59	20	11.4	1.8	47.2	3.6
Lupin white	5	3.0	7.6	1.2	33	9	555.9	228.3	38.3	2.5
Lupin blue	5	3.0	4.8	0.7	39	8	40.2	18.9	32.6	1.6
Rapesseed meal	9	6.6	6.2	1.4	146	11	71.1	4.2	65.5	3.5
Rapesseed meal cake	10	6.4	5.6	1.1	144	31	65.0	4.5	60.9	2.5
Soybean meal	14	3.6	13.2	2.2	187	51	34.9	3.9	52.2	3.4
Soybean meal cake	10	3.8	16.1	2.4	263	90	40.8	6.3	55.2	2.0
Sunflower meal cake	11		33.4	3.0	117	31	32.6	4.1	82.6	6.9
Lineseed meal cake	10	4.2	20.8	3.2	240	116	45.0	6.4	70.7	11.0

they are consequently supplemented to the diet, mainly in forms of oxides, sulphates or chelates. It is however, recognized that several dietary factors can reduce the trace mineral bioavailability and thus modify the optimal level initially assessed with the empirical method. If not quantified, such uncertainties favour the inclusion of safety margins in the quantities of supplemented trace minerals, enhanced with the concern that the supplemented ones would be affected by antagonists, as well. As these feed additives are relatively cheap, the risk of a 'the more the better' is obviously expected. In the EU for example, Zn concentrations in complete feeds for pigs and chickens are mostly closer to the maximal authorized levels than to recommendations (EFSA, 2014). Such situation is of increasing environmental concern, especially in regards to Cu and Zn, as excess trace mineral intake is consequently excreted in the faeces at levels greater than 95% of the dietary supply (Nys *et al.*, 2003; Jondreville *et al.*, 2003). Phytate, the principal storage form of phosphorus (P) in plants, is identified as the major dietary antagonist for trace mineral bioavailability in monogastrics. The hydrolysation of phytate improves P digestibility, but may also improve the digestibility of the trace minerals which are bound to phytate.

It is finally of interest to identify which trace minerals sources are affected by phytate and to quantify their interaction with phytate hydrolysis in order to implement them into trace mineral recommendations. This would be favourable for reducing safety margins, improving the dietary proportion of the mineral retained in the body and thereby reducing excretion in the faeces.

15.2 Phytate as an antagonist to trace mineral bioavailability

The antagonism of phytate on trace mineral bioavailability was mainly conducted under the following circumstances: studies using sodium phytate and trace minerals from supplements; phytate and trace minerals from plants; phytate from plants and trace minerals from plants and from supplements. These are discussed below. Trace mineral bioavailability can be defined as the maximal proportion of ingested trace mineral utilised for the biological and physiological functions by a healthy animal (Windisch, 2001). The optimal conditions to study trace mineral bioavailability are met when the dietary concentrations are not sufficient to allow animal homeostatic regulation and when an adequate parameter is measured which directly responds to the mineral evaluated, such as bone or plasma level for Zn.

15.2.1 Sodium phytate and supplemented trace minerals

Back in the 1960s, O'Dell and Savage (1960) and Oberleas *et al.* (1962) described that the addition of sodium (Na) phytate into a semi-synthetic diet containing 9 and 14 mg Zn/kg reduced growth performance of poultry and pigs. Later, it was demonstrated that Na-phytate strongly limited Zn absorbability in growing rats fed semi-synthetic diets with Zn levels not reaching the homeostatic Zn regulation capacity (Figure 15.1). Phytate forms insoluble complexes with Zn (Davies and Nightingale, 1975) and thus reduces Zn absorbability in the digestive tract. Interestingly, this antagonism occurs with intact phytate (*myo*-inositol 1,2,3,4,5,6-hexakis dihydrogen phosphate, IP_6) or with a lower degree of phosphorylation (IP_5), but not with further phosphorylated IP_4 and IP_3 in rats (Lönnerdal *et al.*, 1989). Studies carried out with Na-phytate showed that the dietary molar ratio of phytate: Zn can be used for evaluating the antagonism of phytate on Zn bioavailability. The bioavailability of Cu, Fe and Mn were also investigated in relation to Na-phytate. The Fe bioavailability was reduced by Na-phytate in rats as shown by Pallauf *et al.* (1999) but this observation was not confirmed by others (Grases *et al.*, 2004; Hayashi *et al.*, 2001; Hunter, 1981;). Sodium phytate did not reduce Cu (Lee *et al.*, 1988) or Mn (Grases *et al.*, 2004) bioavailability in rats.

It should be pointed out here that these studies were all based on a synthetic phytate source, devoid of trace minerals, and the trace minerals were supplemented as salts (oxides,

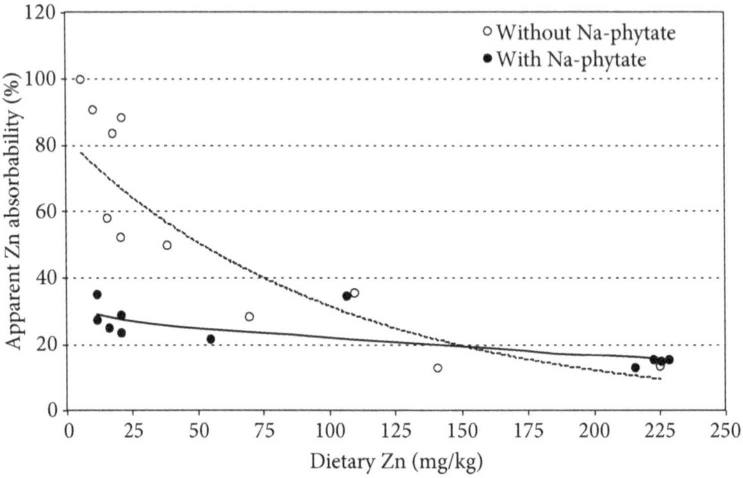

Figure 15.1. *Apparent Zn absorption in growing rats in relation to dietary Zn content and Na-phytate addition in semi-synthetic diets (from Rimbach and Pallauf, 1992; Rimbach* et al., *1995, 1997; Schlegel and Windisch, 2006; Weigand and Kirchgessner, 1980, 1992).*

sulphates). This phytate source contains Na bonds, which are easily dissociated in aqueous solutions. The stronger cation Zn^{2+} replaces Na^{+} to form insoluble phytate – Zn complexes (Davies and Nightingale, 1975; Akter *et al*,. 2015). Other cations, such as Fe^{2+}, Mn^{2+} and Cu^{2+} could theoretically also bind to phytic acid, but Lee *et al.* (1988) demonstrated that in the presence of supplemented Zn, this possibility is limited. In this study, the rats were all supplemented with Zn in the studies revealing no antagonistic effect of Na-phytate on Cu, Fe and Mn, which may have diminished the expected effects of phytate on the additional trace minerals because of the relatively stronger affinity of Zn for phytic acid.

The globally used sentence 'phytates reduce trace mineral bioavailability in monogastrics' is finally based on the above mentioned innovative studies aiming to better understand trace mineral metabolism using Na-phytate in semi-synthetic diets. However, in practical diets, phytates are not present as Na-phytates and trace elements originate from the feedstuffs and from dietary supplements. It is thus of interest to know if plant based phytate modifies trace mineral bioavailability from feedstuffs, from supplements or both.

15.2.2 Phytate and trace minerals from plants

In poultry and swine feed ingredients, it is difficult to avoid the presence of phytate as it is the main storage form of P in the seeds. The chemical form of phytate was characterized in wheat grains (Rodrigues-Filho *et al.*, 2005) and the localisation of phytate and trace elements was determined using X-ray technology in each compartment of rice grains (Iwai *et al.*, 2012; Kyriacou *et al.*, 2014). There is also a strong correlation between phytic P and Mn and Zn, but a weak correlation between phytic P and Cu concentrations in plant based feedstuffs (Table 15.1). These observations suggested that phytic acid is bound to Fe^{2+}, Mn^{2+} and Zn^{2+} but not to Cu^{2+} to form phytate in cereal grains.

To evaluate the effect of plant based phytate on the trace mineral bioavailability in feedstuffs (without trace mineral supplementation) different approaches have been applied, such as: (1) the comparison of non-supplemented trace mineral diets composed of feedstuffs differing in their phytate content; (2) the comparison of genetically modified low-phytate cultivars with traditional cultivars; and (3) the comparison of non-supplemented trace mineral diets with or without hydrolyzation of phytate by microbial or plant based phytases and enhanced by fermentation or liquid feeding. Recent research evaluating Cu in pigs suggests that supplementation of microbial phytase into a cereal based diet without added Cu (7 mg Cu / kg feed) did not improve Cu status (Bikker *et al.*, 2012). These results further indicate, in accordance with the aforementioned findings, that plant based Cu is not naturally bound with phytic acid in grains. Iron absorption was improved when humans consumed wheat without bran compared with those who ate higher bran level (Brune *et al.*, 1989). Iron absorption was also improved when genetically modified low-phytate cultivars were fed to rats (Kruger *et al.*, 2013) or to humans (Mendoza *et al.*, 1998) in comparison to classical cultivars. Finally, blood haemoglobin level was improved in anaemic piglets (Stahl *et al.*, 1999), but not in chickens (Biehl *et al.*, 1997) when microbial phytase was added to a soybean meal based diet without added Fe. Manganese bioavailability was increased in broilers when phytase (1,200 FTU/kg) was supplied to the diets (Mohanna and Nys, 1999). Zinc bioavailability from feedstuffs was improved with genetically modified low-phytate cultivars of cereals in rats (Lönnerdal *et al.*, 2011) and chickens (Linares *et al.*, 2007) and with diets containing ingredients with low phytate contents in chickens and pigs (Mohanna and Nys, 1999; Schlegel *et al.*, 2010). The Zn status was also improved with fermented diets in pigs (Matsui *et al.*, 1998) or with the addition of microbial phytase in chickens (Mohanna and Nys, 1999; Jondreville *et al.*, 2007; Schlegel *et al.*, 2010) and in pigs (Jondreville *et al.*, 2005; Schlegel *et al.*, 2010). Overall, the effects of phytate hydrolysis on plant-based trace mineral availability were lower in broilers than in pigs, especially illustrated when both species were fed the same cereal based diets containing 1.3 and 2.3 g of phytic P/kg with or without microbial phytase (Schlegel *et al.*, 2010). Finally, Schlegel *et al.* (2013) concluded from a meta-analysis that bone Zn content of piglets fed non-supplemented Zn diets (25-38 mg Zn/kg) decreased with phytate in diets devoid of phytase and increased with phytate hydrolysis by phytase. The molar ratio of phytate: Zn is therefore not suitable in practical diets for livestock as phytase activity is not taken into account. Indeed, when evaluating the combination of dietary phytic P and phytase activity, bone Zn concentration decreased linearly as the non-hydrolysed phytic P in the digestive tract increased (Figure 15.2). The non-hydrolysed phytic P was calculated as the difference between total phytic P and the quantity of P released by vegetal and microbial phytase according to regressions issued by Létourneau-Montminy *et al.* (2012). This phenomenon was however not confirmed in chickens thereby creating an intriguing situation as the efficacy of phytase to hydrolyse phytate is similar between the two species (Selle and Ravindran, 2007, 2008). It has been suggested that due to the observed lower pH and higher Zn solubility in the gizzard than in the pig stomach, Zn from feedstuffs would, at least partly, be liberated in the gizzard and thus would be independent from phytate hydrolysis in chickens (Schlegel *et al.*, 2010). This phenomenon would result in a physiologically higher bioavailability of Zn from feedstuffs in chicken than in pigs and would explain their lower dietary Zn requirement (NRC, 1994, 2012).

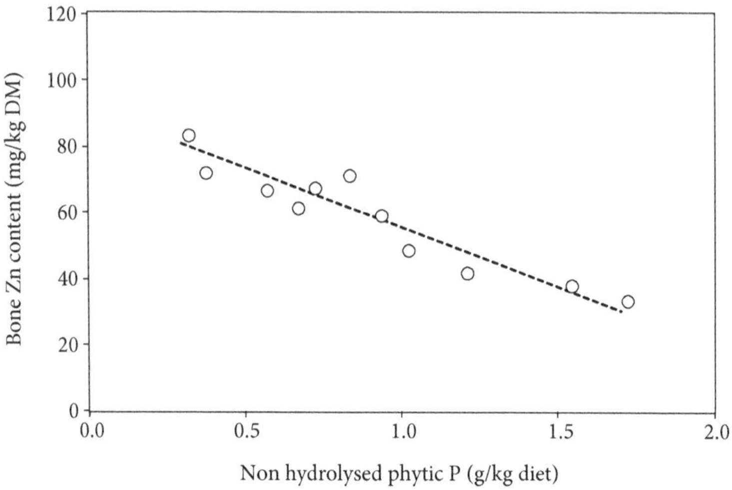

Figure 15.2. *Bone zinc (Zn) in piglets as a function of the non-hydrolysed phytic phosphorus (P) content in diets not supplemented with Zn (Schlegel* et al.*, 2013).*

To conclude, the hydrolysation of phytate from plant-based feedstuffs by phytases or feeding low phytate grains liberated Fe, Mn and Zn from the grains and thus improved their bioavailability. This effect is greater in pigs than in chickens probably due to the lower gastro-intestinal pH allowing higher release of Zn from phytates even when not hydrolysed. It may be deduced from the Cu storage form in grains, that Cu bioavailability of feedstuffs would not be affected by phytate. More data on the phytate to Mn interaction is also required.

15.2.3 Phytate from plants and trace minerals from plants and from supplements

In practical diets, trace minerals are also present as supplemented forms (oxides, sulphates and chelates). Based again on a meta-analysis, Schlegel *et al.* (2013) observed in both broilers and pigs that the increase in bone Zn concentration in response to supplemented $ZnSO_4$ (up to 102 mg Zn/kg) was independent of the phytate or phytase concentrations. Thus, it was concluded that plant based phytates do not interact with supplemented Zn, that supplemental Zn does not require any protection against dietary phytates and that the improved dietary Zn bioavailability from phytate hydrolysis was solely due to the liberated Zn from plants. Similar conclusions regarding supplemented Cu could be advanced as microbial phytase did not modify apparent absorbability of Cu (Adeola, 1995) nor liver Cu stores (Shelton *et al.*, 2004) in pigs fed a Cu supplemented diet (14-20 mg Cu/kg). Further research is required to verify if a same conclusion can be adopted for supplemented Fe and Mn.

With an reverse view, when high levels of ZnO (Augspurger *et al.*, 2004; Walk *et al.*, 2013, 2015) or $CuSO_4$ (Banks *et al.*, 2004) are used as therapeutics for piglets and poultry, calcium and P digestibility and serum levels, and bone mineralisation are expected to be reduced. This phenomenon seems however rather due to an interaction between Ca, non phytic P and the high presence of trace mineral ions in the digestive tract than due to plant based phytates. To conclude, these observations strongly suggest that the bioavailability of supplemented Zn and Cu is not affected by plant based phytates.

15.3 Implications on diet formulations for poultry and swine

Finally, the following is considered as relevant:

- Phytates from plants affect Zn, Fe and possibly Mn bioavailability from plants, but not from supplemented trace mineral sources, at least from Zn.
- The hydrolysis of phytate from plant releases Zn, Fe and possibly Mn from plant origin, but not Cu because it is most probably not bound to plant based phytic acid.
- The lower pH of the gizzard in the growing chicken allows Zn and possibly Fe and Mn to be dissociated from phytic acid and thus become available to the animal, even when phytates are not hydrolysed by phytase.

Therefore the supplemented Zn (as $ZnSO_4$) equivalency of microbial phytase is limited in chickens with 5 to 10 mg Zn/500 FTU (Jondreville *et al.*, 2007; Schlegel *et al.*, 2013). In pigs, the release of P and of Zn from plant based phytic acid are considered as proportional (Jondreville *et al.*, 2005) and the Zn equivalency depends on the presence of non-hydrolysed phytic P from feedstuffs (Schlegel *et al.*, 2013). As an example, microbial phytase was estimated to liberate Zn from plant based phytate at an equivalency of 17, 34 and 48 mg Zn (as $ZnSO_4$) with 250, 500 and 750 FTU/kg diet, respectively when piglets were fed a corn-soybean meal based diet (2.5 g phytic P/kg). With a lower presence of dietary phytate (1.5 g phytic P/kg), when phytic P was limiting for the activity of phytase, the Zn (as $ZnSO_4$) equivalency was estimated at 6, 11 and 13 mg Zn (as $ZnSO_4$) with 250, 500 and 750 FTU/kg diet, respectively. Plant based phytase activity also releases an equivalency of 4-5 mg Zn (as $ZnSO_4$) per 200 FTU/kg in non-soaked diets. We may assume that these equivalencies are underestimated for supplemental Zn sources presenting a lower bioavailability than $ZnSO_4$ (e.g. certain Zn oxides). Higher inclusion levels of microbial phytase (>1,500 FTU/kg, super-dosing) may allow further reduction in Zn supplementation to fulfil pig's requirement as suggested by Zeng *et al.* (2015). Therapeutical levels of Cu (up to 160 mg/kg diet) would not influence this Zn equivalency of phytase (Bikker *et al*, 2012). Calcium has a limited negative impact on Zn availability in poultry and pigs unless fed at extremely excessive levels (EFSA, 2014). A first attempt was made in Switzerland to recommend Zn levels for pigs according to phytic P and phytase activity, based on an initially recommended Zn level of 100 mg/kg diet for piglets (Table 15.2). Iron and Mn equivalencies from phytase in poultry and pigs still need to be assessed.

Table 15.2. *Swiss recommendations of dietary Zn levels for weaned piglets (Schlegel, 2011).*

Native Zn (mg/kg diet)	Phytic P (g/kg diet)	Plant phytase[1] (FTU/kg diet)	Microbial phytase (FTU/kg diet)	Added Zn[2] (mg/kg diet)	Total Zn (mg/kg diet)
30	2.5	0	0	85	115
		0	250	65	95
		0	500	50	85
		0	750	35	65
30	2.0	0	0	70	100 [3]
		100	0	70	100
		200	0	65	95
		0	250	50	80
		0	500	40	70
		0	750	30	60
25	1.5	0	0	65	90
		100	0	65	90
		200	0	60	85
		0	250	50	75
		0	500	40	65
		0	750	35	60

[1] Plant phytase activity in the produced feed.

[2] Added Zn as sulphate or organic (chelate).

[3] Initial Zn recommendation which did not consider effects of phytate and phytase.

15.4 Conclusions

1. When aiming for improved sustainability in pig and poultry mineral nutrition, the conditions in which the antagonist phytate reduces trace mineral bioavailability need to be known and quantified. Otherwise, the chances are low that the actual applied safety margins in trace mineral supplementation are reduced.
2. The phytate-Zn antagonism has been quantified in pigs and chickens and can be implemented into dietary Zn recommendations and the phytate-Cu antagonism seems limited in these species.
3. However, further research is required to explore these effects in heavier pigs and other poultry than growing chickens.
4. Further research is also required to better quantify the phytate-Fe and phytate-Mn antagonism before implementing these interactions into recommendations.

References

Adeola, O., 1995. Digestive utilization of minerals by weanling pigs fed copper- and phytase-supplemented diets. Canadian Journal of Animal Science 75:603-610.

Akter, M., Graham, H. and Iji, P.A., 2015. Interactions between phytase and different dietary minerals in *in vitro* systems. Journal of Food, Agriculture and Environment 13:38-44.

Augspurger, N.R., Spencer, J.D., Webel, D.M. and Baker, D.H., 2004. Pharmacological zinc levels reduce the phosphorus-releasing efficacy of phytase in young pigs and chickens. Journal of Animal Science 82:1732-1739.

Banks, K.M., Thompson, K.L., Jaynes, P. and Applegate, T.J., 2004. The effects of copper on the efficacy of phytase, growth, and phosphorus retention in broiler chicks. Poultry Science 83:1335-1341.

Biehl, R., Emmert, J.L. and Baker, D., 1997. Iron bioavailability in soybean meal as affected by supplemental phytase and 1α-hydroxycholecalciferol. Poultry Science 76:1424-1427.

Bikker, P., van Diepen, J.T.M., Binnendijk, G.P. and Jongbloed, A.W., 2012. Phytase inclusion in pig diets improves zinc status but its effect on copper availability is inconsistent. Journal of Animal Science 90:197-199.

Brune, M., Rossander, L. and Hallberg, L., 1989. Iron absorption: no intestinal adaptation to a high-phytate diet. American Journal of Clinical Nutrition 49:542-545.

Davies, N.T. and Nightingale, R., 1975. Effects of phytate on intestinal-absorption and secretion of zinc, and whole-body retention of zinc, copper, iron and manganese in rats. The British Journal of Nutrition 34:243-258.

EFSA, 2014. Scientific opinion on the potential reduction of the currently authorised maximum zinc content in complete feed. EFSA Journal 12:3668.

Grases, F., Simonet, B.M., Perellò, J., Costa-Bauza, A. and Prieto, R.M., 2004. Effect of phytate on element bioavailability in the second generation of rats. Journal of Trace Element in Medicine and Biology 17:229-234.

Hayashi, K., Hara, H., Asvarujanon, P., Aoyama, Y. and Luangpituksa, P., 2001. Ingestion of insoluble dietary fibre increased zinc and iron absorption and restored growth rate and zinc absorption suppressed by dietary phytate in rats. British Journal of Nutrition 86:443-451.

Hunter, J., 1981. Iron availability and absorption in rats fed sodium phytate. The Journal of Nutrition 111:842-847.

Iwai, T., Takahashi, M., Oda, K., Terada, Y. and Yoshida, K.T., 2012. Dynamic changes in the distribution of minerals in relation to phytic acid accumulation during rice seed development. Plant Physiology 160:2007-2014.

Jondreville, C., Hayler, R. and Feuerstein, D., 2005. Replacement of zinc sulphate by microbial phytase for piglets given a maize-soya-bean meal diet. Animal Science 81:77-83.

Jondreville, C., Lescoat, P., Magnin, M., Feuerstein, D., Gruenberg, B. and Nys, Y., 2007. Sparing effect of microbial phytase on zinc supplementation in maize-soya-bean meal diets for chickens. Animal 1:804-811.

Jondreville, C., Revy, P.S. and Dourmad, J.Y., 2003. Dietary means to better control the environmental impact of copper and zinc by pigs from weaning to slaughter. Livestock Production Science 84:147-156.

Kruger, J., Taylor, J.R.N., Du, X., De Moura, F.F., Lönnerdal, B. and Oelofse, A., 2013. Effect of phytate reduction of sorghum, through genetic modification, on iron and zinc availability as assessed by an *in vitro* dialysability bioaccessibility assay, Caco-2 cell uptake assay, and suckling rat pup absorption model. Food Chemistry 141:1019-1025.

Kyriacou, B., Moore, K.L., Paterson, D., de Jonge, M.D., Howard, D.L., Stangoulis, J., Tester, M., Lombi, E. and Johnson, A.A.T., 2014. Localization of iron in rice grain using synchrotron X-ray fluorescence microscopy and high resolution secondary ion mass spectrometry. Journal of Cereal Science 59:173-180.

Lee D-Y., Schröder, J. and Gordon, D., 1988. Enhancement of Cu bioavailability in the rat by phytic acid. The Journal of Nutrition 118:712-717.

Létourneau-Montminy, M.P., Jondreville, C., Sauvant, D. and Narcy, A., 2012. Meta-analysis of phosphorus utilization by growing pigs: effect of dietary phosphorus, calcium and exogenous phytase. Animal 6:1590-1600.

Linares, L.B., Broomhead, J.N., Guaiume, E.A., Ledoux, D.R., Veum, T.L. and Raboy, V., 2007. Effects of low phytate barley (*Hordeum vulgare L.*) on zinc utilization in young broiler chicks. Poultry Science 86:299-308.

Lönnerdal, B., Mendoza, C., Brown, K.H., Rutger, J.N. and Raboy, V., 2011. Zinc absorption from low phytic acid genotypes of maize (*Zea mays L.*), barley (*Hordeum vulgare L.*), and rice (*Oryza sativa L.*) assessed in a suckling rat pup model. Journal of Agricultural Food Chemistry 59:4755-4762.

Lönnerdal, B., Sandberg, A.-S., Sandström, B. and Kunz, C., 1989. Inhibitory effects of phytic acid and other inositol phosphates on zinc and calcium absorption in suckling rats. Journal of Nutrition 119:211-214.

Matsui, T., Susaki, H., Tamura, A., Yano, H., Nakajima, T., Matsuda, M. and Yano, F., 1998. The improvement of zinc availability in fermented soybean meal in growing pigs. Animal Science and Technology 69:589-591.

Mendoza, C., Viteri, F.E., Lönnerdal, B., Young, K.A., Raboy, V., Brown, K.H., 1998. Effect of genetically modified, low-phytic acid maize on absorption of iron from tortillas. American Journal of Clinical Nutrition 68:1123-1127.

Mohanna, C. and Nys, Y., 1999. Changes in zinc and manganese availability in broiler chicks induced by vegetal and microbial phytases. Animal Feed Science and Technology 77:241-253.

National Research Council (NRC), 1994. Nutrient Requirements of Poultry, 9th revised edition. National Academy Press, Washington, DC, USA.

National Research Council (NRC), 2012. Nutrient Requirements of Swine, 11th revised edition. National Academy Press, Washington, DC, USA.

Nys, Y., Revy, P.S. and Jondreville, C., 2003. Zinc, cuivre et manganèse en aviculture: rôle, disponibilité et risque pour l'environnement. 5e Journées de la Recherche Avicole, Tours, France.

O'Dell, B.L. and Savage, J.E., 1960. Effect of phytic acid on zinc availability. Proceedings of the Society for Experimental Biology and Medicine 103:304-306.

Oberleas, D., Muhrer, M.E. and O'Dell, B.L., 1962. Effects of phytic acid on zinc availability and parakeratosis in swine. Journal of Animal Science 21:57-61.

Pallauf, J., Pippig, S., Most, E. and Rimbach, G., 1999. Supplemental sodium phytate and microbial phytase influence iron availability in growing rats. Journal of Trace Elements in Medicine and Biology 13:134-140.

Rimbach, G. and Pallauf, J., 1992. Effekt einer zulage mikrobieller phytase auf die zinkverfügbarkeit. Zeitschrift für Ernährungswissenschaft 31:269-277.

Rimbach, G., Pallauf, J., Brandt, K. and Most, E., 1995. Effect of phytic acid and microbial phytase on Cd accumulation, Zn status, and apparent absorption of Ca, P, Mg, Fe, Zn, Cu, and Mn in growing rats. Annals of Nutrition and Metabolism 39:361-370.

Rimbach, G., Walter, A., Most, E. and Pallauf, J., 1997. Effect of supplementary microbial phytase to a maize-soya-diet on the availability of calcium, phosphorus, magnesium and zinc: *in vitro* dialysability in comparison with apparent absorption in growing rats. Journal of Animal Physiology and Animal Nutrition 77:7-9.

Rodrigues-Filho, U.P., Vaz, S. Jr., Felicissimo, M.P., Scarpellini, M., Cardoso, D.R., Vinhas, R.C.J., Landers, R., Schneider, J.F., Mc Garvey, B.R., Andersen, M.L. and Skibsted, L.H., 2005. Heterometallic manganese/zinc-phytate complex as a model compound for metal storage in wheat grains. Journal of Inorganic Biochemistry 99:1973-1982.

Schlegel, P. and Windisch, W., 2006. Bioavailability of zinc glycinate in comparison with zinc sulphate in the presence of dietary phytate in an animal model with ^{65}Zn labelled rats. Journal of Animal Physiology and Animal Nutrition 90:216-222.

Schlegel, P., 2011. Zink in der Schweinefütterung. ALP aktuell no.40. Agroscope, Posieux, Switzerland

Schlegel, P., Nys, Y. and Jondreville, C., 2010. Zinc availability and digestive zinc solubility in piglets and broilers fed diets varying in their phytate contents, phytase activity and supplemented zinc source. Animal 4:200-209.

Schlegel, P., Nys, Y. and Jondreville, C., 2013. Bioavailability of zinc sources and their interaction with phytates in broilers and piglets. Animal 7:47-59.

Selle, P.H. and Ravindran, V., 2007. Microbial phytase in poultry nutrition. Animal Feed Science and Technology 135:1-41.

Selle, P.H. and Ravindran, V., 2008. Phytate-degrading enzymes in pig nutrition. Livestock Science 113:99-102.

Shelton, J.L., Southern, L.L., Le Mieux, F.M., Bidner, T.D. and Page, T.G., 2004. Effects of microbial phytase, low calcium and phosphorus, and removing the dietary trace mineral premix on carcass traits, pork quality, plasma metabolites, and tissue mineral content in growing-finishing pigs. Journal of Animal Science 82:2630-2639.

Stahl, C.H., Han, Y.M., Roneker, K.R., House, W.A. and Lei, X.G., 1999. Phytase improves iron bioavailability for hemoglobin synthesis in young pigs. Journal of Animal Science 77:2135-2142.

Walk, C.L., Srinongkote, S. and Wilcock, P., 2013. Influence of a microbial phytase and zinc oxide on young pig growth performance and serum minerals. Journal of Animal Science 91:286-291.

Walk, C.L., Wilcock, P. and Magowan, E., 2015. Evaluation of the effects of pharmacological zinc oxide and phosphorus source on weaned piglet growth performance, plasma minerals and mineral digestibility. Animal 9:1145-1152.

Weigand, E. and Kirchgessner, M., 1980. Total true efficiency of zinc utilization – determination and homeostatic dependence upon to zinc supply status in young rats. Journal of Nutrition 110:469-480.

Weigand, E. and Kirchgessner, M., 1992. Absorption, endogenous excretion, and balance of zinc in growing rats on diets with various sugars replacing starch. Biological Trace Element Research 34:67-77.

Windisch, W., 2001. Homeostatic reactions of quantitative Zn metabolism on deficiency and subsequent repletion with Zn in ^{65}Zn labeled adult rats. Trace Elements and Electrolytes 18:128-133.

Zeng, Z., Li, Q., Tian, Q., Zhao, P., Xu, X.,Yu, S. and Piao, X., 2015. Super high dosing with a novel *Buttiauxella* phytase continuously improves growth performance, nutrient digestibility, and mineral status of weaned pigs. Biological Trace Element Research 168:103-109.

Chapter 16
From floor sweepings to fish flesh – phytase superdosing in the US catfish industry

E. Peatman[1] and B.H. Beck[2]*
[1]School of Fisheries, Aquaculture, and Aquatic Sciences, Auburn University, 211 Swingle Hall / 377 CASIC Bldg., Auburn, AL 36849, USA; [2]United States Department of Agriculture-Agricultural Research Service (USDA-ARS), Aquatic Animal Health Research Unit, 990 Wire Road, Auburn, AL 36832, USA; peatmer@auburn.edu; both authors contributed equally to the manuscript

Abstract

Market competition, rising commodity prices, and a push towards sustainability are dictating higher inclusion rates of plant-based feedstuffs in aquaculture feeds worldwide. Such shifts bring with them many challenges, including recognizing and quantifying losses in digestibility and performance due to anti-nutritional factors. Feed enzymes can help to transform lower cost ingredients into high value nutrition – floor sweepings into fish flesh. The extra-phosphoric effects of phytase superdosing have received relatively little attention in the aquaculture industry. Here we review aspects of phytic acid, both in the cultured organism and in the environment, as they pertain to the catfish industry, the largest segment of aquaculture in the United States. We highlight results of recent research on superdosing in catfish and point out areas in need of additional study.

Keywords: phytate, phytic acid, aquaculture, anaemia

16.1 Background

The consequences of poor nutrition or poorly utilised dietary ingredients in aquaculture differ from those seen in terrestrial livestock agriculture. Whereas in both cases, the animal fails to obtain needed protein, fat, or minerals, excreting them instead to the external environment, in aquaculture there are significant carryon effects. These undigested nutrients can shape the aquatic environment in several lasting, consequential ways, including contributing to algal blooms, increasing the biological oxygen demand of pond bottom sediment, and fostering the growth of pathogenic bacteria. While these consequences are more commonly associated with overfeeding, indigestible elements in plant-based feedstuffs play overlooked and mounting roles (Kumar *et al.*, 2012).

The catfish industry is the largest segment of aquaculture in the United States, albeit a relatively minor contributor on the global stage where aquaculture remains the fastest growing food sector (OECD/FAO, 2014). The industry raises channel catfish (*Ictalurus punctatus*) and hybrid catfish (*I. punctatus* × *Ictalurus furcatus*) predominantly in western Alabama (AL) and eastern Mississippi (MS) in watershed ponds and in the MS Delta

DOI 10.3920/978-90-8686-836-0_16, © Wageningen Academic Publishers 2016

in levee-style ponds. Unlike some other US livestock industries, the catfish industry is only partially integrated and in most cases, hatcheries, food fish producers, feed mills, and processors are unconnected entities. In 2015, the industry produced approximately 316 million pounds of catfish, feeding approximately 400,000 tons of feed (http://www.agecon.msstate.edu/whatwedo/budgets/docs/catfish2014.pdf). These numbers represent a dramatic reduction from the industry peak in 2003 when over 600 million pounds of catfish were produced (Hanson and Sites, 2014). This reduction can be contributed to two primary factors: most significantly, a rise in imported *Pangasius* catfish product from Vietnam, and secondarily, high feed costs linked to higher grain prices.

Catfish diets have changed over the years in response to research on dietary requirements (Eya and Lovell, 1997) and feed ingredient prices (Li and Robinson, 2013). Feed formulations can differ substantially between mills and over the course of the year as commodity prices and availability fluctuate. After the fingerling stage, food fish are usually fed a 28 or 32% protein extruded floating diet. Primary dietary components are soybean meal, corn, cottonseed meal, and wheat middlings. The catfish industry currently is notable among aquacultured species for its minimal inclusion of animal-based ingredients. Fish meal, once a common dietary component, is reserved for fingerling diets and some catfish food fish diets have no animal contribution beyond a fish oil topcoat. While many formulations remain confidential, a clear trend toward lower corn and higher wheat middlings inclusion rates has been in place for the last decade. These changes are predicted to have sizeable consequences on nutrient availability and digestibility.

Phytate, or inositol hexakisphosphate, is a phosphorus (P) storage molecule present in grains and seeds commonly used in aquafeeds. Its structure is composed of six P molecules tethered to a *myo*-inositol (MYO) core. Phytate is widely known to be indigestible to monogastric organisms, such as channel catfish (Andrews, 1973). Inclusion of phytate-containing ingredients in fish diets has been demonstrated to impair growth in a variety of cultured fish species, such as carp, tilapia, trout, and salmon (Francis *et al.*, 2001). While the factors controlling impaired growth are still not fully understood, in its parent form, the phytate anion phosphate groups act to chelate or sequester important divalent and trivalent mineral ions including iron, zinc, magnesium, copper, and calcium and can also form complexes with peptides or proteins thereby reducing protein availability (Hardy, 1998; Helland *et al.*, 2006; Morales *et al.*, in press).

As mentioned above, catfish diets over the last two decades have largely eliminated fishmeal and have reduced inclusion percentages of lower phytate plant sources (soybean meal and corn) in favor of less expensive, but higher phytate sources such as wheat middlings (Li and Robinson, 2013). A recent, in-depth analysis of nutrient composition in poultry feed ingredients, found that phytate levels in wheat middlings were the highest among the tested feedstuffs. These levels, measured from 31 US origin samples, averaged 0.80% phytate-P, 234% of the NRC (1994) value. Interestingly, the same study found increasing levels of phytate-P in soybean meal over the last two decades, likely due to genetic selection for improved nutritional values (nitrogen levels) inadvertently increasing phytate levels as well (Tahir *et al.*, 2012). The effect of rising phytate contents of common feedstuffs was likely at work in a recent study in channel catfish, which reported P apparent availability coefficients (AACs) for soybean meal (36%), cottonseed meal (22%), and wheat middlings (20%) which were significantly lower than historical reports (Buyukates *et al.*, 2000; Li *et al.*, 2015; Lovell, 1978). To our knowledge, there has been no attempt to-date to characterise phytate levels in commercially available catfish diets. Therefore, before initiating the studies described below we surveyed catfish diets from all major catfish feed mills in AL and MS and the phytate levels in these diets ranged from 1.54 to 2.22% (average 1.92%; data not shown).

High and (potentially) increasing levels of phytate in catfish diets could be a contributing factor to several persistent problems faced by US catfish producers. Catfish producers report rising feed conversion ratios (FCRs), degraded water quality associated with low dissolved oxygen (DO) and high ammonia levels, the increasing prevalence and virulence of opportunistic bacterial pathogens, and periodic cases of anaemia. These potential impacts reflect the diverse roles of phytate both within the fish and in the environmental effects highlighted above. With diets averaging over 1.90% phytate, assuming a feeding rate of 14,000 lbs of feed per acre per year, with a 20% digestion rate, and 10-15% of feed going uneaten (Boyd and Tucker, 1998), roughly 200 lbs of phytic acid is predicted to reach the pond sediment each year on a per acre basis. More highly-stocked ponds or overfed ponds would see significantly higher phytic acid accumulation. Given the market pressures on the US catfish industry, regular draining, 'fallowing' and reworking of ponds is increasingly a rare event, particularly in the deeper watershed ponds of western AL and east MS, meaning that these patterns of accumulation can continue over many production cycles.

Although the cycling of orthophosphate monoesters, including phytic acid, are relatively poorly understood, it is known that they resist degradation in the water through strong adsorption to particulate materials, therefore accumulating in the sediment (Cade-Menun *et al.*, 2006; Masuda and Boyd, 1994; Reitzel *et al.*, 2007). There, bacteria and fungi with phytase capabilities can cleave phytic acid resulting in P release and MYO availability. The parameters under which phytic acid dissociation is more likely to occur are tightly linked with soil and water chemistry and conditions optimizing phytase activity (Berry and Hadich, 2013). Deeper sediments in lakes appear to be a sink for phytate-P, in some cases delaying P recycling for years (Jørgensen *et al.*, 2011). In fish ponds, where accumulation of organic matter (faeces, uneaten feed) can significantly increase oxygen demand and establish reducing conditions, anoxia and low pH are associated with release of sediment-bound P and subsequent algal blooms (Kassila *et al.*, 2001; Kassila and Hussenot, 2006). These conditions correspond with known peaks of maximum microbial phytase activity at pH 2.5 and pH 5.0-5.5 (Kumar *et al.*, 2012).

The release of P and MYO from the pond sediments also impacts the bacterial community, potentially triggering or encouraging the growth of pathogens. *Myo*-inositol can be utilised by certain bacteria for growth and has, in some cases, been associated with virulence. These include *Salmonella enterica* serovar Typhimurium, *Clostridium perfigens*, and *Bacillus subitilis* (Kröger *et al.*, 2011; Staib and Fuchs, 2014). In *S. enterica*, MYO-utilising strains have been shown to exhibit a non-virulent bistable phenotype. In the presence of MYO and sufficient levels of CO_2, however, the strain becomes homogeneous and shows more rapid growth (Kröger *et al.*, 2011). Indeed, work by a separate group on *Salmonella* suggests that priming of secretion effectors in MYO-rich environments may then ensure rapid invasion once a host is encountered (Cordero-Alba *et al.*, 2012). The potential role of MYO in the growth and virulence of catfish pathogens has arisen as an important topic for study in light of an emerging virulent strain of *Aeromonas hydrophila* (vAh) in the US catfish industry. This strain appears to be functioning as a primary pathogen, often causing the mortality of >50% of fish within a pond within three days of the first signs of disease (Griffin *et al.*, 2013; Ran *et al.*, 2012). Biochemical and genomic analyses have revealed that: (1) all vAh isolates (but not non-epidemic Ah) were capable of good growth in a minimal medium containing only MYO as a sole carbon source, and (2) that the genomic region predicted to be involved in MYO catabolism is uniquely present in all vAh isolates and is part of the epidemic-associated region in these isolates (Hossain *et al.*, 2013). Ongoing research in our labs as well as in USDA-funded field trials is currently examining the role of phytic acid (and its degradation) in vAh pathogenesis. We hypothesized, based on preliminary studies that MYO-rich and/or anaerobic conditions may change patterns of growth and virulence (Figure 16.1 and 16.2). Recent pilot challenge studies in the lab are also pointing to the importance of growth conditions for triggering vAh virulence (data not shown).

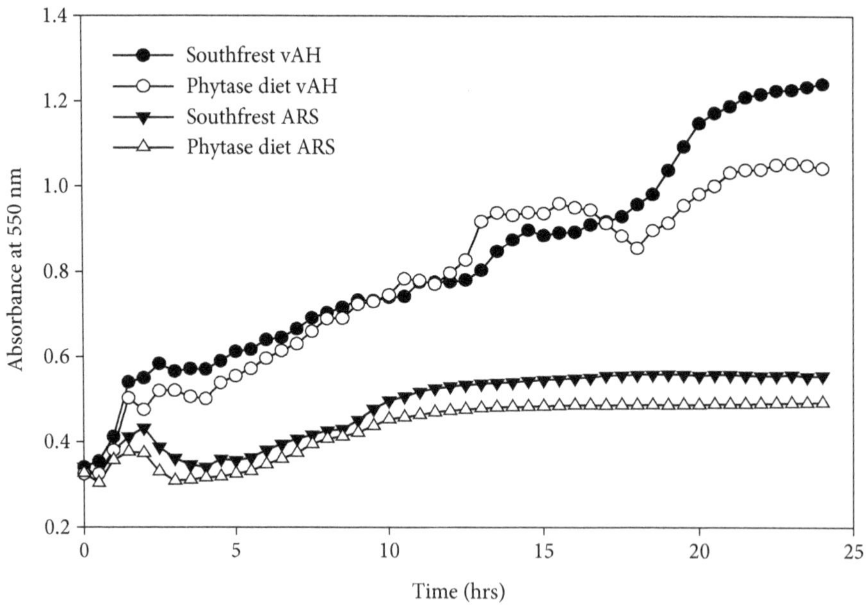

Figure 16.1. *Growth of virulent* Aeromonas hydrophila *(vAh) and non-epidemic* A. hydrophila *(rAh) in a commercial catfish diet homogenate. vAh and rAh isolates were cultured in a diluted homogenate (2% w/v) of a commercial diet (32% protein) at 30 C with 200 rpm shaking. Measurements (OD at a wavelength of 550 nm) were collected every 30 min from six replicate wells per isolate.*

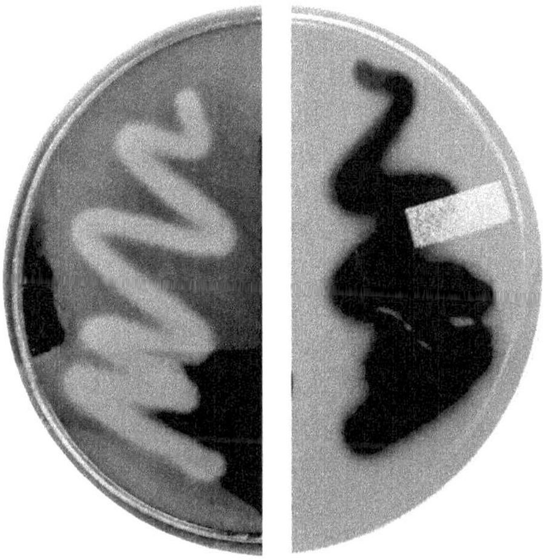

Figure 16.2. *Culture of virulent* Aeromonas hydrophila *(vAh) under aerobic conditions produces α-hemolytic patterns; however when grown anaerobically vAh colonies exhibit a β-hemolytic phenotype, suggesting that vAh possesses an oxygen-labile hemolysin.*

Effects of phytate within the fish also have the potential for impairing growth and negatively impacting health. One of the major impacts of phytate in this regard may come through its ability to tightly bind and thereby limit the absorption of iron both in livestock and humans (Nielsen *et al.*, 2013). Iron is critical for animal health, not only due to its roles in blood parameters, oxygen supply to tissues, and anaemia (Witeska, 2015), but also due to its role in dictating the outcomes of host-pathogen interactions (Parrow *et al.*, 2013). Anaemia remains a persistent and prevalent issue in the catfish industry with an estimated 4.7% of annual cases submitted to the National Warmwater Aquaculture Center in Stoneville, MS from 1994-2012 being anaemia-related (Camus *et al.*, 2014). Whereas the precise mechanisms governing feed-related anaemia are not known, recent findings by Camus *et al.* (2014) indicated sufficient dietary iron levels in examined commercial catfish diets but a 'failure of iron uptake, either as result of poor dietary bioavailability or from an unknown mal-absorptive process.' Intriguingly, administration of parenteral iron to anaemic catfish brought about a recovery from anaemia documented by marked increases in packed cell volume (PCV) and iron stores in the serum and liver (Camus *et al.*, 2014).

Given the potential consequences of high phytate diets in catfish, how can we better turn lemons into lemonade, or more appropriately here 'floor sweepings' into fish flesh? Phytases are phosphatase enzymes that catalyse the hydrolysis of phytate, thereby liberating utilizable inorganic phosphate and *myo*-inositol. In addition to boosting bioavailable P levels *in vivo*, dietary phytase usage has also been linked to improved mineral and nutrient uptake by destroying phytate and its other downstream hydrolysis products (inositol pentakis-, tetrakis-, and tris- phospates; IP_5, IP_4, and IP_3, respectively). Whereas their use has been studied in fish nutrition, including in channel catfish (Jackson *et al.*, 1996; Satoh *et al.*, 1989; Yan *et al.*, 2002), reports of supplementation of phytase in commercially-available fish diets are extremely limited. Phytase studies in channel catfish primarily focused on reducing formulation costs by freeing up P (through hydrolysis of phytate), and thereby eliminating the need for di-calcium phosphate supplementation, as well as reduction in excreted P levels (Li *et al.*, 2004). Li and Robinson (1997) reported that 250 to 750 phytase units (FTU) per kg of a fungal-derived phytase could eliminate the need for P supplementation from inorganic phosphates and improve feed intake, weight gain, and FCR. More recently, second-generation of bacterial phytases have been developed. These phytases are more active at a lower pH range than traditional fungal phytases, as well as having a greater intrinsic resistance to endogenous proteases (Dersjant-Li *et al.*, 2015). Studies in poultry and swine have shown improved performance in diets with high or 'superdosed' inclusion rates of these bacterial phytases (Cowieson *et al.*, 2015; Manobhavan *et al.*, 2016; Walk *et al.*, 2014). Below we highlight results from completed and ongoing studies of phytase superdosing in commercial catfish diets.

16.2 Phytase superdosing in catfish: study highlights

We first wanted to assess the effect of a 2,500 FTU/kg phytase (Quantum Blue, AB Vista, Marlborough, UK) inclusion rate on erythrocyte PCV (hematocrit) and mineral uptake. As shown in Figure 16.3, channel catfish fingerlings fed a phytase-fortified, 28% protein diet (1.9% phytate) had significantly higher hematocrit values ($P<0.05$) beginning six weeks after initiation of feeding, with differences increasing in magnitude at 9 and 12 weeks ($P<0.001$). Analysis of liver and blood showed differences in the uptake of minerals in fish fed the phytase-supplemented diet (Table 16.1). In particular, elevated levels of iron were present in both the blood and liver of fish fed the diet superdosed with phytase. In the liver, phytase-fed animals had a nearly 50% increase in iron stores. At 24-h post-feeding, faeces were analysed for phytate content and fish fed the phytase-supplemented diet had a 70% reduction in phytate concentration in the contents of the distal intestine at a water temperature of 24±2 °C.

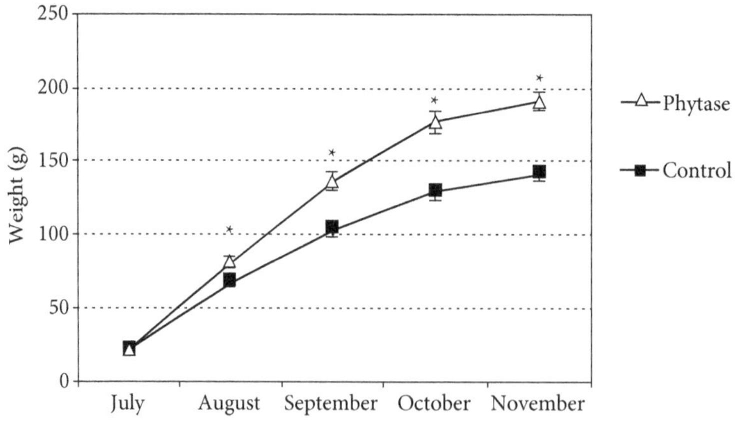

Figure 16.3. *Weight gain of channel catfish over the duration of a pond study where fish were fed a ration containing 0 FTU/kg (control) or 2,500 FTU/kg (phytase; ±SE). Asterisks denote statistically significant differences at* P<0.05.

Table 16.1. *Mineral content (mg/kg) in blood and liver of channel catfish fed for 12 weeks in tanks with an unsupplemented commercial catfish diet (control) and the diet supplemented with 2,500 FTU/kg of diet of phytase (phytase). Values are means (±SE) of three replicates. Means with an asterisk (*) are significantly different (P<0.05) from fish fed the unsupplemented diet.*

Mineral	Blood (control)	Blood (phytase)	Liver (control)	Liver (phytase)
Calcium	547.2±55.4	604.7±5.8	96.1±13.3	96.7±6.2
Phosphorus	5,699.7±121.0	6,541.7±140.4*	7,409.6±311.5	8,869.8±136.4*
Sodium	10,963.3±721.3	12,125.0±89.0	2,059.0±230.5	2,283.5±108.9
Aluminum	ND	ND	ND	ND
Copper	1.1±0.3	1.0±0.2	6.6±0.85	8.5±0.39
Iron	1,072.7±66.8	1,312.0±53.2*	38.7±3.53	73.1±14.5*
Magnesium	427.1±10.0	526.7±13.9*	487.6±23.7	562.2±16.8*
Manganese	ND	ND	5.0±0.47	3.6±0.26*
Potassium	7,945.7±171.7	9,159.7±215.2*	11,771.6±362.3	12,591.5±164.8
Sulfur	2,414.3±30.2	2,563.7±62.4	3,803.3±304.7	4,170.4±136.4
Zinc	100.2±8.0	156.8±34.7	62.5±5.3	73.3±3.1

We then proceeded with a larger, replicated pond study at the USDA-ARS Stuttgart National Aquaculture Research Center utilizing a 32% protein, commercial diet, containing 2.2% phytate, from the same supplier and the same inclusion rate of Quantum Blue phytase (2,500 FTU/kg). Channel catfish fingerlings fed diets containing phytase consumed more feed and gained more weight than fish fed the control ration without supplemental phytase (Figure 16.4). Beginning in August, approximately four weeks after initiation of the study, fish fed the phytase-supplemented diet were significantly (*P*<0.001) larger in mass, and were larger at each sampling time thereafter with final individual weights of control fish averaging 140.8±8.1 g and phytase-fed fish averaging 191.3±9.4 g. Feed conversion ratio

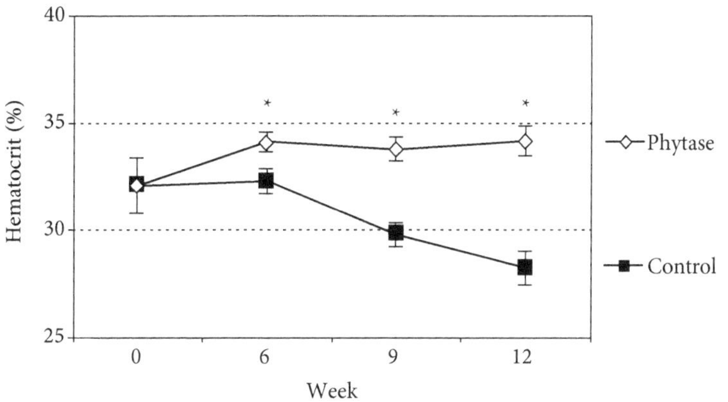

Figure 16.4. *Hematocrit of channel catfish fed a ration containing 0 FTU/kg (Control) or 2,500 FTU/kg (Phytase) in indoor tanks (±SE). Asterisks denote statistically significant differences at* P<0.05.

was different ($P<0.05$) between the two groups and was determined to be 1.76 for the control treatment compared with 1.43 for fish fed supplemental phytase.

Fish fed the diet containing supplemental phytase had increased ($P<0.001$) hematocrit values (Figure 16.5) beginning at 9 weeks post-initiation and continuing until the completion of the trial at 15 weeks. We also performed more comprehensive analyses of the blood including absolute red blood cell counts, which revealed greater red blood cell content in fish fed the phytase-supplemented diet ($P<0.001$). Fish fed the phytase superdosed feed had over 500,000 more red blood cells/μl (an approximate 20% increase) than fish fed the control ration. Hemoglobin values were also higher ($P<0.001$) in the phytase-fed fish (>20% increase). Mean corpuscular volume was higher ($P<0.05$) in fish fed the control diet indicating larger erythrocyte size (Table 16.2). Whereas the red cell compartment was augmented by dietary inclusion of phytase, there were no differences in the absolute number of white blood cells between treatments. Fish fed the diet containing supplemental phytase had increases of iron within the blood and numerous differences ($P<0.05$) in mineral content in the liver including over 40% more iron and substantial increases in P, sodium, copper, magnesium, potassium, sulfur, and zinc (Table 16.3). Whole body analysis of minerals revealed increases ($P<0.05$) in calcium, P, manganese, sulfur and zinc in fish fed the phytase supplement, whereas copper levels were higher ($P<0.05$) in the control fish (Table 16.3). Accordingly, whole body proximate analysis indicated that fish fed the phytase supplemented diet had higher ($P<0.05$) concentration of ash (3.2 vs 2.5%; Table 16.4). Fish did not differ in moisture, lipid, or crude protein content.

Simultaneously, hybrid catfish fingerlings were grown for 15 weeks in replicated 2,650 l flow-through raceway tanks in Auburn, AL, using 32% crude protein commercial feed from the same lot as used in the pond studies. Even at higher densities than achieved in the pond study, similar patterns were revealed. Phytase supplemented fish diets resulted in fish that consumed more feed and gained more weight than control fed fish (Figure 16.6). Final average weights were 311±16 g in the phytase-fed fish and 251±5 g in the

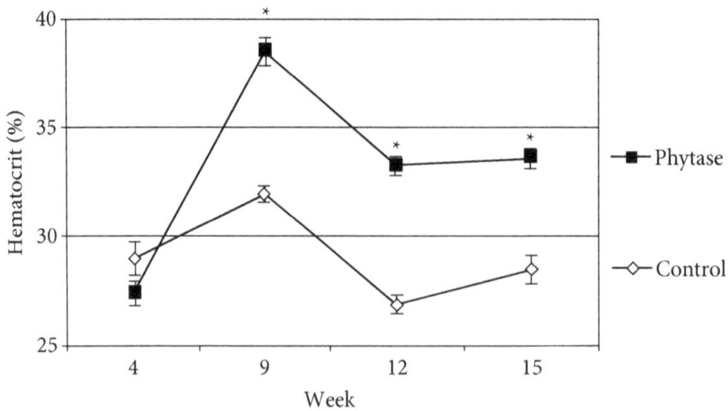

Figure 16.5. *Hematocrit of channel catfish fed a ration containing 0 FTU/kg (Control) or 2,500 FTU/kg (Phytase) in earthen ponds (±SE). Asterisks denote statistically significant differences at* $P<0.05$.

Table 16.2. *Hematological parameters of channel catfish fed for 15 weeks in ponds with a commercial catfish diet (control) and the diet supplemented with 2,500 FTU/kg of diet with phytase (phytase). Values are means (±SE) of three replicates. Means with an asterisk (*) are significantly different (P<0.05) from fish fed the unsupplemented diet.*[1]

Parameter	Treatment	
	Control	Phytase
TCC (10^6/µl)[1]	2.21±0.06	2.70±0.06*
RBC (10^6/µl)	2.01±0.06	2.52±0.05*
WBC (10^5/µl)	1.87±0.09	1.85±0.09
Hemoglobin (g/dl)	7.92±0.18	9.97±0.17*
Hematocrit (%)	28.5±0.65	33.6±0.46*
MCV (fl)	145.6±4.0	135.4±2.8*
MCH (pg)	40.9±1.7	40.1±0.8
MCHC (%)	28.2±0.8	29.9±0.6

[1] MCH = mean corpuscular haemoglobin; MCHC = mean corpuscular hemoglobin concentration; MCV = mean corpuscular volume; RBC = red blood cell; TCC = total cell count; WBC = white blood cell.

Table 16.3. *Mineral content (mg/kg) in blood, liver, and whole body of channel catfish fed for 15 weeks in ponds with an unsupplemented commercial catfish diet (control) and the diet supplemented with 2,500 FTU/kg of diet of phytase (phytase). Values are means (±SE) of three replicates. Means with an asterisk (*) are significantly different (P<0.05) from fish fed the unsupplemented diet.*

	Blood (control)	Blood (phytase)	Liver (control)	Liver (phytase)	Body (control)	Body (phytase)
Ca	586.3±21.5	609.5±27.1	100.14±3.9	105.9±5.8	17,521.5±1342.4	25,345.8±936.3*
P	6,783.5±60.4	6,909.4±83.0	7,809.1±185.2	8,925.5±149.1*	14,743.1±719.4	18,837.9±387.5*
Na	14,420.6±210.3	14,186.9±356.3	2,319.2±83.8	2,675.9±85.8*	3,399.9±73.7	3,604.7±80.6
Al	8.2±2.4	4.8±1.6	20.1±3.6	26.3±6.7	6.0±1.8	6.7±0.7
Cu	1.2±0.2	1.4±0.3	8.21±0.4	9.6±0.4*	1.6±0.0	1.4±0.0*
Fe	1,567.9±21.8	1,680.3±28.2*	87.6±6.9	146.6±8.8*	50.4±0.9	50.1±2.0
Mg	727.8±12.5	723.0±7.8	574.7±17.7	650.2±12.3*	958.0±18.2	1,136.8±63.3
Mn	1.3±0.1	0.9±0.0*	4.0±0.3	4.6±0.6	3.8±0.2	7.6±0.1*
K	7,571.1±197.6	7,608.0±198.8	9,924.4±218.5	10,943.3±202.3*	10,445.4±51.3	10,150.0±162.3
S	4,118.1±57.5	3,958.3±46.6	6,061.6±165.2	6,633.3±197.3*	7,567.4±59.8	6,922.9±47.3*
Zn	140.0±2.8	138.6±2.7	73.6±2.5	87.0±1.9*	63.4±1.7	80.9±1.3*

Table 16.4. *Pond-reared channel catfish whole-body proximate analysis (%). Asterisk denotes statistically significant difference at* P<0.05.

Diet	Moisture	Ash	Lipid	Protein
Control	72.6±0.24	2.5±0.23	7.4±0.17	16.7±0.08
Phytase	72.2±0.12	3.2±0.11*	7.4±0.12	16.3±0.17

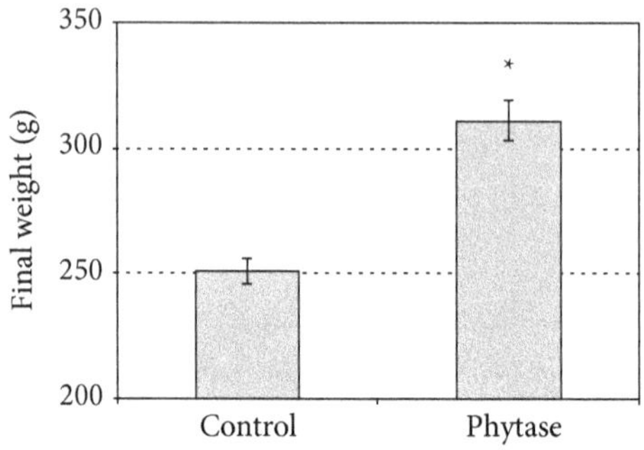

Figure 16.6. *Final average weights of hybrid catfish fed a ration containing 0 FTU/kg (Control) or 2,500 FTU/kg (Phytase) in raceway tanks (±SE). Asterisks denote statistically significant differences at* P<0.05.

control-fed fish (*P*<0.001). Feed conversion ratios were also significantly improved from an average of 1.92 in control fed fish to 1.59 in phytase fed fish (*P*<0.05). Overall yield of phytase-fed fish was 21.5% higher (*P*<0.05) than the corresponding controls. Similar patterns also were observed for hematological parameters with improvements (*P*<0.05) also being realized in hematocrits and red blood cell counts (Figure 16.7). Additional analyses on this experiment are currently in progress.

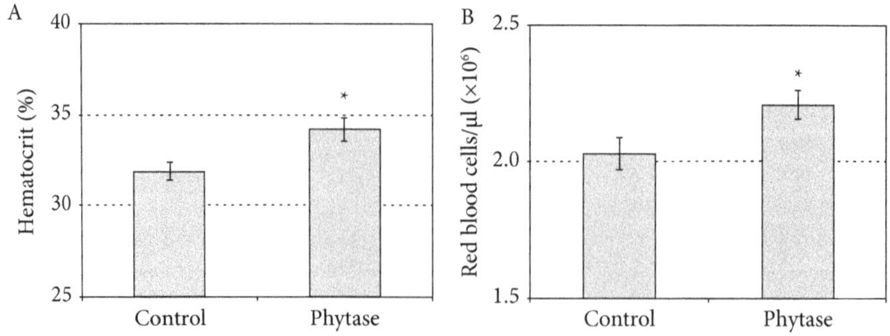

Figure 16.7. *Hematocrit (A) and red blood cell counts of hybrid catfish fed a ration containing 0 FTU/kg (Control) or 2,500 FTU/kg (Phytase) in raceway tanks (±SE). Asterisks denote statistically significant differences at* P<0.05.

16.3 Summary, future directions and key messages

Results of research have indicated that addition of Quantum Blue phytase to a commercial catfish ration significantly improves mineral absorption, hematological parameters, growth, and FCR. Results also indicate that the high phytate content of some plant-based diets may be one contributing factor to feed-related anaemia on commercial catfish farms in the US. The magnitude of these effects is expected to differ depending on differences in feed formulation, dietary phytate levels, and water temperature, which necessitates additional studies that account for these variables.

Current available commercial phytases are unable to withstand the high temperatures of extrusion processing common to the manufacture of floating feeds for catfish and other fish species, necessitating addition of equipment to allow spraying of phytase on finished pellets (Kumar *et al.*, 2012; Lemos and Tacon, in press). When P gains alone have been considered, the challenge of time and money for installation of spray equipment has often been judged by fish feed mills to outweigh the potential benefits of phytase addition. However, our more recent understanding of extra-phosphoric effects of phytase superdosing in addressing anti-nutritional aspects of phytate (Walk *et al.*, 2013) may result in greater phytase utilisation in commercial fish diets in the future. In 2016, AL-based catfish feed mills added post-extrusion spray equipment for phytase, opening the door to industry-wide research and utilisation of not only phytase, but other nutritional enzymes including those targeting non-starch polysaccharides (Selle *et al.*, 2003).

Ongoing and future studies will examine several potential aspects and impacts of phytase superdosing in the US catfish industry by conducting experiments in both the laboratory and field trial environments. Research topics include the role of varying culture temperatures on phytate destruction, the impact of phytase on host disease resistance and pathogen prevalence, and the ability of phytase-fed fish to withstand low dissolved oxygen (hypoxic) conditions.

1. Phytic acid levels in aquaculture feeds are increasing along with increasing utilisation of plant-based feedstuffs.
2. The potential benefits of phytase superdosing extend beyond the cost savings attributable to P liberation from phytic acid.
3. Utilising commercial aquaculture diets for research may provide insights and applications necessary for industry uptake.

Acknowledgments

The authors are grateful for funding from the US Department of Agriculture's National Institute of Food and Agriculture under Grants 2016-68008-25030 and 2015-70007-24276, as well as financial and in-kind support provided by AB Vista.

References

Andrews, J.W., 1973. Interactions of dietary levels of protein and energy on channel catfish (*Ictalurus punctatus*). The Journal of Nutrition 103:1339-1346.

Berry D.F. and Harich, K., 2013. Tetrachlorofluorescein TInsP as a substrate analog probe for measuring phytase activity in surface water: Proof of concept. Journal of Environmental Quality 42:56-64.

Boyd, C.E. and Tucker, C.S., 1998. Pond aquaculture water quality management. Kluwer Academic Publishers, Boston, MA, USA.

Buyukates, Y., Rawles, S.D. and Gatlin III, D.M., 2000. Phosphorus fractions of various feedstuffs and apparent phosphorus availability to channel catfish. North American Journal of Aquaculture 62:184-188.

Cade-Menun, B.J., Navaratnam, J.A. and Walbridge, M.R., 2006. Characterizing dissolved and particulate phosphorus in water with 31P nuclear magnetic resonance spectroscopy. Environmental Science and Technology 40:7874-7880.

Camus, A.C., Wise, D.J., Khoo, L.H., Shi, J. and Berghaus, R.D., 2014. Iron status of channel catfish *Ictalurus punctatus* affected by channel catfish anemia and response to parenteral iron. Diseases of Aquatic Organisms 107:241-248.

Cowieson, A.J., Aureli, R., Guggenbuhl, P. and Fru-Nji, F., 2015. Possible involvement of *myo*-inositol in the physiological response of broilers to high doses of microbial phytase. Animal Production Science 55:710-719.

Cordero-Alba, M., Bernal-Bayard, J. and Ramos-Morales, F., 2012. SrfJ, a Salmonella type III secretion system effector regulated by PhoP, RcsB, and IolR. Journal of Bacteriology 194:4226-4236.

Dersjant-Li, Y., Awati, A., Schulze, H. and Partridge, G., 2015. Phytase in non-ruminant animal nutrition: a critical review on phytase activities in the gastrointestinal tract and influencing factors. Journal of the Science of Food and Agriculture 95:878-896.

Eya, J.C. and Lovell, R.T., 1997. Available phosphorus requirements of food-size channel catfish (*Ictalurus punctatus*) fed practical diets in ponds. Aquaculture 154:283-291.

Francis, G., Makkar, H.P. and Becker, K., 2001. Antinutritional factors present in plant-derived alternate fish feed ingredients and their effects in fish. Aquaculture 199:197-227.

Griffin, M.J., Goodwin, A.E., Merry, G.E., Liles, M.R., Williams, M.A., Ware, C. and Waldbieser, G.C., 2013. Rapid quantitative detection of *Aeromonas hydrophila* strains associated with disease outbreaks in catfish aquaculture. Journal of Veterinary Diagnostic Investigations 25:473-481.

Hanson, T.R., 2016. U.S. catfish feed delivery report. The Catfish Journal, Catfish Farmers of America, Jackson, MS, USA, 30:1.

Hardy, R.W., 1998. Phytate. Aquaculture Magazine 11/12:877-880.

Helland, S., Denstadli, V., Witten, P.E., Hjelde, K., Storebakken, T., Skrede, A., Asgard, T. and Baeverfjord, G., 2006. Hyper dense vertebrae and mineral content in Atlantic salmon (*Salmo salar* L.) fed diets with graded levels of phytic acid. Aquaculture 261:603-614.

Hossain, M.J., Waldbieser, G.C., Sun, D., Capps, N.K., Hemstreet, W.B., Carlisle, K., Griffin, M.J., Khoo, L., Goodwin, A.E., Sonstegard, T.S., Schroeder, S., Hayden, K., Newton, J.C., Terhune, J.S. and Liles, M.R., 2013. Implication of lateral genetic transfer in the emergence of *Aeromonas hydrophila* isolates of epidemic outbreaks in channel catfish. PLoS One 8:e80943.

Jackson, L.S., Li, M.H. and Robinson, E.H., 1996. Use of microbial phytase in channel catfish *Ictalurus punctatus* diets to improve utilization of phytate phosphorus. Journal of the World Aquaculture Society 27:309-313.

Jørgensen, C., Jensen, H.S., Andersen, F.Ø., Egemose, S. and Reitzel, K., 2011. Occurrence of orthophosphate monoesters in lake sediments: significance of *myo*- and scyllo-inositol hexakisphosphate. Journal of Environmental Monitoring 13:2328-2334.

Kassila, J., Hasnaoui, M., Droussi, M., Loudiki, M. and Yahyaoui, A., 2001. Relation between phosphate and organic matter in pond sediments of the Deroua fisheries farm (Béni-Mellal, Morocco): implications for pond management. Hydrobiologia 450:57-70.

Kassila, J. and Hussenot, J., 2006. Heterogeneity of sediments and settling particles in aerated ponds. Aquaculture Research 37:449-465.

Kröger, C., Srikumar, S., Ellwart, J. and Fuchs, T.M., 2011. Bistability in *myo*-inositol utilization by *Salmonella enterica* serovar Typhimurium. Journal of Bacteriology 193:1427-1435.

Kumar, V., Sinha, A.K., Makkar, H.P.S., De Boeck, G. and Becker, K., 2012. Phytate and phytase in fish nutrition. Journal of Animal Physiology and Animal Nutrition 96:335-364.

Lemos, D. and Tacon, A.G.J., in press. Use of phytases in fish and shrimp feeds: a review. Reviews in Aquaculture. DOI: http://doi.org/10.1111/raq.12138.

Li, M.H. and Robinson, E.H., 1997. Microbial phytase can replace inorganic phosphorus supplements in channel catfish *Ictalurus punctatus* diets. Journal of the World Aquaculture Society 28:402-406.

Li, M.H., Manning, L.B. and Robinson, E.H., 2004. Summary of phytase studies for channel catfish. Mississippi Agricultural and Forestry Experiment Station, Mississippi State, MS, USA.

Li, M.H. and Robinson, E.H., 2013. Feed ingredients and feed for channel catfish. Southern Regional Aquaculture Center, 1806, Stoneville, MS, USA.

Li, M.H., Robinson, E.H. and Lucas, P.M., 2015. Apparent phosphorus availabilities of selected traditional and alternative feedstuffs for channel catfish. North American Journal of Aquaculture 77:136-140.

Lovell, R.T., 1978. Dietary phosphorus requirement of channel catfish (*Ictalurus punctatus*). Transactions of the American Fisheries Society 107:617-621.

Manobhavan, M., Elangovan, A.V., Sridhar, M., Shet, D., Ajith, S., Pal, D.T. and Gowda, N.K., 2016. Effect of super dosing of phytase on growth performance, ileal digestibility and bone characteristics in broilers fed corn-soya-based diets. Journal of Animal Physiology and Animal Nutrition (Berl) 100:93-100.

Masuda, K. and Boyd, C.E., 1994. Effects of aeration, alum treatment, liming, and organic matter application on phosphorus exchange between soil and water in aquaculture ponds at Auburn, Alabama. Journal of the World Aquaculture Society 25:405-416.

Morales, G.A., Denstadli, V., Collins, S.A., Mydland, L.T., Moyano, F.J. and Overland, M., in press. Phytase and sodium diformate supplementation in a plant-based diet improves protein and mineral utilization in rainbow trout (*Oncorhynchus mykiss*). Aquaculture Nutrition. DOI: http://doi.org/10.1111/anu.12340.

Nielsen, A.V., Tetens, I. and Meyer, A.S., 2013. Potential of phytase-mediated iron release from cereal-based foods: a quantitative view. Nutrients 5:3074-3098.

NRC., 1994. Nutrient requirements for poultry. 9th rev. ed. National Academy Press, Washington, DC, USA.

OECD/FAO., 2014. OECD-FAO agricultural outlook 2014, OECD Publishing, Paris, France. DOI: http://doi.org/10.1787/agr_outlook-2014-en.

Ran, C., Carrias, A., Williams, M.A., Capps, N., Dan, B.C., Newton, J.C., Kloepper, J.W., Ooi, E.L., Browdy, C.L., Terhune, J.S. and Liles, M.R., 2012. Identification of *Bacillus* strains for biological control of catfish pathogens. PLoS One 7:e45793.

Parrow, N.L., Fleming, R.E. and Minnick, M.F., 2013. Sequestration and scavenging of iron in infection. Infection and Immunity 81:3503-3514

Reitzel, K., Ahlgren, J., DeBrabandere, H., Waldebäck, M., Gogoll, A., Tranvik, L. and Rydin, E., 2007. Degradation rates of organic phosphorus in lake sediment. Biogeochemistry 82:15-28

Satoh, S., Poe, W.E. and Wilson, R.P., 1989. Effect of supplemental phytate and/or tricalcium phosphate on weight gain, feed efficiency and zinc content in vertebrae of channel catfish. Aquaculture 80:155-161.

Selle, P.H., Ravindran, V., Ravindran, G., Pittolo, P.H. and Bryden, W.L., 2003. Influence of phytase and xylanase supplementation on growth performance and nutrient utilisation of broilers offered wheat-based diets. Asian Australasian Journal of Animal Sciences 16:394-402.

Staib L. and Fuchs, T.M., 2014. From food to cell: nutrient exploitation strategies of enteropathogens. Microbiology 160:1020-1039.

Tahir, M., Shim, M.Y., Ward, N.E., Smith, C., Foster, E., Guney, A.C. and Pesti, G.M., 2012. Phytate and other nutrient components of feed ingredients for poultry. Poultry Science 91:928-935.

Walk C.L., Bedford, M.R., Santos, T.T., Paiva, D., Bradley, J.R., Wladecki, H., Honaker, C. and McElroy, A.P., 2013. Extra-phosphoric effects of superdoses of a novel microbial phytase. Poultry Science 92:719-725.

Walk, C.L., Santos, T.T. and Bedford, M.R., 2014. Influences of superdoses of a novel microbial phytase on growth performance, tibia ash, and gizzard phytate and inositol in young broilers. Poultry Science 93:1172-1177.

Witeska, M., 2015. Anemia in teleost fishes. Bulletin of the European Association of Fish Pathologists 35:148-160.

Yan, W., Reigh, R.C. and Xu, Z., 2002. Effects of fungal phytase on utilization of dietary protein and minerals, and dephosphorylation of phytic acid in the alimentary tract of channel catfish *Ictalurus punctatus* fed an all-plant-protein diet. Journal of the World Aquaculture Society 33:10-22.

Chapter 17
Trace minerals – what role should they play in today's poultry industry with respect to fast growth rate and woody breast?

T.W. York[1], M.R. Bedford[2] and C.L. Walk[2]*
[1]AB Vista, Suite 270, 150 S Pine Island Road, Plantation, FL 33324, USA;
[2]AB Vista, Woodstock Court, Blenheim Road, Marlborough Business Park, Marlborough, Wiltshire SN8 4AN, United Kingdom; tara.york@abvista.com

Abstract

Trace minerals, such as zinc (Zn), selenium (Se), and iron (Fe) as well as copper (Cu) and manganese (Mn) play important roles in growth, immunity, gut health, antioxidant status, as well as numerous other roles within the animal. However due to pressure on feed costs, trace minerals are sometimes fed at lower than optimal levels compared to those suggested by the primary breeding companies. New challenges in the form of white striping and woody breast have recently surfaced in today's broiler industry and are reportedly tied to genetic selection for fast growth rate and higher meat yield. With these increased pressures come potentially higher metabolic stressors where the use of additional trace minerals, as well as superdosing of phytase may aid to support the antioxidant capacity of the broiler. Recent results indicated that supplementing broiler diets with additional Zn, Fe, Se, tryptophan and/or ethoxyquin in combination with superdosing levels of phytase reduced the severity of woody breast by 50% while improving average daily gain and feed efficiency. These results may suggest that superdoses of phytase with the provision of minerals to support the antioxidant system of the bird may allow for continued improvements in growth performance without increasing the severity of woody breast.

Keywords: antioxidant, phytase, breast meat quality

17.1 Introduction

Poultry genetics is continuously changing in an effort to maximize broiler performance with respect to growth rate, feed conversion and breast meat yield. As with other aspects of modernisation, broiler and genetic companies are learning that with change comes new challenges. One such challenge is the emergence of woody breast, a muscle myopathy with an unknown cause. The initial suspect is genetics and fast growth rate (Trocino *et al.*, 2015). With fast growth rate comes a series of challenges, all of which could be potentially stressful for the animal and its ability to keep up metabolically. Eliminating woody breast does not appear to be a simple task and, according to the genetic companies, it will take them several years to remove this issue (A. Corzo, personal communication). Although the incidence of woody breast cannot be abolished quickly, research is being conducted

C.L. Walk *et al.* (ed.) **Phytate destruction – consequences for precision animal nutrition**
DOI 10.3920/978-90-8686-836-0_17, © Wageningen Academic Publishers 2016

around the world, including genetic companies, looking for ways to reduce the occurrence and severity of this issue. The biggest challenge for the poultry industry is pushback from the consumer due to the unfavourable impact on meat quality and palatability. Taste panels conducted by researchers at the University of Arkansas suggest as the severity of woody breast increases the acceptance to the consumer decreases (C. Owens-Hanning, personal communication). Therefore, poultry companies who sell further processed breast meat need to find ways to reduce the most severe forms of woody breast or risk losing their customers.

One promising area of current research is the importance of trace minerals and antioxidants. Adding to this promising area of research is the interaction of phytase, an enzyme traditionally used for phosphorus release, and the role it may play in the reduction of woody breast when fed at higher than normal inclusion levels. The intent of this chapter is not to go into detail regarding trace minerals or phytases, differences in their bioavailability, or the antioxidant system. Instead this article is a reminder of the important role trace minerals provide and the interaction between them. It is also intended to challenge the conventional logic of the poultry industry of constantly putting pressure on feed costs, especially considering the desired growth rate and yield expected from today's modern broiler. As an industry the more we expect from the broiler, the more we need to invest to ensure we accomplish our goal, to provide safe, nutritious, high quality poultry meat to the consumer, in an effective manner. Although a novel concept, recent studies evaluating the impact of trace minerals and antioxidants in combination with feeding higher levels of phytase suggest they may help reduce the incidence and severity of woody breast.

17.2 Trace minerals

Trace minerals, despite the important role they play, are not in the forefront of nutritionists' minds when it comes to formulating diets. Increasing the level of trace minerals would add cost to poultry diets and commercial nutritionists are constantly pressured to minimize cost. The recommendation for trace minerals found in poultry nutrition text books or by the primary breeding companies suggest feeding levels above those typically found in today's poultry diets for enhanced performance and bird health (Cobb-Vantress, 2013; Leeson and Summers, 2005). While the recommendations for these trace minerals are higher, it is also advised to feed these levels across all phases. Yet the trace mineral levels in diets of broilers fed today may be lowered from the starter to the withdrawal phase or completely removed from the final feed, presumably for cost savings (T. York, personal observation).

Trace minerals can have very specific tasks, in addition to serving in cross-functional roles when it comes to growth, immunity, oxidative status or bone development (McDonald *et al.*, 1995; Oviedo-Rondon and Ferket, 2005; Surai, 2002). An example is zinc (Zn), which is necessary for immunity, collagen development, as well as the intracellular antioxidant enzyme superoxide dismutase (SOD) (Dibner *et al.*, 2007; Harris, 1992; Surai, 2002). Selenium (Se) is another mineral that can serve in a single or multifunctional capacity and of particular interest is the role Se plays for the antioxidant enzyme glutathione peroxidase (GPx) (Surai, 2002).

Given the stressors imposed on today's broilers it may be time to reconsider the trace mineral and antioxidant components fed within current diets. Especially in light of some of the latest challenges the breeder and broiler industries are facing today, such as feathering, leg health, moving to antibiotic free feeds, seven day livability and most recently white striping and woody breast. The concept of lowering trace minerals in the feed or pulling them completely out as the bird reaches maturity, is not consistent with promoting optimal performance through optimal nutrition.

17.3 Natural defence system

Optimising trace mineral utilisation and inclusion in the diet, specifically Zn, copper (Cu), selenium (Se) and manganese (Mn), should enhance the bird's natural defence system (Kidd, 2004). Supporting the antioxidant system by increasing the availability of natural and synthetic antioxidants (specifically fat soluble vitamin E) in addition to feeding optimal levels of Zn, Mn, Cu and Se (Kidd, 2004; Surai, 2002) would assist in maintaining efficient levels of endogenous antioxidants in the tissues, which are needed to combat free radical production. Free radicals are unstable atoms that can destroy important biological components, such as lipids within the cell walls. If free radical scavenging enzymes cannot keep up with the production and removal, it may lead to greater oxidative stress and tissue damage (Jaeschke, 1995). Due to free radicals being produced and used by the immune system to destroy harmful microrganisms, the bird's body is always under some form of assault (Surai, 2002). Whether the production of free radicals within the animal leads to detriment, rather than good, is ultimately determined by the efficiency of the antioxidant system and its supporting players, such as the trace minerals mentioned previously, as well as the level of environmental and growth stressors imposed on that animal.

When considering the genetic improvements that have resulted in rapid weight gain and improved feed efficiency, the movement towards antibiotic free diets, and the everyday production stressors, it only seems logical that the production of free radicals within the tissues have risen than in years past. Although genetically the bird has been selected for growth it does not appear to have been selected to handle the added internal stress that may result from the accelerated growth rate (Sandercock *et al.*, 2009). The steps today's modern broiler pursues to protect cells from oxidative stress most likely has not adapted and still follows the same line of defence as outlined by Surai (2002). The first line of defence being the prevention of harmful free radicals with the help of the antioxidant enzymes SOD and GPx, which utilize Zn and Se. The second involves vitamin E, which is an important chain breaking antioxidant to stop the formation of new peroxides. Ubiquinol and glutathione also play an important role in this second line of oxidative defence (Surai, 2002). The critical thing to remember is the overall defence system requires trace minerals such as Zn, Cu, Mn and Se to be effective. It is also important to remember there may be a delicate balance between having too much or too little of any particular mineral. For example Cu and Fe, although play important roles also have the ability to act as promoters of lipid oxidation if available in excess (Halliwell, 1999). While diets fed to today's broiler are most likely not deficient for these trace minerals the question that should be asked is are they sufficient? Selenium and vitamin E deficiency have been shown to impair the antioxidant capacity against oxidative stress, especially in the pectoralis major (Avanzo *et al.*, 2001).

17.4 Role of phytase and phytate

Phytic acid and its salts (phytate; IP_6) was first identified over 150 years ago (Hartig, 1855). Since then we have recognized many roles phytate can play in animal nutrition. Although it is a nutrient, it is more known for its anti-nutritive properties and negative impact on animal performance due its ability to bind to and reduce phosphorus, trace mineral and amino acid digestibility (Cowieson *et al.*, 2004; Humer *et al.*, 2015). Another role of phytic acid, which is often overlooked, is its role as an antioxidant. Phytic acid is reported to act as an antioxidant by chelating minerals such as Cu and Fe, which are known pro-oxidants in lipid metabolism (Midorikawa *et al.*, 2001). When investigating methods to prevent cancer in humans, Midorikawa *et al.* (2001) suggested that phytic acid inhibited the Fe-mediated oxidative reaction as well as limited site specific DNA damage due to this antioxidant role. While research would need to be conducted to confirm similar findings in poultry, this suggests a potential benefit for phytic acid within the cell, with respect to lipid oxidation and meat quality.

17.4.1 Superdosing of phytase

The concept of feeding phytase at levels beyond the standard 500 FTU/kg in the diet has been reported in the literature but it was not until recently that the return on investment has made it worthwhile to implement (Nelson *et al.*, 1971; Shirley and Edwards, 2003). Today, the practice of feeding (>1,500 FTU/kg) of phytase has been coined 'superdosing' (SD). Although the definition of this practice may vary depending on the enzyme supplier, for the purpose of this discussion it is defined as feeding 1,500 FTU/kg of an enhanced *Escherichia coli* phytase (Quantum Blue, AB Vista, Marlborough, United Kingdom) with only the mineral matrix applied to the first 500 FTU/kg. When fed at standard levels the primary target for this enhanced phytase is to remove the higher phytate esters such as IP_6 and inositol pentaphosphate (IP_5) for the release of available phosphorus. When SD levels are fed, the objective is to obtain closer to 90% phytate destruction by breaking IP_6 down to inositol triphosphate (IP_3) or inositol diphosphate (IP_2) and thereby allowing for endogenous or microbial phytases to create *myo*-inositol (Figure 17.1). Although the newer generation phytases have this capability, the amount of phytase need to achieve this level of destruction will differ depending on the phytase source. Obtaining phytate destruction beyond inositol tetraphosphate (IP_4) is important for several reasons. The first reason is the recent realisation that the phytate ester IP_4 continues to act as an anti-nutrient and bind minerals (Xu *et al.*, 1992; Persson *et al.*, 1998; Yu *et al.*, 2012) and pepsin (Yu *et al.*, 2012) to interfere with protein digestion. The second reason is for the generation of inositol which has reported antioxidant functions (Dam, 1944). However, in order to produce inositol from phytate in the diet, the phytate has to be broken down to lower phytate esters through exogenous phytase, endogenous phosphatase and bacterial phytases in the gut (Figure 17.1).

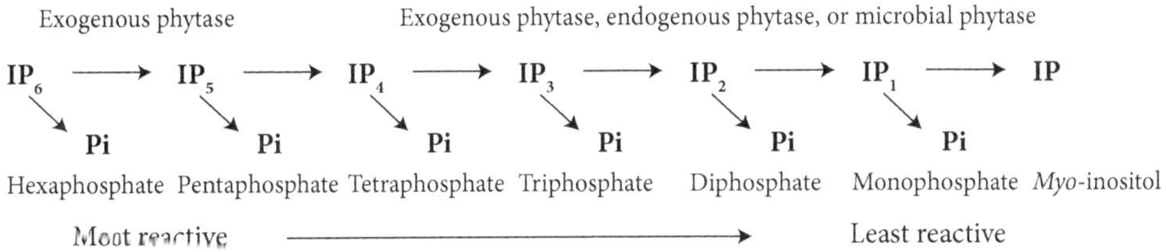

Figure 17.1. *Breakdown of phytate and lower phytate esters to inositol and phosphorus through the action of phytase.*

Once inositol is generated it may combine with phosphate at the cellular level to recreate phytate, which allows it to act as an antioxidant within the cell by binding with minerals (Midorikawa *et al.*, 2001). Recent studies in poultry (Walk *et al.*, 2014) and pigs (Wilcock *et al.*, 2014) have demonstrated a correlation between the generation of inositol in the proximal gastrointestinal tract and improved feed conversion as shown in Figure 17.2. However, with greater destruction of phytate comes a higher availability of trace minerals and amino acids concomitant with less energy loss due to lower phytate esters making their way into the small intestine. This is a partial explanation of why SD may result in improved performance beyond a phosphorus effect. Although it is believed that inositol has an important role, it is one that is less well understood. Specifically, it is the role of inositol and the release of additional trace minerals that are of interest when considering the role of SD phytase and the antioxidant capacity of the animal.

17.4.2 Woody breast

Researchers in Finland were among the first to provide a detailed histological description of woody breast (Sihvo *et al.*, 2013). In their study, breast filets were selected from a processing facility which exhibited the characteristics of woody breast, such as pale meat with noticeable out bulging and hardened ridges. In breast filets demonstrating woody

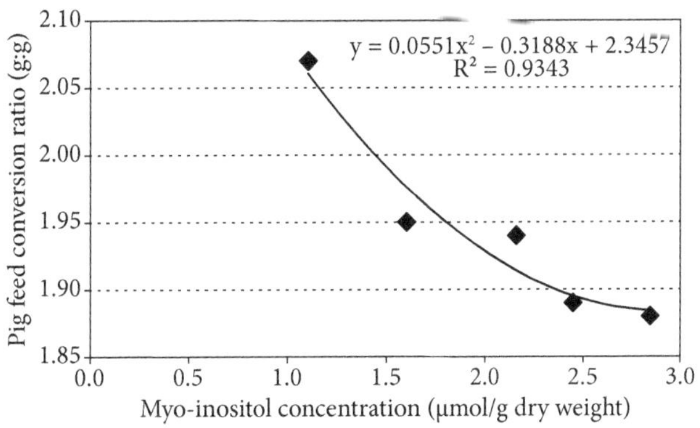

Figure 17.2. *Feed conversion ratio compared with the generation of inositol from superdoses of phytase in the stomach content of 55 kg pigs (Wilcock* et al., *2014).*

breast, the hardness can vary from mild to severe. A severe woody breast is essentially like a board, the meat is very hard and difficult to bend back upon itself. While mild conditions are limited to hardness in either the caudal or cranial region of the breast filet. It is interesting to hypothesise why the variation exists within breast filets of birds raised under similar conditions and body weights. Histologically, woody breast was described as having myofibres with variable diameters, degeneration and necrosis of tissue and infiltration of inflammatory cells, primarily macrophages (Sihvo *et al.*, 2013). The affected areas were found to have a high concentration of connective tissue (fibrosis) separating the muscle fibres. It was noted that perivascular inflammation occurred around the veins.

In 2015, researchers at the University of Bologna (Mazzoni *et al.*, 2015) conducted a pen study to evaluate woody breast in heavy sized broilers. Breast meat was randomly selected and measured for various parameters. In this study, researchers found histological lesions ranging from mild to severe and described similar findings of those from Sihvo *et al.* (2013). Their conclusion was that woody breast was a chronic inflammatory process due to the presence of T-lymphocytes, fibrosis and necrosis and attributed the cause of this chronic inflammation to the genetic pressure placed on increased growth rate of the breast muscle (Mazzoni *et al.*, 2015). As to the true aetiology of woody breast, this remains unknown. One hypothesis put forth by researchers involved in studying woody breast is that the growth of the connective tissue in the muscle does not keep pace with the muscle fibre radial growth and the fibres outgrow the supporting connective tissues leading to muscle damage (Kranen *et al.*, 2000; Mazzoni *et al.*, 2015; Swatland, 1990). This suggests that blood flow to these tissues may be temporarily limited or even nonexistent at times. This also means that nutrients are not being delivered to the tissue and products of cellular metabolism, such as free radicals and immune cells, are not being removed which may lead to the inflammation reported in the veins.

Building on the hypothesis of Mazzoni *et al.* (2015), an article published by medical researchers in Minnesota recently suggested that periods of injury to tissues where blood flow is halted temporarily and then flow is re-established leads to even greater damage (Chamoun *et al.*, 2000). The belief is that the absence of oxygen and nutrients from blood during this ischemic period leads to a condition which results in inflammation and oxidative stress rather than restoring the tissues to normal function. The restoration of blood flow and oxygen to the cells may lead to increased damage to cellular proteins and plasma membranes as a result of a lack of scavenger enzymes to keep up with the oxidation stress occurring. A proposed inflammatory cascade for this ischemic-reperfusion injury, as suggested by Chamoun *et al.* (2000), is outlined in Figure 17.3. The white clots refer to a thrombus. In birds, thrombocytes circulate as mononuclear cells but essentially the formation of a thrombus is the aggregation of the mononuclear cells to form a plug and then a deposition of crosslinked fibrin.

The hypothesis of Mazonni *et al.* (2015) and the description of the ischemic-reperfusion injury by Chamoun *et al.* (2000) would suggest that oxidative stress may be the underlying factor for woody breast development. What initiates or causes this may perhaps be the lack of supportive tissues or oxygen reaching the breast muscle due to the selection for increased growth rate as has been proposed. What is clear is that this is a complex series

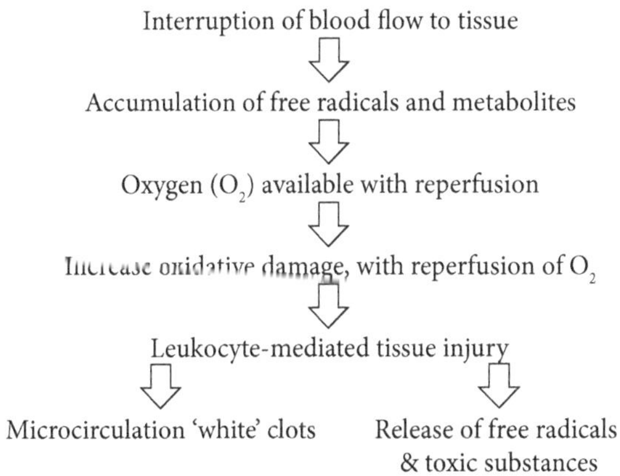

Figure 17.3. *Inflammatory cascade associated with ischemic-reperfusion injury (adapted from Chamoun* et al.*, 2000).*

of events. Supporting the antioxidant system of the bird to a greater extent in order to potentially slow or reduce the damage which may be occurring makes logical sense.

A lack of blood and nutrients to support the developing muscle tissue leads to tissue death and the infiltration of inflammatory cells. Sihvo *et al.* (2013) stated that the main inflammatory cells found around the degenerative fibre were macrophages. The activation of macrophages in stress conditions creates a source of additional free radicals that if not combated by the animals' natural line of defence may lead to more cellular destruction due to lipid peroxidation of the biological membranes compromising their structure and function (Surai, 2002; Zhao *et al.*, 1998). Macrophages use superoxide to destroy pathogens as their first line of defence. However, superoxide is toxic to the cells and needs to be converted to a less toxic form quickly by the enzyme superoxide dismutase (Surai, 2002). This enzyme requires key minerals such as Zn, Mn and Cu.

17.4.3 Superdosing: a benefit for woody breast?

In 2015, Italian researchers evaluated the impact of SD an enhanced *E. coli* phytase and inositol on broiler performance and meat quality in a phosphorus deficient diet (F. Sirri and C. Walk, unpublished). In this study the birds fed the phosphorus deficient diet had the lowest severity of woody breast and also the lowest average daily gain, supporting the idea that woody breast is a result of fast growth rate. However, this study also demonstrated that the incidence of woody breast was lower in Ross 308 male broilers when fed SD levels of phytase even though these treatments resulted in the highest average daily gains (F. Sirri and C. Walk, unpublished; Figure 17.4). The levels of Zn, Mn and Cu (mg/kg) fed were indicative of levels commonly fed to broilers in Europe and were stepped down in each feeding phase.

Woody breast is believed to be attributed to the genetic pressure for increased growth rate of the breast muscle or rate of growth overall. However, a second study was recently

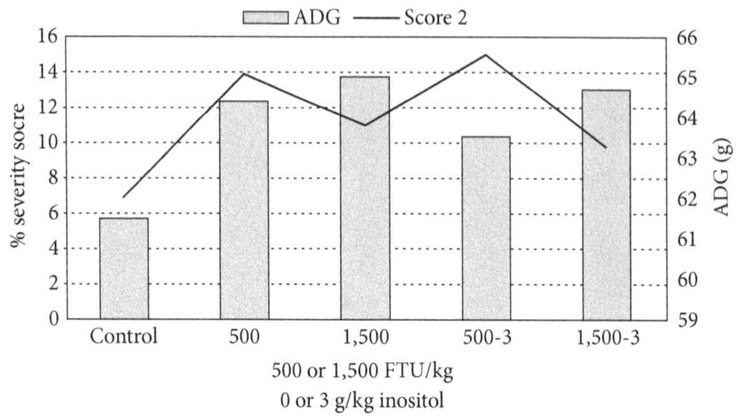

Figure 17.4. *Incidence of severe woody breast and average daily gain in Ross 308 male broilers fed superdoses of phytase or inositol from hatch to 49-days post-hatch (F. Sirri and C. Walk, unpublished).*

conducted which showed a similar trend in that the incidence of woody breast was not increased even though birds fed the SD treatment exhibited some of the best growth rates overall (A. Corzo and M. Bedford, personal communication). Woody breast was present in both of these studies so it poses the question of why SD would allow for lower incidences of or a reduction in the severity of woody breast without reducing growth rate? It could be hypothesised that the increase in trace minerals and amino acids (Coweison *et al.*, 2004), inositol (Walk *et al.*, 2014; Wilcock *et al.*, 2014), and antioxidant levels in the liver (Karadas *et al.*, 2009), and/or the combination of, which occur when birds are fed SD levels of phytase allows for the release of nutrients that support growth while aiding in the reduction of oxidative stress. While there is not enough information in the literature to make firm conclusions, there are a number of studies recently conducted that may support this theory.

The importance of trace minerals such as Zn, Cu, Mn, and Se and their role in supporting the antioxidant enzymes SOD and GPx has been well established. The mineral which has not been discussed in regards to the antioxidant enzymes is Fe. Iron, due to the role it plays in improving the oxygen carrying capacity of blood, is another mineral of interest and one that is negatively impacted by the anti-nutrient phytate (Selle *et al.*, 2011). Phytate, IP_6 and IP_5 have been reported to bind up to 70% of the Fe in the diet, while IP_4 and IP_3 can still bind up to 30% of Fe (Yu *et al.*, 2012). *In vitro* work looking at the impact of increasing levels of phytase on the release of free Fe in soybean meal reported that 1,200 FTU/kg, but not 400 or 800 FTU/kg of phytase, resulted in an increase in the release of free Fe by 239% (Stahl *et al.*, 1999). More recently, when investigating the impact of SD of phytase in catfish, researchers found that feeding increased phytase levels (>2,000 FTU/kg enhanced *E. coli* phytase) resulted in not only an increase in trace minerals in the blood and liver (Figure 17.5), but also an increase in red blood cells, percent hematocrit and g/dl of haemoglobin (Chapter 16; Peatman and Beck, 2016), all parameters which improve the oxygen uptake capacity of the animal (Figure 17.6). It would appear the release of additional Fe may be incorporated into a form that is beneficial to the animal rather than remaining in a free

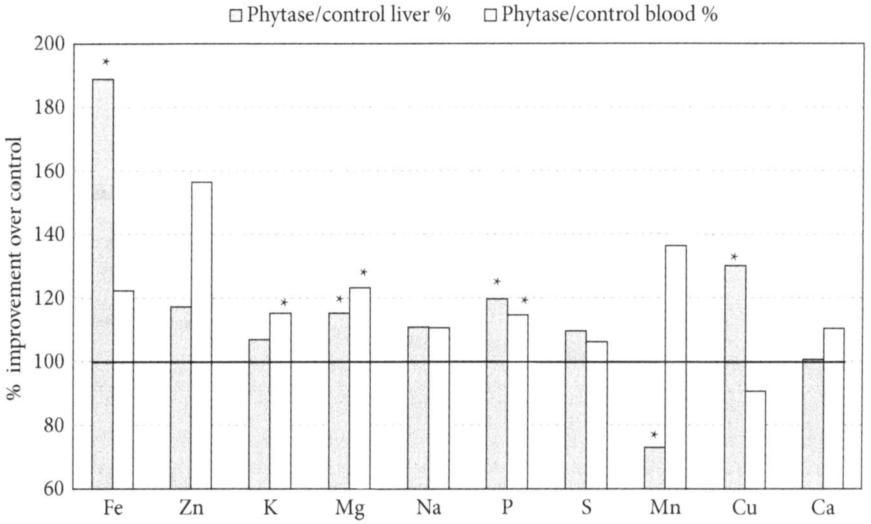

Figure 17.5. *Impact of feeding increased levels of phytase (>2,000 FTU/kg enhanced* Escherichia coli *phytase) on trace mineral levels in the blood and liver of catfish (Peatman and Beck, 2016).*

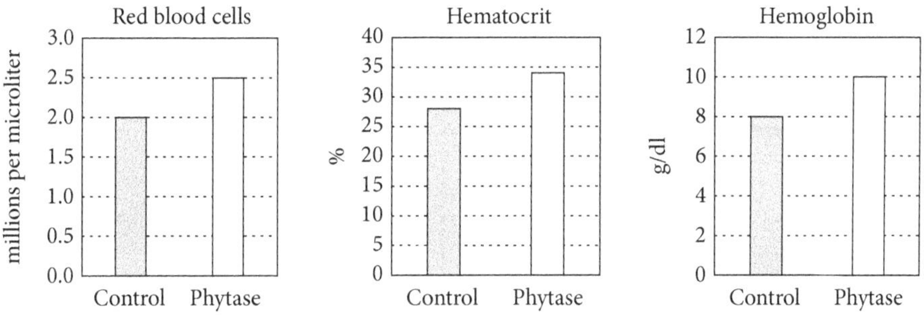

Figure 17.6. *Impact of feeding increased levels of phytase (>2,000 FTU/kg enhanced* Escherichia coli *phytase) on red blood cells, hematorcit and haemoglobin in catfish (Peatman and Beck, 2016).*

form which can become detrimental (Halliwell, 1987). Cu is another mineral which has not been discussed in detail. Cu is often fed at levels beyond requirement to serve as an antimicrobial agent, yet also plays a role in the enzymes systems involved in Fe transport and metabolism, red blood cell formation and immunity (Cohen, 2002).

A study comparing a standard dose (500 FTU/kg) of an enhanced *E. coli* phytase against a diet containing 1,500 FTU/kg in Ross 708 male broilers was recently concluded in the US. In addition to testing the impact of SD, researchers in this study also evaluated parameters which had been reported in the literature as possibly having an impact on the reduction of woody breast or a positive impact on meat quality (Fe, Zn, Se and ethoxyquin, and tryptophan, Trp) either on their own or in combination with the increased phytase treatment. Zinc and Fe were added to the diets in the sulphate form. Sulphates are reported as having higher bioavailability than oxides which are used often in broiler diets to lower the cost of trace mineral premixes. In this study, the total level of Zn and Fe were stepped down as the bird aged, but were well above those of the industry on average. The treatments for this study as well as the amount of Zn, Fe and Trp added are shown in Table 17.1 (unpublished data).

Body weight, feed conversion and breast meat yield were recorded. On day 57, 120 birds per treatment were randomly selected to measure the percent breast meat and to score for the severity of woody breast. All breast filets were scored by the same person using the scale developed by Tijare *et al.* (2015) and shown in Table 17.2.

The percent of breast filets that were scored as either a 2 or 3 on day 57 for the Ross 708 male broilers is depicted in Figure 17.7 (unpublished data). According to taste panels conducted by researchers at the University of Arkansas it is the more severe forms of

Table 17.1. *Experiment treatments designed to test the impact of superdoses of phytase or nutrients (Zn, Fe, Trp and Se and ethoxyquin) on the incidence and severity of wooden breast.*

Dietary treatment	Phytase (FTU/kg)	Added Fe sulfate (mg/kg)	Added Zn sulfate (mg/kg)	Added tryptophan (%)
Control, 500[1]	500	0	0	0
Control, SD[1]	1,500	0	0	0
SD + AOX[2]	1,500	0	70	0
SD + Fe	1,500	80	0	0
SD + Trp	1,500	0	0	0.05
SD + All[3]	1,500	80	70	0.05

[1] Nutrition matrix applied in the diet account for the supplementation of phytase was 0.15% avP, 0.16% Ca, 0.03% Na.

[2] Diet contained 125 mg/kg ethoxyquin (66.6% Mix 6, Novus International, St. Louis, MO, USA) and a standard trace mineral premix with 0.3 mg/kg Se.

[3] Diet included standard trace mineral premix with ethoyxquin and Se, added Zn, Fe and tryptophan.

Table 17.2. *Scoring system for woody breast characteristics in breast fillets of broilers detected by tactile evaluation (adapted from Tijare* et al.*, 2015).*

Woody score		Woody characteristics
Normal	(0)	flexible throughout
Mild	(1)	hardness mainly in cranial region, flexible in caudal region
Moderate	(2)	hardness throughout with some flexibility in mid to caudal region
Severe	(3)	very hard, rigid throughout

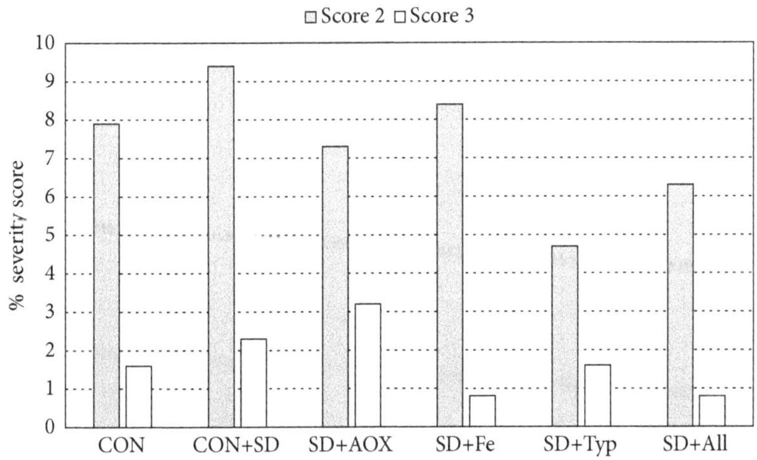

Figure 17.7. *Incidence of moderate and severe woody breast scores in 57-day old broilers fed a control diet with 500 FTU/kg compared with birds fed superdoses of phytase with added nutrients (Zn, Fe, Trp and Se and ethoxyquin) known to influence the antioxidant status of the animal (C. Walk, M. Bedford, C. Wyatt and T. York, unpublished data).*

woody breast that are deemed unacceptable with a score of 3 being the most severe form (C. Owens-Hanning, personal communication). At first glance, the data from Figure 17.7 would suggest that several treatments were successful in lowering the scores of 2 and 3 for woody breast with the SD + Trp or SD + all treatment exhibiting the lowest scores overall. However, upon further evaluation of the data, a different picture evolves. As observed in the previous studies, the treatment which resulted in the lowest form of severe woody breast, yet had the highest average daily gain was the treatment which included SD phytase and all the other dietary components that support the antioxidant capacity of the animal (SD + all treatment, Figure 17.8). This treatment had the highest body weight (P=0.0256), lowest feed conversion ratio and the second highest percent breast meat yield (Table 17.3) compared to all others. While the overall incidence of woody breast in this study was low for reasons beyond the scope of this article, when comparing the control (500 FTU/kg phytase) treatment to the SD + all treatment, the most severe form of woody breast was reduced by 50% (1.6 vs 0.8% respectively; unpublished data). While not found to be statistically significant, the reduction in severity in woody breast would be significant to the industry today.

Researchers from the University of Arkansas recently compared diets where Ross 708 male broilers were fed two levels of phytase (500 vs 1,500 FTU/kg), two levels of ethoxyquin (0 or 125 mg/kg) and two vitamin trace mineral (VTM) premixes, industry standard or enhanced (K. Christensen, C. Walk, D. Koltes, C. Owens-Hanning, C. Wyatt and T. York, unpublished). The industry standard premix provided Zn, Mn, Cu and Fe in the sulfate form, with an inorganic form of Se. The enhanced premix provided Mn and Fe in the sulfate form, with Zn and Cu being supplied as glycinates, Se from an organic source, and contained a higher level of vitamin E. The trace mineral levels varied slightly between the two premixes but were both indicative of current levels fed for broilers in the US. In this

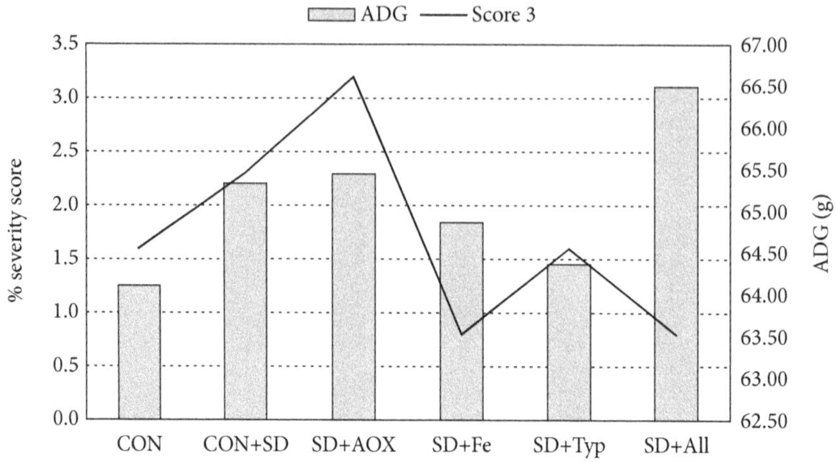

Figure 17.8. *Incidence of severe woody breast scores and the rate of average daily gain of 57-day old broilers fed a control diet with 500 FTU/kg compared with birds fed superdoses of phytase with added nutrients (Zn, Fe, tryptophan and Se and ethoxyquin) known to influence the antioxidant status of the animal (C. Walk, M. Bedford, C. Wyatt and T. York, unpublished data).*

Table 17.3. *Growth performance and breast meat yield of broilers fed a control diet with 500 FTU/kg of phytase compared to birds fed superdoses of phytase with added nutrients (Zn, Fe, tryptophan and Se and ethoxyquin) known to influence the antioxidant status or meat quality of animals (C. Walk, M. Bedford, C. Wyatt and T. York, unpublished data).*

Dietary treatment	BW gain (kg)	FCR[1]	Breast meat yield (%)[2]
Control, 500 FTU/kg	3.914[a]	1.624	32.05
Control + 1,500 FTU/kg (SD)	3.989[ab]	1.600	32.60
SD + AOX	3.996[ab]	1.593	32.66
SD + Fe	3.961[b]	1.598	32.81
SD + Trp	3.930[b]	1.636	32.34
SD + All	4.060[a]	1.564	32.78
Diet[3]	0.0256	0.1064	0.0836

[1] BW corrected using mean experimental body weight and 33 g = 0.01 pt.

[2] Breast meat yield reported as a percent of hot carcass weight.

[3] *P*-value for the t-test which gives pairwise comparisons of the means (all 6 diets).

study, the VTM was kept constant across all feeding phases and not reduced as the bird aged. An overview of the different treatment combinations are shown in Table 17.4.

Body weights, feed conversion, breast meat yields and woody breast scores were measured at two different time periods. The scoring system was the same as the one reported previously (Tijare *et al.*, 2015). Overall, the incidence and severity of woody breast was

Table 17.4. *Experimental treatments designed to test the impact of superdoses of phytase in the presence or absence of ethoxyquin when fed industry standard or enhanced vitamin and trace minerals on the severity of woody breast in male broilers (K. Christensen, C. Walk, D. Koltes, C. Owens-Hanning, C. Wyatt and T. York, unpublished).*

Treatment	Vitamin and trace mineral premix	Phytase (FTU/kg)	Ethoxyquin (mg/kg)[2]
1	standard	500[1]	0
2	standard	500	125
3	standard	1,500[1]	0
4	standard	1,500	125
5	enhanced[3]	500	0
6	enhanced	500	125
7	enhanced	1,500	0
8	enhanced	1,500	125

[1] Matrix applied for the enhanced *Escherichia coli* phytase; 0.15% avP, 0.16% Ca, 0.13 Na.
[2] Mix 6, Novus International, St. Louis, MO, USA, 66.6% ethoxyquin.
[3] Contains Zn and Cu glycinate, BASF, Ludwigshafen, Germany and organic Se (Cypress).

high on both day 44 and 63 with the majority of breast filets scored exhibiting some form of woody breast ranging from mild (score of 1) to severe (score of 3). Although the data is still being validated, when comparing the extremes (treatment 1 vs treatment 8), the birds fed the SD phytase treatment with an enhanced VTM had the highest average daily gain yet exhibited a reduction in the most severe form of woody breast by almost 47% from 3.7% to 1.96%, respectively (Figure 17.9). As in the previous studies, the combination

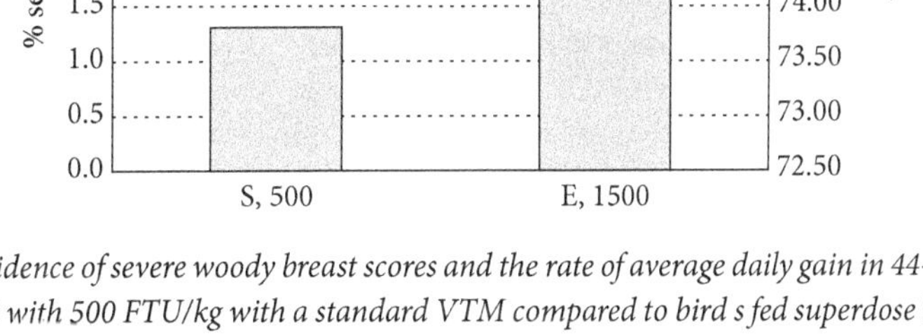

Figure 17.9. *Incidence of severe woody breast scores and the rate of average daily gain in 44-day old broilers fed a control diet with 500 FTU/kg with a standard VTM compared to bird s fed superdose levels of phytase and an enhanced VTM (K. Christensen, C. Walk, D. Koltes, C. Owens-Hanning, C. Wyatt and T. York, unpublished).*

treatments when incorporating the SD of phytase appear to be beneficial in allowing the bird to grow to its maximum potential while aiding in the reduction of the most severe form of woody breast.

17.5 Conclusions

Currently there is a wide variety of studies being conducted by Universities and researchers around the globe trying to determine the cause of the meat quality issue coined 'woody breast'. While there is not a clear answer, the most recent studies presented in this chapter encourage the view that dietary manipulations can reduce the incidence of woody breast without resulting in the need to reduce growth or feed efficiency. From this series of work, some of the key messages are:

1. Supplementing the diet with ingredients which support the animal's internal line of defense and antioxidant capacity aide in the reduction of woody breast.
2. Feeding superdosing levels of an enhanced *E. coli* phytase in combination with nutrients involved in supporting the antioxidant pathway of the animal appears to allow for improvements in average daily gain while not increasing the most severe form of woody breast.
3. Increasing the overall trace mineral levels or implementing practices which aid in their release, while supporting the antioxidant capacity in combination, is showing promise in the reduction of woody breast.

References

Avanzo, J.L., De Mendonca, C.X. Jr. and Pugine, S.M.P., 2001. Effect of vitamin E and selenium on resistance to oxidative stress in chicken superficial pectoralis muscle. Comparative Biochemistry and Physiology. Toxicology and Pharmacology 129:163-173.

Chamoun, F., Burne, M., O'Donnell, M. and Rabb, M., 2000. Pathophysiologic role of selectins and their ligands in ischemia reperfusion injury. Frontiers in Bioscience 5: 103-109.

Cobb Vantress, 2013. Broiler performance and nutrition supplement. Cobb 500. October 1, 2013. Siloam Springs, Arkansas, USA.

Cohen, J., 2002. 'Strategic role' for copper. Feed International, December 2002.

Cowieson, A. J., Acamovic, T. and Bedford, M.R., 2004. The effect of phytase and phytic acid on the loss of endogenous amino acids and minerals from broiler chickens. British Poultry Science 45:101-108.

Dam, H., 1944. Studies on vitamin E deficiency in chicks. Journal of Nutrition 27:193-211.

Dibner, J. J., Richards, J.D., Kitchell, M.L. and Quiroz, M.A., 2007. Metabolic challenges and early bone development. Journal of Applied Poultry Research 16:126-137.

Halliwell, B., 1987. Oxidants and human disease: some new concepts. FASEB Journal 1:358-364.

Halliwell, B., 1999. Antioxidant defence mechanisms: from the beginning to the end (of the beginning). Free Radical Research 31:261-272.

Hartig, T., 1855. Über das Klebermehl. Botanische Zeitung 13:881-885.

Harris, E.D., 1992. Copper as a cofactor and regulator of copper, zinc-superoxide dismutase. The Journal of Nutrition 122:636-640.

Humer, E., Schwarz, C. and Schedle, K., 2015. Phytate in pig and poultry. Journal of Animal Physiology and Animal Nutrition 99:605-625.

Jaeschke, H., 1995. Mechanisms of oxidant stress-induced acute tissue injury. Proceedings of the Society for Experimental Biology and Medicine 209:104-111.

Leeson, S. and Summers, J.D., 2005. Commercial poultry nutrition, 3rd edition. University Books, Guelph, Ontario, Canada, 398 pp.

Karadas, F., Pirgozliev, V., Pappas, A.C., Acamovic, T. and Bedford, M.R., 2009. Effects of different dietary phytase activities on the concentration of antioxidants in the liver of growing broilers. Journal of Animal Physiology and Animal Nutrition 94:519-526.

Kidd, M.T., 2004. Nutritional modulation of immune function in broilers. Poultry Science 83:650-657.

Kranen, R. W., Lambooy, E., Veerkamp, C.H., Kuppevelt, T.H. and Veerkamp, J.H., 2000. Histological charcteristization of hemorrages in muscle of broilers. Poultry Science 79:110-116.

Mazzoni, M., Petraci, M., Meluzzi, A., Cavani, C., Clavenzani, P. and Sirri, F., 2015. Relationship between pectoralis major muscle histology and quality traits of chicken meat. Poultry Science 94:123-130.

McDonald, P., Edwards, R.A., Greenhalgh, J.F.D. and Morgan, C.A. (eds.), 1995. Animal Nutrition Fifth Edition. Longman Scientific & Technical, Essex England, pp 607.

Midorikawa,K., Murata, M., Oikawa, S., Hiraka, Y. and Kawanishi, S., 2001. Protective effect of phytic acid on oxidative DNA damage with reference to cancer chemoprevention. Biochemical and Biophysical Research Communications 288:552-557.

Nelson, T.S., Shieh, T.R., Wodzinski, R.J. and Ware, J.H., 1971. Effect of supplemental phytase on the utilization of phytate phosphorus by chicks. Journal of Nutrition 101:1289-1294.

Oviedo-Rondon, E.O. and Ferket, P.R., 2005. Nutritional factors that affect leg problems in meat poultry: A review. In: Proceedings for the 32nd Annual Carolina Poultry Nutrition Conference. October 27th, 2005. Research Triangle Park, NC, USA, pp. 58-88.

Peatman, E. and Beck, B.H., 2016. From floor sweepings to fish flesh: phytase superdosing in the US catfish industry. In: Walk, C.L., Kühn, I., Stein, H.H., Kidd, M.T. and Rodehutscord, M. (eds.) Phytate destruction – consequences for precision animal nutrition. Wageningen Academic Publishers, Wageningen, the Netherlands, pp. 237-250.

Persson, H., Turk, M., Nyman, M. and Sandberg, A.S., 1998. Binding of Cu, Zn and Cd to inositol tri-, tetra-, penta- and hexaphosphates. Journal of Agriculture and Food Chemistry 46:3194-3200.

Sandercock, D.A., Barker, Z.E., Mitchell, M.A. and Hocking, P.M., 2009. Changes in muscle cell cation regulation and meat quality traits are associated with genetic selection for high body weight and meat yield in broilers chickens. Genetic Selection Evolution 41. DOI: http://doi.org/10.1186/1297-9686-41-8.

Selle, P.H., Ravindran, V., Cowieson, A.J. and Bedford, M.R., 2011. Phytate and phytase. In: Bedford, M.R. and Partridge, G.G (eds) Enzymes in Farm Animal Nutrition, 2nd edition, CAB International, United Kingdom, p. 160-205.

Shirley, R.B. and Edwards, H.M., Jr., 2003. Graded levels of phytase past industry standards improves broiler performance. Poultry Science 82:671-680.

Sihvo, H.K., Immonen, K. and Puolanne, E., 2013. Myodegeneration with fibrosis and regeneration in the pectoralis major muscle of broilers. Veterinary Pathology 51:619-623.

Stahl, C.H., Han, Y.M., Roneker, K.R., House, W.A., and Lei, X.G., 1999. Phytase improves iron bioavailability for haemoglobin synthesis in young pigs. Journal of Animal Science 77:2135-2142.

Surai, P.F., 2002. Natural antioxidants in avian nutrition and reproduction. Nottingham University Press, Thrumpton, Nottingham, UK, 615 pp.

Swatland, J. J., 1990. A fibre-optic probe for muscle composition in poultry. Food Science and Technology Journal 23:239-241.

Tijare, V.V., Yang, F.L., Alvarado, C.Z., Coon, C. and Owens, C.M., 2015. Meat quality of broilers breast fillets with white striping and woody breast muscle myopathies. Poultry Science (E. Suppl): M36 Abstract.

Trocino, A., Piccirillo, A., Birolo, M., Radaelli, G., Bertotto, D., Filiou, E., Petracci, M. and Xiccato, G., 2015. Effect of genotype, gender and feed restriction on growth, meat quality and the occurrence of white striping and wooden breast in broiler chickens. Poultry Science 94:2996-3004.

Walk, C.L., Santos, T.T. and Bedford, M.R., 2014. Influence of superdoses of a novel microbial phytase on growth performance, tibia ash, and gizzard phytate and inositol in young broilers. Poultry Science 93:1172-1177.

Wilcock, P., Bradley, C.L., Chewning, J.J. and Walk, C.L., 2014. The effect of superdosing phytase on inositol and phytate concentration in the gastrointestinal tract and its effect on pig performance. Journal of Animal Science 92:E-supplement 2, pp 383.

Xu, P., Price, J., Wise, A. and Aggett, P.J., 1992. Interaction of inositol phosphates with calcium, zinc, and histadine. Journal of Inorganic Biochemistry 47:119-130.

Yu, S., Cowieson, A.J., Gilbert, C., Plumstead, P. and Dalsgaard, S., 2012. Interactions of phytate and *myo*-inositol phosphate esters (IP_{1-5}) including IP_5 isomers with dietary protein and iron and inhibition of pepsin. Journal of Animal Science 90:1824-1832.

Zhao, W., Han, Y., Zhao, B., Hirota, S., Hou, J., and Xin, W., 1998. Effect of carotenoids on the respiratory burst of rat peritoneal macrophage. Biochimica et Biophysica Acta 1381:77-88.

www.ingramcontent.com/pod-product-compliance
Lightning Source LLC
Chambersburg PA
CBHW061204240926
51846CB00006B/1111
* 9 7 8 9 0 8 6 8 6 2 9 0 0 *